International Conference e-Learning and Digital Learning (ELDL2025), and Sustainability, Technology and Education (STE2025)

Held at the Multi-Conference on Computer Science and Information Systems (MCCSIS 2025)

Lisbon, Portugal
23-25 July 2025

ISBN: 979-8-3313-2770-5

Printed from e-media with permission by:

Curran Associates, Inc.
57 Morehouse Lane
Red Hook, NY 12571

International Association for Development of the Information Society (IADIS)
Rua Sao Sebastiao Da Pedreira
No. 100, 30, 1050-209
Lisbon, Portugal

Phone: 351 21 3151373
Fax: 351 21 3151244

secretariat@iadis.org

Additional copies of this publication are available from:

Curran Associates, Inc.
57 Morehouse Lane
Red Hook, NY 12571 USA
Phone: 845-758-0400
Fax: 845-758-2633
Email: curran@proceedings.com
Web: www.proceedings.com

INTERNATIONAL CONFERENCES

E-LEARNING AND DIGITAL LEARNING 2025

AND

SUSTAINABILITY, TECHNOLOGY AND EDUCATION 2025

part of the

MULTI CONFERENCE ON COMPUTER SCIENCE

AND INFORMATION SYSTEMS 2025

PROCEEDINGS
OF THE
INTERNATIONAL CONFERENCES

E-LEARNING AND DIGITAL LEARNING 2025

AND

SUSTAINABILITY, TECHNOLOGY AND EDUCATION 2025

Lisbon, Portugal

JULY 23 - 25, 2025

Organised by

iadis

international association for development of the information society

As a member of Crossref (a non-profit membership organization for scholarly publishing working with the purpose to make content easy to find, link, cite and assess) each published paper in this book of proceedings will be allocated a DOI (Digital Object Identifier) number for its fast and easy citation and indexation.

Volume Editors:
Miguel Baptista Nunes, Pedro Isaías, Tomayess Issa and Theodora Issa

Computer Science and Information Systems Series Editors:
Piet Kommers and Pedro Isaias

Associate Editor: Luís Rodrigues

TABLE OF CONTENTS

FOREWORD ix

PROGRAM COMMITTEE xiii

KEYNOTE LECTURE xvii

FULL PAPERS

E-LEARNING AND DIGITAL LEARNING

LEARNING OUTCOMES AND QUALITY OF EDUCATIONAL SERVICE AS PREDICTORS OF ACADEMIC SATISFACTION 3
Olger Gutierrez Aguilar, Aleixandre Duche Pérez, Anani Gutierrez Aguilar and Sandra Chicana Huanca

EXPLORATION ON THE LEARNING PROCESS: A COMPARATIVE STUDY OF HIGH AND LOW PRE-TEST PERFORMANCE 11
Wannisa Matcha

HYBRID LEARNING 2.0: DESIGNING SUSTAINABLE MODELS FOR POST-PANDEMIC HIGHER EDUCATION IN JORDAN 22
Maha S. Yaseen

AN AI-ASSISTED WORKFLOW FOR MULTILINGUAL ACCESSIBLE VIDEO PRODUCTION: THE EDUNEXT CASE STUDY 29
Alice Pierluigi, Marina Linacero Martín, Annamaria De Santis, Claudia Bellini and Tommaso Minerva

INTEGRATING SLA AND GENAI IN A PRACTICE-ORIENTED INTRODUCTION TO INFORMATION SYSTEMS COURSE: A MIXED METHODS STUDY 37
Cherifa Boudia, Amina Houari and Kamal Benhaoua

PROBLEM SOLVING AND ARTIFICIAL INTELLIGENCE: A BLENDED INTENSIVE PROGRAMME EXPERIENCE 48
Vittoria Bonanzinga

ADVANCING INTERNATIONAL ACCOUNTING EDUCATION WITH ARTIFICIAL INTELLIGENCE 55
Marja-Liisa Tenhunen

ENHANCING GEOMETRIC CONSTRUCTIONS IN ELEMENTARY EDUCATION USING GEOGEBRA: A CONSTRUCTIVIST APPROACH 63
Lilla Korenova, Angelika Schmid and Zuzana Gemrot

PEDAGOGICAL INTENTION AND LEARNER PERCEPTION OF TRANSVERSAL SKILLS 71
Marc Beutner and Jan-Phillip Lüttke

THE MEDIATING ROLE OF CREATIVITY IN ACADEMIC PERFORMANCE THROUGH THE USE OF CHATGPT FOR UNIVERSITY STUDENTS 81
Olger Gutierrez Aguilar, Aleixandre Duche Pérez, Anani Gutierrez Aguilar and Sandra Chicana Huanca

RETHINKING EDUCATION: DESIGNING MINDS FOR INDUSTRY 5.0 89
Suzana Sampaio, Kathryn Cormican, Pablo Alvarez and Deirdre Green

GENERATIVE ARTIFICIAL INTELLIGENCE IN SCIENCE EDUCATION AND TEACHER TRAINING: DEVELOPMENT OF A DATA COLLECTION INSTRUMENT 97
Juliana C. C. Monteiro, Margarida M. Marques and Lúcia Pombo

FORESTED: INTERACTIVE TOOL FOR COMPETENCY CENTERED CURRICULUM VISUALIZATION 105
Sorana Cimpan, Romain Ferrer and Hinata Yokoyama

INTEGRATING AI-BASED SEARCH FEATURES IN LMS: DESIGN AND IMPLEMENTATION NEEDS AND ISSUES 113
Andrea Molinari

GEOMETRIC FIGURES IN STEREO BY CONICAL AXONOMETRIC PERSPECTIVE USING GEOGEBRA 2D SOFTWARE 121
Karla Bastos Guedes, Ronaldo da Silva Tavares, José Geraldo Franco Méxas and Marcelle de Sá Guimarães

EDUCATIONAL ASSESSMENT IN THE AGE OF AI: A NARRATIVE REVIEW ON DEFINITIONS AND ETHICAL-LEGAL PRINCIPLES FOR TRUSTWORTHY AUTOMATED SYSTEMS 129
Giannangelo Boccuzzi, Flavio Manganello and Alberto Nico

UNDERSTANDING LEARNERS' PERCEPTIONS ON A GENERATIVE AI-ENABLE SYSTEM IN PERSONALIZED CONTEXTUAL VOCABULARY LEARNING 137
Mohammad Nehal Hasnine, Mirai Yamada, Hiroshi Ueda and Kensuke Akao

EXPLORING THE PERCEIVED EFFECTS AND PERCEPTIONS OF USING NO-CODE AI EDUCATIONAL TOOLS IN AI LITERACY COURSES FOR NON-IT MAJOR UNDERGRADUATE STUDENTS 145
Hyunjin Cha

DIGITAL TECHNOLOGIES AND HEALTHY LIFESTYLES: MOBILE APPLICATIONS FOR PHYSICAL ACTIVITY 152
Lilla Korenova, Marian Merica, Tomas Barot, Lucia Hlatka and Katarina Zilkova

USE OF GENERATIVE AI BY PHD STUDENTS: IMPACT ON PRODUCTIVITY AND SKILLS DEVELOPMENT 159
Bruno F. Gonçalves

INFORMAL LEARNING AND AFFINITY SPACES: LESSONS LEARNED FROM THE ANALYSIS OF THE CURIOSITY, INTEREST AND ENGAGEMENT CONSTRUCT OBSERVED IN PROJECT NOOA 167
Juliana C. C. Monteiro, Carla Morais and Miguel Carvalhais

DESIGN OF A SYNTAX–ANALOGY-BASED LEARNING TOOL FOR ENJOYABLE HARMONIC EDUCATION USING NATURAL LANGUAGE STRUCTURES 175
Yoshitsugu Kohjiba, Hinata Nakamura, Soma Arai and Takuro Shibayama

ASPECTS OF COMPUTATIONAL THINKING DEVELOPMENT IN FIRST-YEAR HEALTH SCIENCE STUDENTS 183
Edita Butrime and Virginija Tuomaite

SUSTAINABILITY, TECHNOLOGY AND EDUCATION

FRAMEWORK FOR NET ZERO PATHWAY FOR INDIAN CHEMICAL INDUSTRY 191
Mukundan Ramakrishnan, Wilson Kia Onn Wong and Suchismita Das

INTEGRATED INTRUSION DETECTION AND PREVENTION MODEL FOR MOODLE LEARNING MANAGEMENT SYSTEM 199
Olayinka Anthony Ojo, Celestine Iwendi and Adedapo Paul Aderemi

TOWARDS SUSTAINABLE LEATHER GOODS. INNOVATION IN MATERIALS 208
Dalia Gallico

SHORT PAPERS

E-LEARNING AND DIGITAL LEARNING

ENHANCING EDUCATIONAL FEEDBACK WITH LLM-POWERED MULTI-AGENT SYSTEMS 219
Lorenzo Pozzi

EXPLORING AI DIGITAL TEXTBOOK DESIGN FOR PERSONALIZED DIAGNOSTICS IN SPECIAL EDUCATION 224
Seol-Hee Lee, Yedana Lee and Ga-Young Lee

ENHANCING DIGITAL COMPETENCY IN PRE-SERVICE EARLY CHILDHOOD TEACHERS: A PROGRAM DEVELOPMENT STUDY 229
Minkyung Suh and Wooyong Jeun

CONCEPTUAL MODEL OF THE TRIPARTITE APPROACH TO IMPROVING LEARNING OUTCOMES OF THE STUDY PROGRAM 234
Ljerka Luic and Ana Mulovic Trgovac

FROM THE PERSPECTIVE OF 3C-3I INTEGRATION: RESEARCH ON THE CHALLENGES AND TRANSFORMATION STRATEGIES OF DIGITAL TEACHING IN MIDDLE SCHOOLS 239
Xixi Lu

SUSTAINABILITY, TECHNOLOGY AND EDUCATION

RAIL TRAFFIC RESEARCH IN GERMANY: EVALUATION OF GUIDELINES, REGULATIONS AND DESIGN VALUES WITH RESPECT TO CLIMATE CHANGE 244
Jens Lehnen, Michael Eble, Carina Herrmann and Frauke von den Driesch

FEASIBILITY STUDY ON THE PROCESSING OF RECYCLED HDPE/PLA MIXED FILAMENT FOR FDM 3D PRINTING 249
Peng-Ching Chiu, Jui-Hung Cheng and Chang-Zhe Lin

IDENTIFICATION AND ANALYSIS OF DATA SOURCES ON THE IMPACT OF RIVER FLOODING ON TRANSPORT INFRASTRUCTURE IN GERMANY 253
Michael Eble, Jens Lehnen, Frauke von den Driesch and Carina Herrmann

CULTURAL HERITAGE SUSTAINABILITY IN THE DESTINATION BRANDING OF SRI LANKA AND THE MALDIVES 258
Adriaan De Man

TRAINING SKILLS FOR SUSTAINABILITY: THE 3RPLAY METHODOLOGY 263
Sara Nosari

REFLECTION PAPERS

E-LEARNING AND DIGITAL LEARNING

THE VALUE, APPLICATION AND CHALLENGES OF ARTIFICIAL INTELLIGENCE IN INFORMATION TECHNOLOGY COURSES IN PRIMARY AND SECONDARY SCHOOLS 271
Zhiqi Miao

AN OVERVIEW OF THE VILLAGE PROJECT AND INSTRUCTIONAL DESIGN MODELS OPEN ACCESS DATABASE 275
Fırat Sarsar, Özge Andiç Çakır, Christos Vasilakis, Stella Tsovou, Stavros Theocharis, Miša Mojca Cajnko, Matic Pavlin, Blaž Likozar, Saqib Rasool Chaudhry, Linda Pavitola, Valdis Priedols, Nida Şensoy Mercan, Berk Deniz, Sima Mammadova and Urfan Taghiyev

POSTERS

E-LEARNING AND DIGITAL LEARNING

PIL@B – INTERACTIVE LABORATORY PROGRAM 281
Alice Melo Ribeiro, Jamila Santos Khalifa, Samara Pereira Brito and Anna Luiza Moreira Lana de Almeida

SUSTAINABILITY, TECHNOLOGY AND EDUCATION

STRATEGIC OPTIONS ON STANDARDS ISSUES FOR ACHIEVING BUSINESS GOALS 284
Byoungho Jun

AUTHOR INDEX

FOREWORD

These proceedings contain the papers and posters of the: 19th International Conference on e-Learning and Digital Learning (ELDL 2025) and 13th International Conference on Sustainability, Technology and Education (STE 2025), which was organised by the International Association for Development of the Information Society, 23 - 25 July, 2025. These conferences are part of the 19th Multi Conference on Computer Science and Information Systems 2025, 23 - 25 July, which had a total of 608 submissions.

The e-Learning and Digital Learning (ELDL) 2025 conference aims to address the main issues of concern within e-Learning. This conference covers both technical as well as the non-technical aspects of e-Learning.

The conference accepted submissions in the following seven main areas: Organisational Strategy and Management Issues; Technological Issues; e-Learning Curriculum Development Issues; Instructional Design Issues; e-Learning Delivery Issues; e-Learning Research Methods and Approaches; e-Skills and Information Literacy for Learning.

The above referred main submission's areas are detailed below:

Organisational Strategy and Management Issues
- Higher and Further Education
- Primary and Secondary Education
- Workplace Learning
- Vocational Training
- Home Schooling
- Distance Learning
- Blended Learning
- Change Management
- Educational Management
- Continuous Professional Development (CPD) for Educational and Training Staff
- Return on e-Learning Investments (ROI)

Technological Issues
- Learning Management Systems (LMS)
- Managed Learning Environments (MLEs)
- Virtual Learning Environments (VLEs)
- Computer-Mediated Communication (CMC) Tools
- Social Support Software
- Architecture of Educational Information Systems Infrastructure
- Security and Data Protection
- Learning Objects
- XML Schemas and the Semantic Web
- Web 2.0 Applications

e-Learning Curriculum Development Issues
- Philosophies and Epistemologies for e-learning
- Learning Theories and Approaches for e-learning
- e-Learning Models
- Conceptual Representations
- Pedagogical Models
- e-Learning Pedagogical Strategies
- e-Learning Tactics
- Developing e-Learning for Specific Subject Domains

Instructional Design Issues
- Designing e-Learning Settings
- Developing e-Learning Pilots and Prototypes
- Creating e-Learning Courses
- Collaborative Learning
- Problem-based Learning
- Inquiry-based Learning
- Blended Learning
- Distance Learning
- Designing e-Learning Tasks
- E-learning Activities
- Online Groupwork
- Experiential Learning
- Simulations and Modelling
- Gaming and Edutainment
- Creativity and Design Activities
- Exploratory Programming
- Learning Analytics

e-Learning Delivery Issues
- e-Delivery in different contexts
  - Higher and Further Education
  - Primary and Secondary Schools
  - Workplace Learning
  - Vocational Training
  - Distance Learning
- Online Assessment
- Innovations in e-Assessment
- e-Moderating
- e-Tutoring
- e-Facilitating
- Leadership in e-Learning Delivery
- Networked Information and Communication Literacy Skills
- Participation and Motivation in e-Learning

e-Learning Research Methods and Approaches
- Action Research
- Design Research
- Course and Programme Evaluations
- Systematic Literature Reviews

- Historical Analysis
- Case Studies
- Meta-analysis of Case Studies
- Effectiveness and Impact Studies
- Evaluation of e-Learning Technologies
- Evaluation of Student and Tutor Satisfaction
- Learning and Cognitive Styles
- Ethical Issues in e-Learning

e-Skills and Information Literacy for Learning
- Teaching Information Literacy
- Electronic Library and Information Search Skills
- ICT Skills Education
  - in schools and colleges
  - for business, industry and the public sector
  - in adult, community, home and prison education
  - informal methods (peer groups, family)
- Education for Computer-mediated Communication Skills
  - Netiquette
  - Online safety for children and vulnerable users
  - Cybercrime awareness and personal prevention
- Student Production of Online Media
  - Web design
  - Digital storytelling
  - Web 2.0 tools
  - etc.
- Digital Media Studies

The International Conference on Sustainability, Technology and Education 2025 (STE 2025) has the purpose to address the main issues which occur by evaluating the relationship between Sustainability, Education and Technology. The main areas of interest for this conference are: Sustainability and Leadership, Sustainability and Green IT, Sustainability and Education.

These events received 165 submissions from more than 23 countries. Each submission has been anonymously reviewed by an average of four independent reviewers, to ensure that accepted submissions were of a high standard. Consequently, only 26 full papers were approved, which meant an acceptance rate of 16%. A few more papers were accepted as short and reflection papers, as well as posters. An extended version of the best papers will be selected for publishing in the Interactive Technology and Smart Education (ITSE) journal (ISSN:1741-5659), IADIS International Journal on WWW/Internet (ISSN: 1645-7641) and also in the IADIS International Journal on Computer Science and Information Systems (ISSN: 1646-3692). Other outlets may also receive extended versions of the best papers.

A successful conference requires the effort of many individuals. We would like to thank the members of the Program Committee for their hard work in reviewing and selecting the papers that appear in this book. We are grateful to all authors who submitted their papers to this conference and to the presenters who provided the substance of the meeting. We wish to thank all members of our organizing committee. We are especially grateful to Professor Dr. Piet Kommers, UNESCO Professor of Educational Technology (The Netherlands), for serving as the keynote speaker at MCCSIS 2025.

Last but not least, we hope that everybody enjoyed the presentations, and we invite all participants for next years´ edition of these conferences.

Miguel Baptista Nunes, Xi'an Jiaotong-Liverpool University, China
Pedro Isaias, Universidade Aberta (Portuguese Open University), Portugal
e-Learning and Digital Learning 2025 Conference Program Co-Chairs

Tomayess Issa, Curtin University, Perth, Australia
Theodora Issa, The University of Western Australia, Australia
Sustainability, Technology and Education 2025 Conference Program Co-Chairs

Piet Kommers, UNESCO Professor of Educational Technology, The Netherlands
Pedro Isaias, Universidade Aberta (Portuguese Open University), Portugal
MCCSIS 2025 General Conference Co-Chairs

July 2025

PROGRAM COMMITTEE

E-LEARNING AND DIGITAL LEARNING 2025
CONFERENCE PROGRAM CO-CHAIRS

Miguel Baptista Nunes, Xi'an Jiaotong-Liverpool University, China
Pedro Isaias, Universidade Aberta (Portuguese Open University), Portugal

SUSTAINABILITY, TECHNOLOGY AND EDUCATION 2025
CONFERENCE PROGRAM CO-CHAIRS

Tomayess Issa, Curtin University, Perth, Australia
Pedro Isaias, Universidade Aberta (Portuguese Open University), Portugal
Theodora Issa, The University of Western Australia, Australia

MCCSIS 2025 GENERAL CONFERENCE CO-CHAIRS

Piet Kommers, UNESCO Professor of Educational Technology, The Netherlands
Pedro Isaias, Universidade Aberta (Portuguese Open University), Portugal

E-LEARNING AND DIGITAL LEARNING 2025

CONFERENCE COMMITTEE MEMBERS

Adamantios Koumpis, University Hospital Cologne, Germany
Airina Volungeviciene, Vytautas Magnus University, Lithuania
Alejandro Sánchez, Universidad de Alicante, Spain
Ali Fawaz Shareef, The Maldives National University, Maldives
Aminath Zifna, The Maldives National University, Maldives
Ana Barata, ISEP-GILT, Portugal
Andreas Bollin, Klagenfurt University, Austria
Andreas Papasalouros, University of the Aegean, Greece
Angélica De Antonio, Universidad Politecnica de Madrid, Spain
Anthony Abrome, University of Bologna, Italy
Antoanela Naaji, Vasile Goldis West University of Arad, Romania
Antonio Navarro, Universidad Complutense de Madrid, Spain
Antonio Panaggio, Italian Ministry of Education, Italy
Apostolos Gkamas, University of Ioannina, Greece
Araz Yusubov, ADA University, Azerbaijan
Bertil P. Marques, Instituto Superior De Engenharia Do Porto (ISEP), Portugal
Bruno F. Gonçalves, Instituto Politécnico de Bragança, Portugal
Carolina Lorenzo, University of Alicante, Spain
Charalampos Karagiannidis, University of Thessaly, Greece
Christos Bouras, University of Patras, Greece
Copelia Mateo Guillén, Universidad de Alicante, Spain
Danielle Allessio, University Of Massachusetts, College Education, USA
Dara Tafazoli, University of Newcastle, Australia
Demetrios Sampson, University of Piraeus, Greece
Dessislava Vassileva, Sofia University "st. Kliment Ohridski", Bulgaria
Dimitra Pappa, National Center for Scientific Research - Demokrit, Greece
Egle Butkeviciene, Kaunas University of Technology, Lithuania
Eliza Stefanova, Sofia University, Bulgaria

Elliot Soloway, University of Michigan, USA
Esma Aimeur, University of Montréal, Canada
Esteban Vázquez Cano, Spanish National University of Distance Education, Spain
Eugenia Kovatcheva, University of Library Studies and Information Tech, Bulgaria
Eva Jereb, University of Maribor, Slovenia
Francesca Pozzi, Instituto Tecnologie Didattiche – CNR, Italy
Geoffrey Lautenbach, University of Johannesburg, South Africa
George Tsihrintzis, University of Piraeus, Greece
Giovanni Bonaiuti, University of Cagliari, Italy
I Putu Indra Kusuma, Universitas Pendidikan Ganesha, Indonesia
Ibrahim Ahmed, University of Bahrain, Bahrain
Igor Bernik, University of Maribor, Slovenia
Jan-Torsten Milde, Hochschule Fulda, Germany
Jesús Sánchez Allende, CUNEF Universidad, Spain
Joan Ballantine, University of Ulster, United Kingdom
Jose Bidarra, Universidade Aberta, Portugal
Juan M. Santos-Gago, atlanTTic - University of Vigo, Spain
Khalid El Khadiri, Chouaib Doukkali University, Morocco
Larbi Esmahi, Athabasca University, Canada
Luh Indrayani, Universitas Pendidikan Ganesha, Indonesia
Luis Álvarez-González, Universidad Austral de Chile, Chile
Made Hery Santosa, Universitas Pendidikan Ganesha, Indonesia
Maiga Chang, Athabasca University, Canada
Malinka Ivanova, Technical University of Sofia, Bulgaria
Manuel Caeiro-Rodríguez, University of Vigo, Spain
Marina Rui, University of Genoa, Italy
Martin Wessner, Hochschule Darmstadt, Germany
Martín Llamas-Nistal, atlanTTic Research Center - University of Vigo, Spain
Martina Benvenuti, University of Bologna, Italy
Michail Kalogiannakis, University of Thessaly, Greece
Nicola Capuano, University of Salerno, Italy
Patrick Blumschein, Teacher University of Freiburg, Germany
Paula Escudeiro, ISEP/GILT, Portugal
Paula Miranda, School of Technology, Setubal Polytechnic University, Portugal
Piedade Carvalho, ISEP/GILT, Portugal
Rafael Morales Gamboa, Universidad de Guadalajara, Mexico
Ritesh Chugh, CQUniversity, Australia
Rosabel Martínez Roig, University of Alicante, Spain
Roza Ibrahim, The Maldives National University, Maldives
Rubén Edel Navarro, Universidad Veracruzana, Mexico
Sangeetha Kutty, Cquniversity, Australia
Shelia Anjarani, Universitas Muhammadiyah Purwokerto, Indonesia
Tamilla Mammadova, ADA University, Azerbaijan
Tarek Assassi, University of Mohamed Kheider, Algeria
Telmo Zarraonandia, Universidad Carlos III de Madrid, Spain
Thomas Zarouchas, Computer Technology Institute and Press "diophantu, Greece
Vahid Norouzi Larsari, Charles University, Czech Republic
Valentina Dagiene, Vilnius University, Lithuania
Vassilis Triantafyllou, University of Peloponnese, Greece
Verónica Onrubia, Universidad de Alicante, Spain
Veronika Makarova, University of Saskatchewan, Canada
Yannis Psaromiligkos, Technological Education Institute of Piraeus, Greece

SUSTAINABILITY, TECHNOLOGY AND EDUCATION 2025 CONFERENCE COMMITTEE MEMBERS

Agnieszka Szplit, Jan Kochanowski University, Poland
Anita Pipere, Daugavpils University, Latvia
Cecilia Haskins, Norwegian University of Science and Technology, Norway
Dzintra Ilisko, Daugavpils University, Institute of Sustainable Ed, Latvia
Eleftheria Alexandri, Hellenic Open University, Greece
Ilona Fjodorova, Daugavpils University, Latvia
Khaled Galal Ahmed, United Arab Emirates University, United Arab Emirates
Lindita Bande, United Arab Emirates University, United Arab Emirates
Maatouk Khoukhi, United Arab Emirates University (UAEU), United Arab Emirates
Madalina Calbureanu, University of Craiova, Romania
Mahesh (Michael) Raisinghani, TWU College of Business, USA
Mahmoud Khalifa, ALM College for Higher Education, UK
Maria Bostenaru, University of Architecture and Urbanism, Romania
Martin Dennis Scoppa, United Arab Emirates University, United Arab Emirates
Maziar Asefi, Toronto Metropolitan University, Canada
Mike Burbridge, Curtin University, Australia
Mohamed Mahgoub, United Arab Emirates University, United Arab Emirates
Nena Vukelić, University of Rijeka, Croatia
Pankaj Upadhyay, Indian Institute of Technology Guwahati, India
Peter Strachan, Aberdeen Business School, United Kingdom
Richie Moalosi, University of Botswana, Botswana
Shobhana Madhaven, University of Westminster, United Kingdom
Sorin-Codrut Florut, Politehnica University of Timisoara, Romania
Umut Turker, Eastern Mediterranean University, North Cyprus
Walid Oueslati, Université de Carthage, Tunisia
Wolfgang Lohmann, Randstad Digital Germany AG, Deutschland

KEYNOTE LECTURE

A.I. FOR STUDENTS' METACOGNITIVE SKILLS AND CREATIVE THINKING

Prof. Dr. Piet Kommers
UNESCO Professor of Educational Technology,
The Netherlands

ABSTRACT

Similar to the advancements made in ICT three decades ago, the emergence of AI today signals to educators that the human learning process may be at risk. This keynote lecture will provide an additional perspective on this issue. It is crucial to recognize that the WWW supplies all the information that AI currently presents. The main difference lies in its rhetoric concerning establishing a specific viewpoint. In other words, AI now acts as a powerful catalyst for refining criteria for student assessments, concentrating on discovering unique creative solutions rather than merely elaborating on existing ones. The foundation of creative thinking is metacognition: the awareness of what one already knows, one's preferred thinking style, and an evolving intuition about how learning builds on prior knowledge. This lecture will outline a roadmap for innovative didactic, instructional, and curricular approaches in the future. Previous research into conceptual representations, ontologies, and metaphoric analogies has supported education in expanding the fields of knowledge elicitation and conceptual thinking. Once AI is integrated, we will witness a growing need among young people to navigate complexities, such as sources of information that do not rely on empirical and validated scientific evidence. This keynote will highlight new aspirations that have emerged and how to seamlessly incorporate them into 'standing education' at this very moment. Topics such as teacher education, student selection, and assessment will be discussed.

Full Papers

LEARNING OUTCOMES AND QUALITY OF EDUCATIONAL SERVICE AS PREDICTORS OF ACADEMIC SATISFACTION

Olger Gutierrez Aguilar, Aleixandre Duche Pérez, Anani Gutierrez Aguilar
and Sandra Chicana Huanca
Universidad Católica de Santa María
Arequipa, Perú

ABSTRACT

This research aims to explore the quality of educational services and learning outcomes as predictors of academic satisfaction in university students. It investigates how these dimensions contribute to academic satisfaction, considering the imperative need to optimise teaching processes and support infrastructure and the competencies developed during the student's education. The methodology used is a quantitative explanatory approach using a non-experimental sectional design. We surveyed 555 undergraduate students from various degree programmes at a private university in Arequipa, Peru. We used an instrument designed to include digital competencies, social responsibility, professional and personal competencies, quality of teaching and academic resources and support. The most important result shows that professional and personal competencies have a positive and significant relationship with academic satisfaction, highlighting the critical importance of these competencies in higher education. Likewise, the quality of academic resources and teaching had the highest coefficients in the analysis, highlighting their substantial influence on student satisfaction. It is concluded that the quality of educational services and learning outcomes are strong predictors of academic satisfaction, generating motivation and perception of relevance in the training received both for the professional environment and for society in general. These results reinforce the need for universities to adopt a holistic approach that fosters diverse competencies and continuously improves the quality of educational services to optimise the student experience and satisfaction.

KEYWORDS

Quality of Service, Learning Outcomes, Teaching, Digital Competences, Social Responsibility, Personal Competences

1. INTRODUCTION

Professional training and academic satisfaction are fundamental aspects of the university experience, directly impacting student development and institutional effectiveness. Recent research, such as that of Soliba et al. (2025), has shown that the quality of the educational service and learning outcomes play a crucial role in student satisfaction. This study seeks to delve deeper into how these dimensions interact and contribute to academic satisfaction, addressing the imperative need to optimize both teaching processes and the support infrastructure and the skills developed during student training, as noted in the works of Abdolalipour et al. (2024).

In the university context, the quality of the educational service covers a wide range of factors, from the effectiveness of teaching and the availability of academic resources to institutional support and the integration of digital and social skills in the curriculum, as suggested by Rulinawaty et al. (2024). Furthermore, learning outcomes are considered to reflect the effectiveness of study programs and the capacity of institutions to prepare professionals who respond to the contemporary demands of the labour market and society at large (Hirsch & Rubach, 2024; Rodriguez-Gomez et al., 2024). This multidimensional approach to educational service quality and learning outcomes provides a valuable framework to explore their correlation with academic satisfaction, an area that has captured the interest of numerous scholars.

In conceptual terms, educational service quality is broken down into several key dimensions, including the quality of teaching, recognized by Khan (2024) as a primary factor, and the availability of resources and institutional support, highlighted by Meyer et al. (2024). This study aims to analyze how these categories interact with each other and with learning outcomes to form an educational environment that not only meets students' academic and professional expectations but also fosters high student satisfaction and engagement. By integrating these various dimensions, the study aims to offer more profound and applicable insights to improve the quality of higher education and its impact on students.

2. THEORETICAL BACKGROUND

2.1 Quality of Teaching

Teaching quality refers to the effectiveness of teaching practices in transmitting content, promoting active participation, facilitating student involvement, and adapting pedagogical strategies to the characteristics of each group (Smit et al., 2024). According to Benarroch et al. (2024), the professional development of teachers – which includes individual satisfaction and collaboration with peers – directly affects how classroom activities are planned and executed.

Along these lines, the work of Jurkowski and Abramczyk (2024) highlights that evidence-based continuing education, when accompanied by a collaborative component, improves self-efficacy beliefs and the quality of teaching practice. Such collaboration, for example, can enhance the implementation of cooperative learning methodologies and be reflected in better student performance and satisfaction with the educational process (Jurkowski & Abramczyk, 2024).

2.2 Academic Resources and Institutional Support

Regarding the provision of resources and institutional support, the literature indicates that physical infrastructure, teaching resources and, increasingly, virtual teaching and learning environments are essential to ensure the effectiveness of academic activity (Rulinawaty et al., 2024). The use of learning management platforms (LMS) and the availability of technological support directly influence students' perception of the institution's quality, strengthening their satisfaction and motivation to act independently (Rulinawaty et al., 2024).

In addition, adequate academic guidance and mentoring positively impact the development of competencies and the achievement of training objectives. For example, in the teaching of obstetrics and gynaecology, Meyer et al. (2024) emphasize the need for feedback and supervision structures that guarantee the acquisition of practical skills, aligning the quality of the service with competency-based education standards.

On the other hand, a safe learning environment that provides socio-emotional support is considered a determining factor for well-being and academic satisfaction. According to Shikuku et al. (2024), in contexts with high practical demand such as the training of health professionals not only adequate infrastructures but also consistent support policies for the development of clinical practice and the training of future graduates are required.

2.3 Learning Outcomes as a Quality Framework

The concept of learning outcomes encompasses objectives and competencies that students must achieve at the end of their training, thus cementing the relevance of studies on teaching and assessment strategies aimed at such purposes (Smit et al., 2024). These results can be divided into two large analytical categories:

2.3.1 Professional and Personal Competences

Professional and personal competencies refer to the knowledge, skills and attitudes that enable future graduates to perform effectively in their curricular field (Rodriguez-Gomez et al., 2024). These competencies include the ability to reflect on decision-making, the ability to work collaboratively and the responsibility to manage their learning process autonomously (Getenet et al., 2024).

The development of professional competence involves the reflective application of acquired knowledge through dynamic methodologies that promote critical thinking and adaptation to changing contexts (Benarroch et al., 2024). In addition, integrating self-assessment and reflection in teaching practice itself – for those training as future teachers – is essential to adjust teaching strategies and promote learning effectiveness (Rodriguez-Gomez et al., 2024). Self-regulation and proactivity are emphasised as key traits since they offer the possibility of planning, monitoring and evaluating academic progress on an ongoing basis (Rodriguez-Gomez et al., 2024). Thus, higher education institutions that promote strengthening these competencies tend to foster an environment that favours students' intrinsic motivation and academic satisfaction.

2.3.2 Digital Competences and Social Responsibility

The increasing digitalisation of educational environments has highlighted the need for future graduates to master technological tools and understand their use's ethical and social implications (Getenet et al., 2024; Hirsch & Rubach, 2024). According to Hirsch and Rubach (2024), teaching strategies that combine traditional and advanced methodologies allow students to boost their confidence in their digital capabilities, strengthening their professional development in a world characterised by rapid technological evolution.

Likewise, training in social responsibility, understood as recognising cultural diversity and equity and including the perspective of sustainability in the academic sphere, promotes the construction of more fair and inclusive educational spaces (Ulbricht et al., 2024). To this end, university programs must incorporate content that allows for reflection on the social role of the future professional so that the student becomes an active agent within his or her community.

In contexts where service quality is mediated by digital education and social responsibility, it has been found that implementing appropriate platforms and training oriented towards community service increases student motivation and satisfaction (Khan, 2024; Rulinawaty et al., 2024). In particular, the availability of ongoing training in both digital and citizenship skills favours adaptability, resilience, and active participation in the construction of collaborative teaching-learning environments (Hirsch & Rubach, 2024).

2.4 Academic Satisfaction and its Link to Quality and Learning Outcomes

Academic satisfaction represents, in general terms, the degree of well-being that students experience about the quality of teaching, the content offered, and the perceived usefulness of their training (Soliba et al., 2025). As Abdolalipour et al. (2024) point out, the factors that promote academic satisfaction include teaching characteristics, the curriculum's relevance, and the set of social and institutional supports.

This construct is interrelated with other aspects, such as motivation, mental health, and emotional connection with the teaching staff (Soliba et al., 2025). The higher the quality of the educational experience, the greater the likelihood that students recognize value in their training process. Therefore, when examining academic satisfaction, it is key to integrate perceptions of how the institution meets student needs and how the study programs promote key competencies for professional insertion and civic life (Wilkie et al., 2024).

From a continuous improvement perspective, evaluating academic satisfaction provides relevant information for institutional decision-making, the optimization of teaching planning and the redesign of curricula (Meyer et al., 2024). In turn, a high level of satisfaction may reflect either the effectiveness of the institution in providing adequate infrastructure and teaching or the ability of the student to self-regulate their learning process, provided that they have the necessary support (Soliba et al., 2025).

2.4.1 Theoretical Relationship Between Quality of Educational Service, Learning Outcomes and Academic Satisfaction

The literature suggests that the correlation between educational service quality and learning outcomes can largely determine academic satisfaction (Abdolalipour et al., 2024; Soliba et al., 2025). If the student perceives that the training environment offers real possibilities for professional growth and encourages the mastery of transferable skills, this positive impact translates into higher satisfaction levels (Rodriguez-Gomez et al., 2024).

In terms of teaching, it has been proven that experimenting with evidence-based methodologies, critical reflection and collaboration contribute to developing skills that reinforce students' self-confidence and motivation (Jurkowski & Abramczyk, 2024). This is associated with both the dimension of service quality (due to the methodology and support offered) and the dimension of learning outcomes (by consolidating transferable skills).

On the other hand, in evaluating service quality and learning outcomes, a technological component appears that can improve the implementation of curricula and enhance ongoing training (Rulinawaty et al., 2024). From an institutional perspective, the effective use of an LMS and the availability of timely advice increase the perception of quality (Getenet et al., 2024). In turn, developing students' digital skills and social responsibility promotes greater participation and commitment to their educational process (Ulbricht et al., 2024).

Therefore, the literature converges on the idea that the quality of teaching, the provision of adequate resources, and the acquisition of professional, personal, digital, and social skills are integrated to impact student satisfaction. This binomial – represented by the quality of the service and the learning outcomes – is positioned as a solid predictor of academic satisfaction by generating motivation, a sense of belonging, and the perception that the training received is relevant both for the professional field and for society in general (Hirsch & Rubach, 2024; Rulinawaty et al., 2024; Soliba et al., 2025). Therefore, this theoretical model maintains that the synergy between the quality factors of the educational service (teaching and resources) and the learning outcomes (professional, personal, digital skills and social responsibility) directly promotes academic satisfaction, constituting the articulating axis of the student's educational experience. Based on the preceding theoretical considerations, the following hypotheses are proposed:

H1: Digital skills and social responsibility positively affect students' academic satisfaction.

H2: A positive association exists between professional and personal skills and students' academic satisfaction.

H3: The quality of teaching is positively associated with students' academic satisfaction.

H4: Academic resources and support quality is positively associated with academic satisfaction.

3. METHODOLOGY

The research adopted an explanatory quantitative approach through a non-experimental sectional design. The sample for the study consisted of 555 university students from the professional careers of Industrial Engineering 260 (46%); Veterinary Medicine 103 (19%); Obstetrics and Childcare 55 (10%); and Advertising and Multimedia 137 (25%). The distribution by sex was 40% men (N = 219) and 60% women (N = 336), with ages between 16 and 25 years, with an average of 18.5 years (Standard Deviation = 1.78). The research was carried out at a private university in Arequipa, Peru.

The data were obtained through a survey conducted between November and December 2024. Before applying the instrument, the participants' authorization was previously obtained through a consent form. The instrument was specifically designed for the study. The final version consists of 4 predictor variables: Digital competencies and Social Responsibility (DCSR), Professional and Personal Competencies (PPC), Quality of Service: Teaching (QST) and Quality of Service: Academic Resources and Support (QSARS) and as an endogenous variable, Academic Satisfaction (SAT). To measure the variables, a 5-point Likert scale was used to evaluate the responses: (1) Strongly Disagree (2) Disagree (3) Neutral (4) Agree (5) Strongly Agree. SmartPLS (v: 4.1.0.9) for data processing and analysis, applying the multiple regression analysis module.

4. RESULTS

Table 1. Summary coefficients

	Unstandardized coefficients	Standardized coefficients	SE	T value	P value	2.5 %	97.5 %
Digital Competences and Social Responsibility (DCSR)	0.027	0.027	0.049	0.543	0.588	-0.070	0.123
Professional and Personal Competences (PPC)	0.275	0.275	0.064	4.329	0.000	0.150	0.400
Quality of Service: Academic Resources and Support (QSARS)	0.361	0.361	0.056	6.430	0.000	0.251	0.472
Quality of Service: Teaching (QST)	0.264	0.264	0.057	4.630	0.000	0.152	0.376
Intercept	0.000	0.000	0.030	0.000	1.000	-0.059	0.059

The results in Table 1 indicate that digital competencies and social responsibility (DCSR) do not significantly impact academic satisfaction, reflected by an unstandardized coefficient of 0.027 and a p-value of 0.588. This finding points out that, although these competencies are relevant in the modern educational context, they are not determining factors in the perception of satisfaction among students within the framework of this study. This lack of significance could indicate the need to integrate these competencies more effectively into academic activities or review how they are valued in the educational experience.

In contrast, professional and personal competencies (PPC) show a positive and significant relationship with academic satisfaction, with an unstandardized coefficient of 0.275 and a significant p-value of 0.000. This result underlines the critical importance of these competencies in higher education, highlighting that students value leadership, effective communication, and adaptability, which are directly related to greater academic satisfaction. The strength and significance of this link indicate that interventions to improve these competencies could be a strategic approach to elevate the quality of the educational experience.

Consequently, the quality of educational service, specifically in terms of academic resources and support (QSARS) and quality of teaching (QST), presents the highest and most significant coefficients in the analysis, with values of 0.361 and 0.264 respectively, both with p-values of 0.000. These results highlight the substantial influence of educational service quality on student satisfaction. The availability and quality of academic resources, along with teaching methods that are viewed as effective and empathetic., are identified as the main drivers of satisfaction, indicating that policies and practices that strengthen these aspects can directly and positively impact students' overall perception of their education.

Table 2. Summary ANOVA

	Sum square	df	Mean square	F	P value
Total	304.000	303	0.000	0.000	0.000
Error	82.430	299	0.276	0.000	0.000
Regression	221.570	4	55.392	200.925	0.000

The results obtained in Table 2 are explained through an analysis of variance (ANOVA) that validates the significance of the multiple regression model used in this research. The analysis reveals that the model is statistically significant, with an F-value of 200.925 and a p-value of 0.000, indicating that the independent variables (professional and personal skills, digital skills and social responsibility, and quality of educational service) have a substantial and collective effect on students' academic satisfaction. This high F-value shows that the model has a powerful predictive capacity, highlighting the relevance of improving and focusing educational interventions in these areas to optimize the learning experience and student satisfaction.

Furthermore, the decomposition of the total variability, shown in the regression sum of squares (221.570) and the error sum of squares (82.430), emphasizes that a large proportion of the variation in academic satisfaction can be explained by the variables selected in the model. This confirms the adequacy of the model and the importance of each variable included in the regression. These findings are fundamental for formulating educational policies and teaching strategies that specifically focus on the areas identified as most influential, thus ensuring that educational institutions can effectively improve academic satisfaction and learning outcomes in contemporary educational environments.

Table 3. Correlations

	(DCSR)	(PPC)	(QSARS)	(QST)	(SAT)
Digital Competences and _Social Responsibility (DCSR)	1.000	0.787	0.622	0.607	0.628
Professional and Personal Competences (PPC)	0.787	1.000	0.767	0.775	0.778
Quality of Service: Academic _Resources and Support (QSARS)	0.622	0.767	1.000	0.816	0.804
Quality of Service: Teaching (QST)	0.607	0.775	0.816	1.000	0.788
Satisfaction (SAT)	0.628	0.778	0.804	0.788	1.000

Table 3 shows the correlations between the investigated variables, which are crucial to understanding how the different variables interact with each other and their relationship with academic satisfaction (SAT). Correlations between Digital competencies and Social Responsibility (DCSR) and other variables: With Professional and Personal competencies (PPC): 0.787, indicating a strong correlation, suggesting that students who score high in digital competencies and social responsibility also tend to have high personal and professional competencies. With Quality of Service: Academic Resources and Support (QSARS): 0.622, a moderate correlation indicates that a positive perception of digital competencies and social responsibility is moderately associated with the perception of the quality of academic resources and support. With Quality of Service: Teaching (QST): 0.607, also a moderate correlation, showing a positive relationship between the perception of digital competencies and the quality of teaching. With Satisfaction (SAT): 0.628, a moderate correlation, suggests that digital competencies and social responsibility are moderately related to academic satisfaction.

Professional and Personal Competences (PPC) correlations: With Quality of Service: Academic Resources and Support (QSARS): 0.767, a strong correlation implies that higher personal and professional competencies are related to a better perception of academic resources and support. With Quality of Service: Teaching (QST): 0.775, equally strong, suggesting that personal and professional competencies are closely linked to the perceived quality of teaching. With Satisfaction (SAT): 0.778, a strong correlation, indicating that personal and professional competencies are important predictors of academic satisfaction.

Correlations for Quality of Service: Academic Resources and Support (QSARS) and Quality of Service: Teaching (QST): A robust correlation between QSARS and QST: 0.816, highlighting that perceptions about academic resources and teaching quality are highly interrelated. QSARS with Satisfaction (SAT): 0.804 and QST with SAT: 0.788 are both strong correlations, highlighting that both the quality of academic resources and the quality of teaching are key determinants of academic satisfaction.

These results reflect that all variables have significant correlations with academic satisfaction, personal and professional skills, and educational service quality (both resources and teaching), which show the strongest correlations. This implies that these areas are vital to improving academic satisfaction and should be the priority focus of intervention and improvement in educational environments.

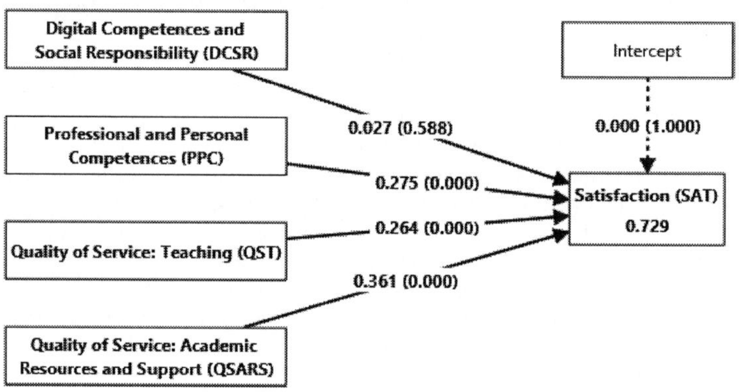

Figure 1. Regression model with standardised coefficients and significance level p-values

Figure 1 illustrates a regression model assessing the influence of various variables on academic satisfaction. The results highlight that professional and personal competencies (PPC) significantly and positively impact satisfaction, with a standardized coefficient of 0.275 and a p-value of 0.000, indicating that these competencies are essential elements to improve students' perception of satisfaction. Similarly, quality of teaching (QST) and academic resources and support (QSARS) also show positive and significant effects on satisfaction, with coefficients of 0.264 and 0.361, respectively. This finding underlines the critical importance of providing high-quality teaching and robust support resources as determining factors in students' educational experiences.

On the other hand, digital competencies and social responsibility (DCSR) present a coefficient of 0.027 with a p-value of 0.588, indicating that they do not significantly impact academic satisfaction in this study. The intercept of the model is zero and not statistically significant, suggesting that, without the influence of the independent variables, predicted academic satisfaction would be zero. This analysis highlights priority areas where interventions and improvements can be most effective in raising academic satisfaction, focusing mainly on improving personal and professional skills and the quality of educational support.

5. CONCLUSIONS

This study has revealed significant findings regarding the determinants of academic satisfaction in a university context. Professional and personal competencies were found to have a direct and robust relationship with student satisfaction, underlining their crucial role in higher education. These results imply that competencies facilitating effective project management, efficient communication, and adaptability are essential for student well-being and satisfaction, highlighting the importance of integrating these aspects into educational curricula.

On the other hand, the quality of academic resources and support and the quality of teaching emerged as the strongest predictors of academic satisfaction, both with significant and positive influences. These findings suggest that investments in the quality of teaching and the provision of adequate resources and support are fundamental for academic satisfaction and strategic for long-term institutional development. Interestingly, digital competencies and social responsibility did not significantly impact academic satisfaction, which could indicate the need for a more strategic alignment of these competencies with students' academic and professional goals. These findings reinforce the need for universities to adopt a holistic approach that fosters various competencies and continuously improves the quality of educational services to optimize student experience and satisfaction.

However, the study presents limitations that must be considered when interpreting the results. The absence of a significant impact of digital competencies and social responsibility on academic satisfaction suggests that these elements might be underutilized or poorly integrated in the current curriculum context. Furthermore, the methodology used is based on a cross-sectional design, which limits the ability to establish definitive causal relationships between the variables studied. These limitations indicate areas for future research, especially in longitudinal design and in exploring how digital competencies and social responsibility can be effectively incorporated to improve student satisfaction.

In terms of practical applications, the study's results can guide educational institutions in optimizing their pedagogical and administrative strategies. Implementing improvements in the quality of teaching and the provision of resources can be an effective strategy to increase satisfaction and, consequently, student retention. Likewise, skills development programs must be reviewed and strengthened, ensuring that they address the comprehensive needs of students and prepare graduates for professional and personal challenges. This approach enhances the educational experience and prepares students to contribute effectively in their future work environments and communities.

ACKNOWLEDGEMENT

We thank the Vice-Rectorate of Research of the Catholic University of Santa María de Arequipa for its support and the academic and administrative facilities it provided for this research.

REFERENCES

Abdolalipour, S., Namdar-Areshtanab, H., Ghaffarifar, S., Ghaffari, R., & Mirghafourvand, M. (2024). Exploring the experiences of medical students regarding the factors affecting academic achievement: a qualitative study [Article]. BMC Medical Education, 24(1), Article 1292. https://doi.org/10.1186/s12909-024-06294-1

Benarroch, A., Briceño-Martínez, J. J., & Bernal-Ballen, A. (2024). Analysis of Conceptions and Beliefs, Argumentative Teaching Practices, and Reflection of University Teachers': A Case Study on Teachers' Professional Development [Article]. SAGE Open, 14(3). https://doi.org/10.1177/21582440241281352

Getenet, S., Haeusler, C., Redmond, P., Cantle, R., & Crouch, V. (2024). First-year Preservice Teachers' Understanding of Digital Technologies and Their Digital Literacy, Efficacy, Attitude, and Online Learning Engagement: Implication for Course Design [Article]. Technology, Knowledge and Learning, 29(3), 1359-1383. https://doi.org/10.1007/s10758-023-09724-z

Hirsch, A. K., & Rubach, C. (2024). Exploring the Interplay Between Teaching Strategies and Digital Competencies Beliefs Among Pre-Service Teachers: A Longitudinal Study [Article]. Education Sciences, 14(12), Article 1342. https://doi.org/10.3390/educsci14121342

Jurkowski, S., & Abramczyk, A. (2024). Collaboration in in-service teacher training - the missing link between empirical evidence and practice? [Article]. Journal of Education for Teaching, 50(4), 550-563. https://doi.org/10.1080/02607476.2024.2327062

Khan, I. A. (2024). Innovative Pedagogies and Role of Teachers for Quality Teacher Education in Saudi Arabia [Article]. Education and Self Development, 19(3), 28-43. https://doi.org/10.26907/esd.19.3.03

Meyer, B., Riedel, F., Amann, N., Graf, A., Stuehrenberg, A., Ritter, V., Wallwiener, M., Heublein, S., Recker, F., Weiss, M., & Riedel, M. (2024). Exploring the current state of clinical and practical teaching in obstetrics and gynecology in the era of competency-based education: a nationwide survey among German teaching coordinators [Article]. BMC Medical Education, 24(1), Article 165. https://doi.org/10.1186/s12909-024-05138-2

Rodriguez-Gomez, D., Muñoz-Moreno, J. L., & Ion, G. (2024). Empowering Teachers: Self-Regulated Learning Strategies for Sustainable Professional Development in Initial Teacher Education at Higher Education Institutions [Article]. Sustainability (Switzerland), 16(7), Article 3021. https://doi.org/10.3390/su16073021

Rulinawaty, Samboteng, L., Purwanto, A. J., Kuncoro, S., Jasrial, Tahilili, M. H., Efendi, Y., & Karyana, A. (2024). Investigating the influence of the updated DeLone and McLean information system (IS) success model on the effectiveness of learning management system (LMS) implementation. Cogent Education, 11(1), 2365611. https://doi.org/10.1080/2331186X.2024.2365611

Shikuku, D. N., Mwaura, C., Nandikove, P., Uyara, A., Allott, H., Waweru, L., Nyaga, L., Tallam, E., Bashir, I., Ndirangu, E., Bedwell, C., Bar-Zeev, S., & Ameh, C. (2024). An evaluation of the effectiveness of an updated pre-service midwifery curriculum integrated with emergency obstetric and newborn care in Kenya: a cluster randomised controlled trial [Article]. BMC Medical Education, 24(1), Article 1562. https://doi.org/10.1186/s12909-024-06581-x

Smit, E., Tuithof, H., & Béneker, T. (2024). Portraying the developing PCK of Dutch pre-service geography teachers [Article]. International Research in Geographical and Environmental Education, 33(3), 177-192. https://doi.org/10.1080/10382046.2023.2281652

Soliba, B. L., Rivera, M. L. A. D., Marquez, A. J. M., Marquez, A. M. B., Ocumen, J. P., Ottao, B. B., Pedro, R. B., Posot, M. T., Ramos, K. E., & Rivera, G. R. E. (2025). Perceived Academic Satisfaction of Fourth-Year Pre-Service Teachers [Article]. SAGE Open, 15(1). https://doi.org/10.1177/21582440241293030

Ulbricht, J., Schachner, M. K., Civitillo, S., & Juang, L. (2024). Fostering Culturally Responsive Teaching Through the Identity Project Intervention: A Qualitative Quasi-Experiment with Pre-Service Teachers [Article]. Identity, 24(4), 307-330. https://doi.org/10.1080/15283488.2024.2361890

Wilkie, B., Santana Cáceres, P. J., Marchena, J. M., & Jordan, A. (2024). Is the development of physical literacy ubiquitous in high-quality physical education? [Article]. European Physical Education Review, 30(1), 36-50. https://doi.org/10.1177/1356336X231179344

EXPLORATION ON THE LEARNING PROCESS: A COMPARATIVE STUDY OF HIGH AND LOW PRE-TEST PERFORMANCE

Wannisa Matcha
Department of Computer and Informatics for Management, Faculty of Communication Sciences, Prince of Songkla University
181 Jaroenpradith Road, Rusamilae, Muang, Pattani 94000, Thailand

ABSTRACT

This paper presents the exploration of the high and low performers captured by using Multiple Choices-based Questions (MCQs) and students' learning processes. The class was conducted in a face-to-face setting, but all learning materials were provided in the Learning Management System (LMS). By using the mean score of the pre-tests as a threshold, students are divided into two groups i.e., those who scored below average and above average. Both groups received the same treatment. The process mining technique was used to examine these two groups' learning processes. The results revealed that those who scored below average during the pre-test focused more on additional learning and practical materials. On average, they also obtained higher mid-term and final scores. Meanwhile, those who score above the average on their pre-tests tend to obtain lower scores in their midterm and final exams. They paid more attention to the solutions of the assignment. However, the statistical tests suggested no significant differences between the two groups in terms of the midterm test and final exam scores.

KEYWORDS

Educational Data Mining, Learning Analytics, Learning Process, Mining Process, Prior Knowledge, Self-Regulated Learning

1. INTRODUCTION

Learning is a process of constructing new knowledge by connecting new information with the existing one. In Thailand's higher educational context, students will be given a lecture – usually face-to-face. They will be provided with some learning materials that are usually made available online via LMS. Students are also expected to spend some time regulating their own learning after the class. Researchers and educators have long studied to understand the learning process. Among the educational theories that try to understand the student's learning process is Self-Regulated Learning (SRL). SRL is described as the repetitive process of learning where learners need to make judgments on their learning, set goals and expectations, select the appropriate learning tactics and strategies, and monitor and control their own learning (Winne & Hadwin, 1998; Zimmerman, 2008). Based on the SRL point of view, learners direct their learning by considering their cognitive ability as well as the environment they are in. This is known as internal and external conditions in the SRL model (Winne & Hadwin, 1998). Examples of internal cognitive conditions are prior knowledge, self-efficacy, motivation, goal orientation, beliefs, and any construct related to the cognitive process (Greene & Azevedo, 2007). The external conditions are time constraints, communication cues, available learning materials, social context and other external cognitive conditions. Taking into account these conditions, students make decisions on what learning tactics and strategies they can apply in a given situation. Hence, learning products such as essays, reports, or grades can be produced. Students then evaluate their results against their expectations and update the conditions, standards, and learning goals.

As mentioned, the prior knowledge is of the cognitive conditions that are often reported to have an association with learning performance (Brod, 2021; van Kesteren et al., 2014). That is, those students who score higher in their prior knowledge assessment tend to have a higher score in academic performance as compared to those who reported having a lower level of prior knowledge. Similarly, Hailikari, (2009) found that prior knowledge is a predictive factor of final grade. McCarthy & McNamara, (2021) also found that prior knowledge affects text comprehension in various dimensions. Therefore, research strongly suggested that the student's prior knowledge should be measured before teaching some new content (Brod, 2021).

A pre-test is often used to measure the students' level of prior knowledge and skills before they are exposed to the course (Delucchi, 2014). Results from the pre-tests could help the teachers to identify what students have already know, and what they need to learn. Hence, teachers can plan the course and identify the intervention needed to support learning. Not only do pre-tests help the teachers in planning the instruction but it can also help students to be motivated and reflect on their own learning. The pre-test can also be used as a priming strategy. By priming the pre-test, students will be aware of what they are expected to learn in the particular course.

Yet much is unknown regarding how students reflect on the pre-tests (Brod, 2021). Some students used the pre-test to drive their learning. Yet, there are some students who may not recognize the value of prior knowledge and pre-tests. Hence, there was a mixed report on the effect of using pre-tests to capture prior knowledge. This study aims to add more insights into the research by examining the student's learning process based on the pre-tests. That is, students were divided into two groups based on their pre-tests' scores designed by using the multiple-choice questions (MCQs) to capture the knowledge before attending the course. Division of students into two groups based on the scores enables them to compare the learning process.

2. LITERATURE REVIEW

The following section reviews the related works relevant to the study on pre-test and prior knowledge as well as learning process. The paper also discusses existing studies that have employed process mining techniques to analyze and gain insights into the learning process. These works illustrate how process mining can uncover hidden patterns of students' learning behaviors.

2.1 Pre-Tests and Prior Knowledge

Pre-tests are often suggested to be included when teachers design the course (Brod, 2021). A pre-test is an instrument that can be used to capture the domain's prior knowledge (Delucchi, 2014). Additionally, lecturers use pre-tests to classify the groups of students to plan the course. Some use pre-tests together with post-tests to measure how much knowledge is gained after the students participated in learning (Delucchi, 2014; Gouldthorpe & Israel, 2013). Some also used pre-tests as a priming strategy. Priming strategy refers to the teaching technique where teachers give a hint of what students are expected from the course (Schreibman et al., 2000). For instance, Gulboy & Yucesoy-Ozkan, (2022) primed the video content to help Autism Spectrum Disorder (ASD) students learn. They found that by priming the video, the learning performance was improved. This is because priming gives some idea to the students on what is expected to happen; hence, students are aware of the situation and can prepare themselves. Priming the domain knowledge by using pre-tests also could offer similar insights. Yet not much research explores the role of pre-test as a priming initiator.

The most common purpose of using pre-tests is to measure prior knowledge. Based on Winne & Hadwin's COPES model describing the SRL process and its influential factors, prior knowledge plays a significant role in students' knowledge development process (Winne & Hadwin, 1998). In fact, many scholars highlighted that when students acquire new knowledge, they do it by connecting the new information to the prior knowledge they have (Brod, 2021; Simonsmeier et al., 2022). Hence, prior knowledge is the essential foundation for learning.

The benefits of pre-tests to capture prior knowledge are well-recognized and practiced by teachers and instructors. The most common instrument used to capture the pre-test is the multiple-choice-based questions (MCQs). MCQ is a test that is comprised of questions and choices of answers. Students are usually asked to select one correct answer. However, the design of MCQ by its nature offers the possibility of guessing (Lee et al., 2017). Guessing has been classified into three types of guessing, including:

- Completely random guessing: refers to making a decision without any specific knowledge or rationale (Foley, 2016). This is the most unreliable and inaccurate measurement of prior knowledge. It can be implied that by using this type of guessing, students lack prior knowledge.

- Eliminated guessing: This refers to the process of eliminating the unlikely options by drawing on prior knowledge (Maguya, 2022). Students rule out choices that they are certain of incorrect or do not align with their understanding. This increases the chance of selecting the right answer.
- Informed guessing: Informed guessing involves using some available information or clues to make a more educated estimate of the correct answer. It might not be fully certain but is guided by some relevant knowledge (Maguya, 2022).

Even though using MCQs might offers the chance to use guessing tactics. Yet, its flexibility and convenience of assessment make the MCQ as one of the most frequently used instruments (Maguya, 2022). Therefore, it is important to acknowledge and understand both drawbacks and potentials. Hence, this research offers another insight into the use of MCQ as a pre-test instrument. This research also examines if pre-tests can be used to inform the students of their own learning performance, help them to reflect on their own learning process, hence, choose the right learning strategies. These questions are among the research questions that capture the attention of educators.

2.2 Learning Process

The application of learning tactics and strategies is the central process of SRL (Srivastava et al., 2022). Learning tactics and strategies are the main drivers in the learning process. Students applied the learning tactics and strategies that they perceived as effective to acquire knowledge and produce the work as required in the course (Greene & Azevedo, 2007; Winne & Hadwin, 1998). Consequently, it contributes to the level of performance (Blasiman et al., 2017; Broadbent, 2017). Hence, several research studies found a connection between learning tactics and strategies with the level of performance. For instance, research studies found that those who applied deep approaches to learning, i.e., intensive efforts and use of multiple tactics and learning resources, tended to have a higher level of performance as compared to those who applied the surface learning approach (Diseth & Martinsen, 2003). Similarly, those who focused on using practical-based learning strategies such as distributed practice, spaced practice, and others were more likely to achieve a higher score as compared to those who use passive learning strategies such as rereading, note taking, and highlighting (Dunlosky, 2013; Fincham et al., 2018). Hence, selecting the appropriate learning strategies is the core of the learning process.

Not all learning tactics and strategies are equally effective. (Bjork et al., 2013) posit that students often perceive the less effective learning strategies as the effective ones. This resulted in a faulty judgment of their learning and made an ineffective decision on the learning tactic and strategy application. For instance, based on the score received from the MCQs, some students might perceive that they acquire sufficient knowledge and select ineffective learning tactics and strategies that might have resulted in the learning performance. Yet, much is not known regarding the association of students' pre-tests and their associated use of learning tactics and strategies to drive the learning process. This research aims to fulfill that gap by exploring the process of learning based on the recorded data. To explore the differences in the learning process, students were divided into two groups by using average pre-test score as a threshold. This method was modified from Ahadi et al., (2015)'s to identify at-risk and non-at-risk students. Hence, two research questions (RQs) were formulated to guide this study, including,

- **RQ1:** Were there any differences in students' online learning processes when comparing between two groups identified based on level of pre-test scores?
- **RQ2:** Does the level of pre-test score associate with the learning performance?

2.3 Mining the Learning Process

Learning is a process of acquiring knowledge through the application of learning tactics and strategies. Previously, the process of learning was studied by using self-report instruments such as questionnaires, think-aloud protocol, or by observing the learning. Both self-report and observation are discrepancies to biases and interventions that might affect the actual learning process. For instance, the thinking-aloud protocol increases the cognitive load as the students need to learn at the same time speaking out loud to elaborate their rationale for the action taken. The questionnaire is more likely to represent the perception of their behaviors rather than the actual action. (Zhou & Winne, 2012) found that the log data, which automatically records the student's interaction when participating in online activities, were more reflective of the actual learning behaviors when compared to the use of questionnaires.

Log data has been increasingly used by researchers to understand the learning process. However, log data is huge and semi-structured data. understanding such data needs data analytics techniques, data mining, or machine learning algorithms to mine and uncover the potential insights hidden in the data source. For instance, Maldonado-Mahauad et al., (2018) used log data to examine the different patterns of the learning process. As a result, they found the learning engagement patterns students used when participating in the online course. Saqr et al., (2023) studied the longitudinal cross-course online tactics, strategies, and trajectories by using the log data. Saint et al., (2021) explored the use of various process mining algorithms to detect the micro-level SRL process from trace data. The exploration revealed that process mining is a useful tool for understanding the process, especially when combining the algorithms which offer different insights. As presented thus far, log data and process mining have the potential to reveal the actual learning process. Hence, log data could be used to extract the students' actual learning behaviors.

3. METHOD

This research aims to investigate the learning process of students who have different levels of MCQ scores. The research was conducted in the Quantitative Analysis in Management course offered for the third-year Information Communication Technology (ICT) related program at one of a Thai university. Participants in this research are those who registered for the course (N=26). The course was running in 2022 for 17 weeks. Students were required to study nine learning topics. Each week they will have two hours of face-to-face class and two hours of tutorial sessions. Students were given an assignment for each learning topic to be completed on their own time. After the due date, the lecturer explained the solution to the problems in the assignments and uploaded the solution together with the explanation for the student's own revision. Additional practical exercises were also provided. All learning materials were uploaded via the university's LMS. Students can also view the scores obtained from the tests via the LMS.

The pre-test was designed to capture students' knowledge that they expected to obtain before attending this course, such as the prior knowledge of business management and basic mathematics and statistics. These two fields are the foundation of this course in which students were expected to learn during the first two years of the ICT program. Students were asked to answer the MCQ-based pre-test during the first week of the course. Prior to taking the pre-test, students were informed of the purpose of the pre-test, and that the pre-test would not account for the final score. During the ninth week, students had the midterm test which accounted for 25 percent of the final score. The final exam was held in week 18 of the semester which accounted for 35 percent of the final grade. The remaining 40 percent were assignments (20 percent), tutorials (15 percent) and class participation and attendance (5 percent).

The collected pre-test, midterm, and final exam scores were analyzed by using descriptive statistics including mean, median, first quartile, third quartile, and standard deviation. By examining the pre-test scores, students were divided into two groups based on their scores captured by using the MCQ-based pre-test. That is, by using the average score as a threshold, students were divided into two groups namely, i) above average group, and ii) below average group.

As mentioned, the learning materials were provided via LMS which automated the recording of students' interactions. This is known as log data. Log data represents the students' learning footprint, it can be used to examine the students' learning process hence, answering RQ1. However, such recorded data is often huge and semi-structured. Hence, a data analytic technique is needed to process the information. Process mining is considered a data analytic method that can be used to discover the ordering of events that reflect the process. Several contemporary research studies applied process mining to understand the student's learning patterns (Bannert et al., 2014; Reimann et al., 2009; Saint et al., 2021). In this study, the process of learning was formulated by using the BupaR R package. That is, the Directly-Follows Graph (DFG) was used to automatically generate the process of students' learning. DFG is the most commonly used algorithm to generate the process (van der Aalst, 2011). It computes the process based on the event. Hence, the data required for this computation included process instance or case, activity, and timestamp. Some other variables are optional to compute the process. In this study, the process of learning was formulated by using the student anonymized id as a case id, type of LMS content that each student visited as an activity, the timestamp, and the learning topic.

The contents in LMS include announcements, course syllabus, textbook, home page, course plan, lecture slides, pre-test, solution, additional information, and pro-gram. Table 1 provides a brief description of the relevant contents that were made available for the students on this course.

Table 1. Content types and description

Content	Description
Additional	contain the additional practical questions. Students were not required to complete this additional learning and practice materials
Announcement	contain the course announcement
Assignment	contain the assignment information, questions, and link to submit the assignment.
Course syllabus	contain the syllabus, course structure, plan, and schedule
Home	course home page displaying a list of learning topics, relevant activities and materials
Lecture	include the lecture slides and reading materials used during the lecture section
Pretest	contain the pretest questions
Solution	contain the solutions and explanation on the tutorials and assignment questions
Other	other information including open-source programs, links to useful information

By using DFG algorithm via bupaR R package library, the learning process of each group identified based on the level of pre-test scores can be illustrated. In general, each process consisted of nodes, numbers, and flows (arrows). The node denotes the content type students interacted with as presented in Table 1. The flows demonstrated by using the arrows to illustrate the direction from source to next learning content.

To answer RQ2 regarding the significant difference between the two groups, descriptive and inferential statistics were used. Prior to exploring the significant differences in terms of scores, the Shapiro-Wilk test was first applied to explore if the collected dataset was normally distributed. The Shapiro-Wilk test is suitable for working with a small-size dataset. Hence, it was used in this study. The result obtained from the Shapiro-Wilk test suggested that the data was normally distributed. Hence, the Welch two-sample t-test was used to investigate the normally distributed with different variances dataset.

4. RESULTS

A total number of 26 students enrolled in this course. 53 percent were male students. The pre-test was implemented at the beginning of the course to capture the students' prior knowledge relevant to the domain knowledge of the course. On average, students score 4.181 over 10 (S.D. = 2.207). By using the average number as the threshold, students were divided into two groups, including,

- Above average group: a total number of 11 out of 26 students (42.31%) were able to score above the average score on their pre-test. When looking into details, they were able to obtain about 6.10 (S.D.= 1.128) out of 10 on average.
- Below average group: a total number of 15 out of 26 students (57.69%) were able to score below average score. That is, students in this group scored approximately 2.77 (S.D. =1.666) out of 10 on average.

4.1 RQ1: Were there any Differences in Students' Online Learning Processes When Comparing between two Groups Identified Based on Pre-Test Scores?

Process mining was applied to examine the learning process based on the types of learning materials that students interacted with. Figure 1 presents 90% of all learning interactions of students who score below average on the pre-test. The number indicates the number of times students accessed those particular learning materials. The arrow indicates the direction of the interaction flow. They usually go back and forth between the nodes.

Students who scored below average in the pre-test began their learning by accessing the home page on the LMS. The majority of students in this group accessed the assignments after accessing the home page and went back to visit the home page (Start → home → assignment → home), and half of them visited the lecture slides right after accessing the home page (home → lecture) and accessed the assignments after viewing the lecture slides (lecture → assignment). Some of the students also access additional learning materials after visiting the home page (home → additional) and lecture slides (lecture → additional). Most of the time after accessing additional learning materials, students revisited the assignment (additional → assignment).

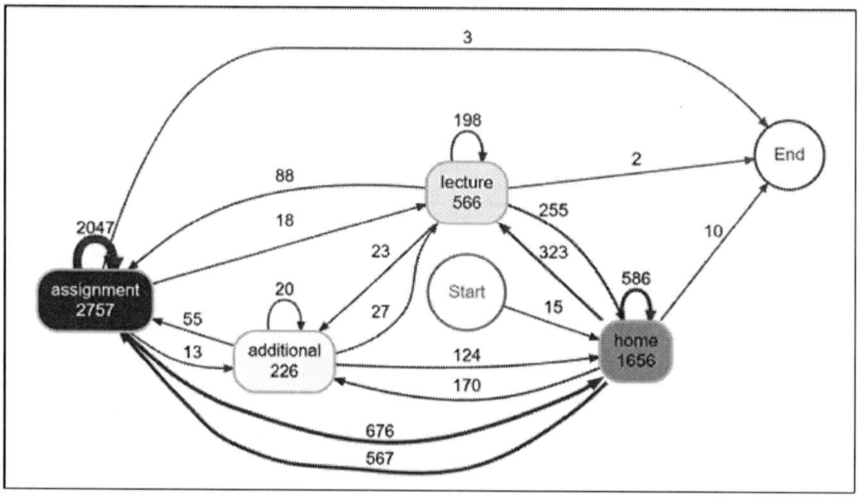

Figure 1. Learning sequence of students who score below average on pre-test score

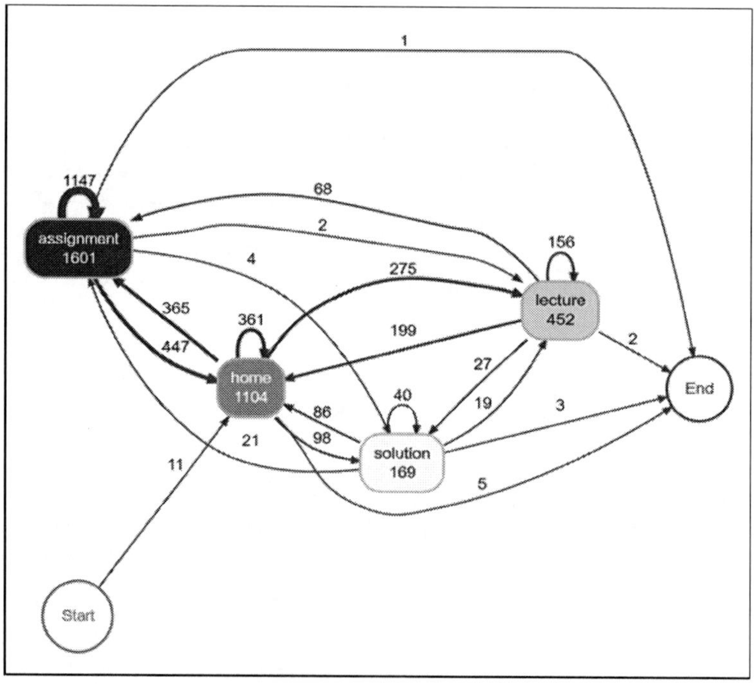

Figure 2. Learning sequence of students who score above average on pre-test score

Figure 2 shows the learning process of students who score above average on pre-test scores. The number of interactions and access to each learning element are much lower. Similar to the below-average group, those who scored higher on the pre-test began their learning by visiting the homepage, followed by assignment access (Start → home → assignment → home). Similarly, half of them also visited lecture slides after accessing the home page (home → lecture). However, rather than focusing on the additional materials, students in this group focused on reviewing the solution given by the lecturer.

4.2 RQ2: Does the Level of Pre-Test Score Associate with the Learning Performance?

Welch two samples t-test was used to examine if a significant difference was observed between the two groups based on their pre-test score. The result showed that there was a significant difference between the two groups at the standard p-value < 0.05 (t = -5.7171, p-value = 0.0000). In terms of the midterm score, it can be seen that those below average were able to achieve higher scores on midterm tests as presented in Figure 3.

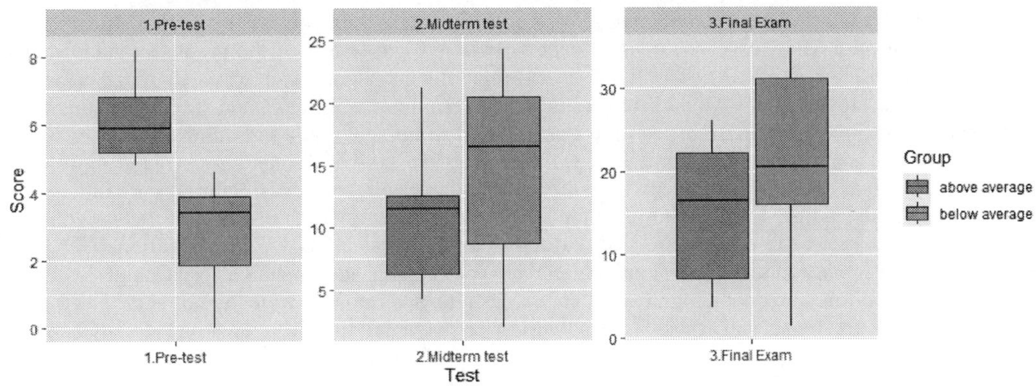

Figure 3. Boxplot of Pre-test (a) and Midterm test (b) of above and below average groups

Figure 3 presents the box plots of pre-test, midterm test, and final exam scores. The edge of the box presents the range of scores between the first (Q1) and third (Q3) quartiles. The line in the box presents the average score. It can be seen that those who have lower prior knowledge performed better during the midterm test. That is, students who scored below average (cyan color) on the pre-test obtained about 14.40 over 25 marks (S.D. = 7.134), median (Q1, Q3) was 16.50 (8.75, 20.50). Whereas those who initially scored above average obtained lower marks as compared to the first group. As presented in Figure 3, the median (Q1, Q3) is equal to 11.50 (6.25, 12.50), and the average score was 10.88 over 25 marks (S.D. = 5.424).

In terms of the final score, the below-average group showed a high average score. That is, the average score was 20.66 (S.D. = 10.53) over 35 and the median (Q1, Q3) was 20.46 (16.02, 31.10). The above-average pre-test group obtained a lower average score. They scored approximately 14.86, S.D. equal to 8.60. The median (Q1, Q3) was 16.42 (7.07, 22.21).

The Welch Two Sample T-test was also applied in order to investigate the significance between the two groups based on their midterm and final exam scores. Even though the box plots and average scores were different but there was no significant difference observed for both exams. That is, the Welch t-test for midterm between the two groups was 1.3712, p-value = 0.183. The t-test for final exam score was 1.5447 and the p-value was 0.1357. That is, the differences in students' prior knowledge derived from the pre-test did not significantly associate with the midterm and final exam scores.

5. DISCUSSION

This research was conducted to examine if priming the pre-test might contribute to the students' learning behaviors. A pre-test is usually used to capture the students' knowledge prior to exposure to the learning.

Pre-tests have been reported on their association with prior knowledge (Dochy et al., 2002), which is one of the factors that influence learning performance (Dochy et al., 2002; Shapiro, 2004; Simonsmeier et al., 2022). Not only pre-tests can reflect on the prior knowledge, but they can also be used to inform the students of the expected learning topic (Delucchi, 2014). The priming technique allows students to be aware of what they are expected to learn before the actual learning begins (Gulboy & Yucesoy-Ozkan, 2022). Hence, it is expected that priming the pre-test would enable students to be aware of their knowledge status, aware of the expected learning knowledge. By actuating awareness, students would be motivated to put in some effort, focus on the learning contents, and apply effective learning strategies when participating in the class. In this study, by priming the pretest, students who scored lower during their pre-test were able to obtain a higher score during a midterm test and final exam. Contrary to several research studies that suggested those who have a higher prior knowledge tend to perform better (Delucchi, 2014; Hailikari, 2009; Kennedy et al., 2015), this study showed that those who obtained higher scores during the pre-test scored lower in their midterm test and final exam. Yet, no significant difference was observed during the midterm and final exams. This can be inferred that even though students scored lower than the other group, the score was not suggestively different.

One of the interesting questions that arise when examining the relationship between the pre-test, midterm, and final exam scores is what makes the students in the 'below average' group score higher on the midterm test and final exam than the 'above average' group. This notion is in contrast with other research works that reported that prior knowledge has a significant association and is one of the predictive factors of learning performance (Delucchi, 2014). That is, it is expected that students who obtained higher pre-test scores (i.e., 'above average group') would achieve higher midterm tests and final exam scores. But this study found otherwise. The investigation of the learning process shed some light on these findings.

It can be observed from the process of learning that the 'above average's group paid more attention to solutions whereas another group focused on the additional learning materials. Solutions are the explanation of the answer to the assignment which was uploaded by the lecturer after the due date of the assignment and the explanation in face-to-face sessions. The additional learning materials were an additional set of tutorials, challenge questions, and some additional materials. Accessing the additional materials suggested that students put additional effort into practicing the challenge questions and tutorials which are uploaded as the additional materials. Whereas those who accessed the solution were expected to review what had already been explained by the lecturer in class. Both revision and practice are considered as necessary learning strategies. According to Dunlosky, (2013), not all learning strategies are equally effective. For instance, practice-oriented learning strategies such as distributed practice, and interleaved practice are considered as highly effective learning strategies, whereas rereading is considered as a less effective learning strategy (Dunlosky et al., 2013). Hence, if students review the solution without practice, this learning strategy might not be as effective as doing the practical exercises. Therefore, learning performance might be lower than those who apply the practice-oriented learning strategies.

However, there are limitations of this study that must be addressed. This study employed MCQs to assess the students' prior knowledge. Even though the MCQ-based pre-test is useful and most frequently used for assessing the students' prior knowledge, there are some limitations that are often discussed in research articles. That is, the use of MCQs offers the opportunity for the students to use the guessing tactic (Foley, 2016). It can be done by simply picking a completely random choice, eliminating the potentially incorrect answers and selecting the most appropriate one, or by inferring from the previous knowledge and selecting the most likely corrected answer (Lee et al., 2017). The last two techniques indicate that students used the previously learned knowledge to make a judgment to answer the questions. Hence, it reflects prior knowledge. However, if students were employed simply picking the choice technique, it would inflate the score. Consequently, it will be impossible to determine the actual level of prior knowledge.

Another limitation is that the students only received feedback in terms of the scores they obtained. Hence, they need to make a judgment on their next steps of learning by themselves. Implementing timely elaborative feedback would be beneficial to students. As highlighted by Bjork et al., (2013) students are prone to make faulty judgments about their learning. In this case, those who obtain the lower score might judge their performance as bad and put in extra effort to learn better. Those with high pre-test scores might presume a sufficient body of knowledge they have already obtained and, hence, make a poor decision to proceed with their learning such as the use of less effective learning strategies (i.e., reviewing) rather than the effective one (i.e., practicing). This study also suggested that feedback on the score does not provide actionable feedback. Several research studies suggested that elaborative personalized feedback is much more effective (Lim et al., 2021; Pardo, 2018; Price et al., 2010). Similarly, Hattie & Timperley, (2007) also stated that effective feedback should help students to be aware of their current state and suggest to them the actions or learning tactics and strategies they should take in order to improve learning and reach the learning outcome of the course.

It is important to highlight that there are several factors influencing learning. Since learning is a dynamic movement of students, it is influenced by both internal and external cognitive conditions. Some of the internal conditions are motivation, prior knowledge, self-efficacy, beliefs, etc (Winne, 2021). The external factors are such as the instructional approach, communication cues, time constraints, social economic status and etc. These factors influence the student's decision to apply what they perceive as 'effective learning strategies'. These dynamic decision-making and actions consequently contribute to the learning outcome and learning performance. For instance, in the case of below-average students' choices to focus more on additional learning and practical materials which might be driven by their internal factors such as motivation. Further exploration of this factor is needed. Therefore, research in students' learning needs to consider these factors when explaining the results acquired from classroom research.

One of the important suggestions derived from the observation of the learning process is the role of timely feedback. That is, after the pre-test, students should get feedback on their performance. However, feedback on the score does not prove to be actionable feedback. The growth of Artificial Intelligence (AI) and the fast adoption of generative AI such as ChatGPT could be used to facilitate the feedback provision. For instance, Woodrow et al., (2024) suggested that AI and Large Language Model (LLMS) algorithms are powerful tools to enable timely feedback. They further explored the use of LLMS to generate style timely feedback and found that those who received and opened the feedback generated by AI tend to perform better in a code writing style for Computer Science course.

Another suggestion for future research is that the use of data mining techniques is powerful. It offers insights into students' learning process. It can be integrated with generative AI or any other AI technology to enhance education.

6. CONCLUSION

This study presents the exploration of the use of pre-tests, students' learning process, and students' performance. The results suggested that even though students did not perform well during the pre-test, they could achieve high scores if they used effective learning strategies. One of the implications of this study is that using the pre-test not only helps the lecturer to be aware of the student's prior knowledge but also helps students to reflect on their learning situation. Hence, the results of the pre-test could facilitate the decision to adopt suitable learning strategies or focus their attention on the learning materials. Priming the pre-test also helps students to be aware of what they can expect to learn in the course. Priming is usually used to direct the attention of learners. Hence, it is advisable to apply for a pre-test in the class.

REFERENCES

Ahadi, A., Lister, R., Haapala, H., & Vihavainen, A. (2015). Exploring machine learning methods to automatically identify students in need of assistance. ICER 2015 - Proceedings of the 2015 ACM Conference on International Computing Education Research, 121–130. https://doi.org/10.1145/2787622.2787717

Bannert, M., Reimann, P., & Sonnenberg, C. (2014). Process mining techniques for analysing patterns and strategies in students' self-regulated learning. Metacognition and Learning, 9(2), 161–185. https://doi.org/10.1007/s11409-013-9107-6

Bjork, R. a., Dunlosky, J., & Kornell, N. (2013). Self-Regulated Learning: Beliefs, Techniques, and Illusions. Annual Review of Psychology, 64(1), 120928131529005. https://doi.org/10.1146/annurev-psych-113011-143823

Blasiman, R. N., Dunlosky, J., Rawson, K. A., Blasiman, R. N., Dunlosky, J., The, K. A. R., Blasiman, R. N., Dunlosky, J., & Rawson, K. A. (2017). The what, how much , and when of study strategies : comparing intended versus actual study behaviour actual study behaviour. 8211.

Broadbent, J. (2017). Comparing online and blended learner's self-regulated learning strategies and academic performance. The Internet and Higher Education, 33, 24–32. https://doi.org/10.1016/j.iheduc.2017.01.004

Brod, G. (2021). Toward an understanding of when prior knowledge helps or hinders learning. Npj Science of Learning, 6(1), 2–4. https://doi.org/10.1038/s41539-021-00103-w

Delucchi, M. (2014). Measuring Student Learning in Social Statistics: A Pretest-Posttest Study of Knowledge Gain. Teaching Sociology, 42(3), 231–239. https://doi.org/10.1177/0092055X14527909

Diseth, Å., & Martinsen, Ø. (2003). Approaches to Learning, Cognitive Style, and Motives as Predictors of Academic Achievement. Educational Psychology, 23(2), 195–207. https://doi.org/10.1080/01443410303225

Dochy, F., de Rijdt, C., & Dyck, W. (2002). Cognitive Prerequisites and Learning: How Far Have We Progressed since Bloom? Implications for Educational Practice and Teaching. Active Learning in Higher Education, 3(3), 265–284. https://doi.org/10.1177/1469787402003003006

Dunlosky, J. (2013). Strengthening the Student Toolbox. American Educator, 37(3), 12–21.

Dunlosky, J., Rawson, K. A., Marsh, E. J., Nathan, M. J., & Willingham, D. T. (2013). Improving Students' Learning With Effective Learning Techniques Promising Directions From Cognitive and Educational Psychology. Psychological Science in the Public Interest, 14(1), 4–58. https://doi.org/10.1177/1529100612453266

Fincham, O. E., Gasevic, D. V., Jovanovic, J. M., & Pardo, A. (2018). From Study Tactics to Learning Strategies: An Analytical Method for Extracting Interpretable Representations. IEEE Transactions on Learning Technologies, 1382(c), 1–14. https://doi.org/10.1109/TLT.2018.2823317

Foley, B. P. (2016). Getting Lucky: How Guessing Threatens the Validity of Performance Classifications. Practical Assessment, Research, and Evaluation, 21, 3. https://doi.org/10.7275/1g6p-4y79

Gouldthorpe, J. L., & Israel, G. D. (2013). Capturing Change: Comparing Pretest-Posttest and Retrospective Evaluation Methods. Edis, 2013(1), 1–4. https://doi.org/10.32473/edis-wc135-2013

Greene, J. A., & Azevedo, R. (2007). A Theoretical Review of Winne and Hadwin's Model of Self-Regulated Learning: New Perspectives and Directions. Review of Educational Research, 77(3), 334–372. https://doi.org/10.3102/003465430303953

Gulboy, E., & Yucesoy-Ozkan, S. (2022). The effects of video priming on facilitating the transition of children with autism spectrum disorder between activities and settings. Journal of Theoretical Educational Science, 15(2), 373–394.

Hailikari, T. (2009). Assessing University Students' Prior Knowledge: Implications for Theory and Practice. https://doi.org/10.1177/0033688206076156

Hattie, J., & Timperley, H. (2007). The Power of Feedback. Review of Educational Research, 77(1), 81–112. https://doi.org/10.3102/003465430298487

Kennedy, G., Coffrin, C., de Barba, P., & Corrin, L. (2015). Predicting success: how learners' prior knowledge, skills and activities predict MOOC performance. Proceedings of the Fifth International Conference on Learning Analytics And Knowledge - LAK '15, 136–140. https://doi.org/10.1145/2723576.2723593

Lee, A. T., Shen, P., Tsai, C., & Lee, T. (2017). Applying Web-Enabled Problem-Based Learning and Self-Regulated Learning to Add Value to Computing Education in Taiwan's Vocational Schools. International Forum of Educational Technology & Society, 11(3), 13–25.

Lim, L., Bannert, M., van der Graaf, J., Molenaar, I., Fan, Y., Kilgour, J., Moore, J., & Gašević, D. (2021). Temporal Assessment of Self-Regulated Learning by Mining Students' Think-Aloud Protocols. Frontiers in Psychology, 12(November), 1–18. https://doi.org/10.3389/fpsyg.2021.749749

Maguya, A. S. (2022). Quantifying the Effects of Guessing, Position Bias and Prior Knowledge in Multiple Choice Exams. Tanzania Journal of Science, 48(4), 760–772. https://doi.org/10.4314/tjs.v48i4.4

Maldonado-Mahauad, J., Pérez-Sanagustín, M., Kizilcec, R. F., Morales, N., & Munoz-Gama, J. (2018). Mining theory-based patterns from Big data: Identifying self-regulated learning strategies in Massive Open Online Courses. Computers in Human Behavior, 80, 179–196. https://doi.org/10.1016/j.chb.2017.11.011

McCarthy, K. S., & McNamara, D. S. (2021). The Multidimensional Knowledge in Text Comprehension framework. Educational Psychologist, 56(3), 196–214. https://doi.org/10.1080/00461520.2021.1872379

Pardo, A. (2018). A feedback model for data-rich learning experiences. Assessment & Evaluation in Higher Education, 43(3), 428–438. https://doi.org/10.1080/02602938.2017.1356905

Price, M., Handley, K., Millar, J., & O'Donovan, B. (2010). Feedback : all that effort , but what is the effect ? Assessment & Evaluation in Higher Education, 35(3), 277–289. https://doi.org/10.1080/02602930903541007

Reimann, P., Building, A., & Work, S. (2009). Using Process Mining to Identify Models of Group Decision Making in Chat Data. 9th International Conference on Computer Supported Collaborative Learning, 98–107.

Saint, J., Fan, Y., Singh, S., Gasevic, D., & Pardo, A. (2021). Using process mining to analyse self-regulated learning: A systematic analysis of four algorithms. LAK21: 11th International Learning Analytics and Knowledge Conference (LAK21), April, 333–343. https://doi.org/10.1145/3448139.3448171

Saqr, M., López-Pernas, S., Helske, S., & Hrastinski, S. (2023). The longitudinal association between engagement and achievement varies by time, students' profiles, and achievement state: A full program study. Computers and Education, 199. https://doi.org/10.1016/j.compedu.2023.104787

Schreibman, L., Whalen, C., & Stahmer, A. C. (2000). The Use of Video Priming to Reduce Disruptive Transition Behavior in Children with Autism. Journal of Positive Behavior Interventions, 2(1), 3–11. https://doi.org/10.1177/109830070000200102

Shapiro, A. M. (2004). How including prior knowledge as a subject variable may change outcomes of learning research. American Educational Research Journal, 41(1), 159–189. https://doi.org/10.3102/00028312041001159

Simonsmeier, B. A., Flaig, M., Deiglmayr, A., Schalk, L., & Schneider, M. (2022). Domain-specific prior knowledge and learning: A meta-analysis. Educational Psychologist, 57(1), 31–54. https://doi.org/10.1080/00461520.2021.1939700

Srivastava, N., Fan, Y., Rakovic, M., Singh, S., Jovanovic, J., Van Der Graaf, J., Lim, L., Surendrannair, S., Kilgour, J., Molenaar, I., Bannert, M., Moore, J., & Gasevic, D. (2022). Effects of Internal and External Conditions on Strategies of Self-regulated Learning: A Learning Analytics Study. ACM International Conference Proceeding Series, 392–403. https://doi.org/10.1145/3506860.3506972

van der Aalst, W. M. P. (2011). Process Mining. In Process Mining. Springer Berlin Heidelberg. https://doi.org/10.1007/978-3-642-19345-3

van Kesteren, M., Rijpkema, M., Ruiter, D. J., Morris, R. G. M., & Fernandez, G. (2014). Building on prior knowledge: Schema- dependent encoding processes relate to academic performance. Journal of Cognitive Neuroscience, 10(26), 2250–2261. https://doi.org/10.1162/jocn

Winne, P. H. (2021). Cognition, Metacognition, and Self-Regulated Learning. Oxford University Press. https://doi.org/10.1093/acrefore/9780190264093.013.1528

Winne, P. H., & Hadwin, A. F. (1998). Studying as Self-Regulated Learning. Metacognition in Educational Theory and Practice, 93, 277–304.

Woodrow, J., Malik, A., & Piech, C. (2024). AI Teaches the Art of Elegant Coding: Timely, Fair, and Helpful Style Feedback in a Global Course. SIGCSE 2024 - Proceedings of the 55th ACM Technical Symposium on Computer Science Education, 1, 1442–1448. https://doi.org/10.1145/3626252.3630773

Zhou, M., & Winne, P. H. (2012). Modeling academic achievement by self-reported versus traced goal orientation. Learning and Instruction, 22(6), 413–419. https://doi.org/10.1016/j.learninstruc.2012.03.004

Zimmerman, B. J. (2008). Investigating Self-Regulation and Motivation: Historical Background, Methodological Developments, and Future Prospects. American Educational Research Journal, 45(1), 166–183. https://doi.org/10.3102/0002831207312909

HYBRID LEARNING 2.0: DESIGNING SUSTAINABLE MODELS FOR POST-PANDEMIC HIGHER EDUCATION IN JORDAN

Maha S. Yaseen

Department of English Language, Al-Ahliyya Amman University, Jordan

ABSTRACT

The COVID-19 pandemic served as a substance for a remarkable transformation in global higher education, encouraging institutions to rapidly adopt digital learning solutions. In Jordan, this sudden shift exposed both the potential and the limitations of existing educational infrastructures, revealing deep-rooted challenges in digital readiness, pedagogical flexibility, and institutional support. As the nation transitions into a post-pandemic educational landscape, there is an urgent need to move beyond emergency remote teaching and toward more resilient and integrated models of instruction. This study proposes a context-sensitive and pedagogically grounded Hybrid Learning 2.0 Framework tailored to Jordanian universities. Applying a mixed-methods approach, including stakeholder interviews, document analysis, and comparative case studies, this research identifies critical barriers and enablers to hybrid learning in the Jordanian context. The findings highlight the importance of equitable access to technology, continuous faculty development, and institutional commitment to digital transformation. The proposed framework emphasizes inclusivity, flexibility, and sustainability, offering practical recommendations for policymakers, university leaders, and educators. Ultimately, this study contributes to the growing body of knowledge on post-pandemic educational reform and provides a strategic proposal for advancing hybrid learning in Jordan's higher education system.

KEYWORDS

Hybrid Learning, Post-Covid Education, Sustainable Model, E-Learning, Jordanian Higher Education

1. INTRODUCTION

The COVID-19 pandemic marked a turning point in the path of the global higher education, necessitating the universities around the globe to move overnight at an ease to online learning centers. In Jordan, like in so many different countries, such a fast transformation exposed the opportunities of digital pedagogy and the strain of digital pedagogy in conventional tertiary institutions. In the past, most universities in Jordan had been slow to adopt online or blended instruction at large scale owing to infrastructure limitations, staff preparedness, and dominant culture that supported traditional, in-person instruction.

The pandemic, however, was a disturbance of the norms. Due to nationwide lockdowns and health-related restrictions, Jordanian universities had to apply emergency remote teaching solutions. Although these initial attempts tended to be both reactive and often lacked pedagogical coherence, institutions were able to test digital tools, delivery platforms, and alternate delivery modalities. As the crisis took place, it was evident that higher education in Jordan could no longer simply resume its status quo ante pandemic. It came as wake-up to the institutional leaders, faculty members, and policymakers to reconsider old assumptions about teaching, learning, and the role of technology within academic experience.

Moving out of the pandemic period and into a post-pandemic world, the primary question facing the Jordanian Universities is no more about integrating digital learning or not, but about doing it efficiently and sustainably. There is increased realization that ad hoc solutions or short-term solutions are not helpful. Rather, institutions need to create strong, context-adapted instructional models that bridge the advantages to face-to-face teaching with the mobility and scalability of digital delivery. Under this understanding, hybrid learning as the intentional combination of online and face-to-face modalities come forth as a promising solution.

In fact, hybrid models have been applauded for supporting student engagement, self-directedness, and enabling access for the student base on a global scale. The actual building of such models in the Jordanian higher education fabric, however, comes with unique challenges. These include the fact that rural areas lack access to effective digital infrastructure compared to urban areas; the fact that faculty members have marked diversity in terms of digital competencies; institutional restrictions on investment in instructional design; and sociocultural resistance to other than traditional teaching modalities and assessment. Furthermore, various faculty and student bodies alike still equate quality education with physical presence in the classroom, further blurring efforts to normalize hybrid modalities.

So far, limited research in Jordan has focused on the long-term adaptation to online learning platforms, particularly concerning the technical and logistical dimensions of e-learning in the post-pandemic era. For example, Sa'di et al. (2023) offer significant recommendations regarding the future of e-learning in Jordanian higher education institutions, emphasizing the need for faculty training, strategic planning, and improved digital infrastructure. However, a notable gap remains in the literature regarding the long-term institutionalization of hybrid learning models that are not only pedagogically effective but also aligned with Jordan's educational, cultural, and technological context.

This paper aims at filling that gap by suggesting a model of Hybrid Learning 2.0 designed for Jordanian universities. Based on the result of previous research and obtained lessons from the best global practices. This model will seek to offer a scalable, sustainable and context-specific approach towards integrating hybrid learning into the core teaching and learning practices of Jordan's higher education institutions. It explores guiding principles, essential resources, and the institutional changes required for successful implementation, while also considering stakeholder perspectives, national education policy, and local constraints.

Ultimately, the study asks: *What constitutes an effective hybrid learning ecosystem for the post-COVID Jordanian university?* Through this inquiry, the paper contributes to the growing body of research seeking to shape the future of higher education in Jordan and the broader Arab region.

2. LITERATURE REVIEW

The emergence of hybrid learning as a dominant pedagogical approach after the COVID-19 pandemic has garnered significant scholarly attention worldwide. Hybrid education, characterized by a deliberate integration of face-to-face and online instruction (Garrison & Vaughan, 2008; Boelens et al., 2017), has been widely discussed in higher education for its potential to enhance learner autonomy, foster engagement, and accommodate cognitive diversity.

In the Arab world, the pandemic was a trigger of digital transformation; however, this shift was uneven across the region. Existing research has highlighted several major challenges in the adoption of online and hybrid learning modalities in Arab higher education institutions. These include widespread infrastructural deficiencies, limited digital competencies among both faculty and students, and broader institutional shortcomings (Abu-Alruz et al., 2021; Almahasees, Mohsen, & Amin, 2021). Nevertheless, the crisis also opened a window of opportunity to reimagine traditional higher education models and to integrate digital learning more strategically and sustainably.

Sa'di et al. (2023) made a timely contribution in presenting e-learning future-oriented proposals for traditional Jordanian tertiary institutions. They highlighted the significance of investment in digital infrastructure, faculty training, and redrawing curricula as the launching points for sustainable e-learning integration. However, they were mostly obsessed with e-learning in the remote or online version and there was a gap when it came to the long-term institutionalization of hybrid learning as a mainstream pedagogical model in the region.

Moreover, much of the existing research in the Arab context has concentrated on the immediate transition to online teaching during the pandemic, often labeled "emergency remote teaching" (Al-Shallakh, M. 2024; Badrakhan et al. 2024; and Hodges et al., 2020), rather than on the design and implementation of deliberate, pedagogically grounded hybrid systems. There is a noticeable scarcity of studies that propose hybrid models specifically tailored to the socio-cultural, economic, and educational contexts of Arab universities.

This paper seeks to address this gap by developing a context-sensitive hybrid learning framework, termed *Hybrid Learning 2.0*, that reflects both international best practices and local realities. Through a critical synthesis of literature and stakeholder-informed analysis, the study aims to offer a model that is not only technologically feasible but also pedagogically effective and institutionally sustainable within Arab higher education.

3. METHODOLOGY

This study adopts a qualitative, exploratory design aimed at proposing a sustainable and context-sensitive hybrid learning model—referred to as Hybrid Learning 2.0—for Jordanian higher education institutions in the post-COVID era. Given the complexity of educational reform and the diversity of institutional capacities across Jordan, this research relies on a multi-source approach, combining document analysis, stakeholder consultation, and comparative review of international models.

The design adopted for the study is design-based research (DBR), apt for solving real-life educational issues among multiple stages of enhancing the model. Through DBR, it becomes possible to build fundamentally theory-based solutions, which are sensitive to the realities of the world (Reeves, 2006). This approach is aligned with the aim of the study to suggest the hybrid learning model that is not only pedagogically correct but also is suited to the environment of education and socio-cultural conditions in Jordan.

Three primary data sources informed the development of the model:

1. Document and Policy Analysis: A review was conducted of strategic documents, regulations, and digital education policies issued by Jordan's Ministry of Higher Education and Scientific Research (MoHESR), as well as individual universities' post-COVID strategic plans (where available). This allowed for mapping institutional readiness, regulatory constraints, and national priorities.

2. Semi-structured Interviews: A total of 12 stakeholders, comprising faculty members, instructional designers, IT staff, and academic leaders from different academic Jordanian institutions, were interviewed. Participants were selected using purposive sampling to ensure a diverse representation of institutional types and experiences. The interviews focused on participants' experiences with digital teaching during and after the pandemic, perceptions of hybrid learning, and institutional support systems.

3. Comparative Case Review: The study examined documented hybrid learning models implemented in higher education systems with comparable socioeconomic conditions (e.g., Egypt, Morocco, Turkey) as well as successful frameworks from global contexts (e.g., the UK, Canada, and Singapore). These cases were analyzed for transferable practices, challenges, and success factors.

Data from interviews were analyzed using thematic analysis, following Braun and Clarke's (2006) six-step process: familiarization, coding, theme development, review, definition, and reporting. Documents and case studies were subjected to qualitative content analysis, with attention paid to recurring patterns in policy orientation, technological integration, and pedagogical structure. Findings from these analyses were synthesized and used to inform the design of the proposed Hybrid Learning 2.0 framework.

To enhance credibility, triangulation was employed by comparing findings across different data sources (interviews, documents, and case studies). Member checking was also conducted by sharing preliminary findings with selected participants to verify interpretations. Peer debriefing sessions with educational technology experts helped refine the model's components and ensure its practical relevance.

As an exploratory and design-oriented study, the research is limited in scope and does not empirically test the implementation of the proposed model. Instead, it aims to provide a theoretically and contextually informed framework that can serve as a foundation for future pilot studies and institutional experimentation.

4. FINDINGS AND MODEL DESIGN: HYBRID LEARNING 2.0 FRAMEWORK

4.1 Key Findings

Through a combination of interviews, policy analysis, and international case studies, several key findings emerged regarding the state of hybrid learning in Jordanian higher education institutions:

1. **Faculty Readiness and Training Gaps:** Interviews with faculty members revealed that while many instructors were able to quickly adapt to emergency remote teaching during the pandemic, they lacked long-term training in hybrid teaching pedagogies. A significant portion of the academic staff expressed a need for continuous professional development to effectively design and deliver hybrid courses. While some institutions have started offering training, this effort remains inconsistent across universities.

2. **Technological Infrastructure and Accessibility:** Although there was a rapid development of digital tools during the pandemic, the quality and availability of these tools varied significantly across Jordan's universities. Public universities, particularly, reported challenges with limited access to high-speed internet, outdated infrastructure, and difficulties in accessing digital resources. Conversely, private universities, particularly those with higher tuition fees, had more advanced infrastructure and better access to e-learning platforms.

3. **Student Engagement and Equity Issues:** A major concern raised by students was the lack of engagement in online classes compared to the interactive nature of face-to-face sessions. This challenge was further worsened by socio-economic disparities, as students from lower-income backgrounds often lacked access to appropriate technology and reliable internet connectivity. Universities also reported that specific groups, such as students residing in rural areas, were disproportionately affected by the digital divide, which significantly hindered their ability to participate effectively in hybrid learning environments.

4. **Cultural Resistance to Non-Traditional Learning:** Despite the rapid digital transformation triggered by the pandemic, some universities faced cultural resistance to non-traditional learning methods, particularly in more conservative and traditional institutions. Faculty and students expressed concerns about the loss of face-to-face interaction and the perceived value of online learning. This resistance was most notable in disciplines that traditionally rely on in-person teaching methods (e.g., medicine, engineering, and law).

5. **Policy and Strategic Gaps:** Although there have been efforts at the national level to integrate e-learning into the education system, most universities in Jordan have not developed comprehensive, long-term strategies for hybrid learning. Policies tend to focus on short-term solutions without addressing the long-term infrastructure, training, and cultural changes required for successful hybrid learning implementation.

4.2 Hybrid Learning 2.0 Framework

Based on these findings, the **Hybrid Learning 2.0** framework was designed to address the unique challenges and opportunities within the Jordanian context. The framework consists of the following key components:

4.2.1 Pedagogical Approach

Blended Pedagogy with Contextual Sensitivity: Hybrid Learning 2.0 should integrate both face-to-face and online learning in a way that aligns with Jordanian students' learning preferences and cultural contexts. Instructors need to be trained in blended learning skills that focus on active learning, collaborative online tools and synchronous learning sessions. They should also value the role of direct interaction with their students for any practical hands-on material that is required for practical learning.

Personalized Learning Pathways: The frame permits the use of adaptive learning technologies in which learners are provided with the opportunity to choose the personalized learning paths that will allow all students to learn at their individual pace. This consists of asynchronous activities such as recorded lectures, assignments and peer reviewed discussions as well as live virtual classes and physical sessions where necessary.

4.2.2 Technology and Infrastructure

Investment in Robust Digital Infrastructure: One of the major pillars of Hybrid Learning 2.0 is ensuring that universities have infrastructure to both support hybrid learning and online learning process. This covers reliable internet, high quality learning management systems (LMS) and cloud-based tools. Public universities, in particular, should focus on the investment of the digitization tools and networks to ensure that technological gap is minimized among their institutions.

Digital Access for All Students: To address the issue of the digital divide, Hybrid Learning 2.0 highlights the importance of equal access to students' technology. This may include subsidizing the students from lower socio-economic settings to acquire devices or obtain affordable internet services. University institutions could also make partnerships with private companies in order to allow the students to afford technological solutions.

Integration of Collaborative Tools: The framework calls for the integration of collaborative tools like discussion boards, video conferencing, and collaborative document editing tools among students. We should integrate such tools within the course design as a means of promoting the interaction of our students, peer learning, as well as group projects that make up part of the hybrid experience.

4.2.3 Faculty Development and Support

Ongoing Faculty Training: One of the core characteristics of the Hybrid Learning 2.0 framework is the necessity for undergoing continuous professional development among the faculty members. Some of these include regular training on the pedagogical principles of hybrid teaching, e-learning platforms and interactive interesting online modules which are to be used for hybrid and e-learning. Methods of managing hybrid classrooms and students' engagement in online and in-person classes should also be of concern to teaching faculty.

Support Systems for Faculty: To ensure that faculty are not overwhelmed, universities need to pair up instructional designers, technical support staff and support peer collaboration. Such support can assist instructors in developing effective hybrid courses, getting technical problems fixed and planning new tactics of teaching online environments.

4.2.4 Student-Centered Design

Flexible Learning Options: The Hybrid Learning 2.0 should be flexible to allow for various students' needs. This includes providing students with synchronous as well as asynchronous learning selections, with a clear message about when they should interact with content and how the interaction should take place. The flexibility should also include ways of assessment where instead of traditional testing, students could perform project-works, on line quizzes or even digital file work.

Enhanced Student Engagement and Support: To deal with issues related to student engagement, Hybrid Learning 2.0 includes the methods to create community and engagement in online learning environments. This covers synchronous check-ins, virtual office hours and a peer mentoring program. Universities should also invest in student support services where students can get comprehensive support for their online and in-person learning. These support services should also be available both online and in person

4.2.5 Institutional and Policy Support

Long-term Strategic Planning: The framework calls for a shift from short-term, reactive strategies to comprehensive, long-term plans for hybrid learning. Universities should design institutional strategies, which integrate hybrid learning into the larger mission and vision of an institution. This involves aligning hybrid learning with accreditation requirements and ensuring that faculty and students are appropriately supported through policies and resources.

National and Institutional Policy Alignment: The Jordanian Ministry of Higher Education and Scientific Research (MoHESR) should provide guidelines and incentives to encourage the universities to adopt the hybrid learning models. These policies should give equal access to technology, match accreditation standards for hybrid practices, and encourage assessing how effective hybrid models are in a regional setting.

5. CONCLUSION

The proposed Hybrid Learning 2.0 Framework provides a future-oriented approach that considers the local situation in Jordanian institutions of higher education and addresses the lessons obtained from the COVID-19 pandemic implementation of hybrid learning. As higher education in Jordan and other Arab countries moves into the post-pandemic era, the return to a hybrid launches possibilities for improvements, and gives rise to numerous issues that should be addressed with responsibility and caution in mind.

This framework highlights the necessity for a balanced, sustainable, and pedagogically sound approach to hybrid learning by combining face-to-face and online approaches in a flexible, inclusive, and adapting system to the various needs of Jordanian students. If there is one thing that Hybrid Learning 2.0 does very well, it is to address what were seen to be the gaps left in the findings: faculty readiness, technological infrastructure, student equity, and cultural aversion to non-traditional learning. With these issues addressed properly, Hybrid Learning 2.0 leads the way for more involvement from students, better learning outcomes, and a caliber of students prepared to face a rapidly evolving global job market.

One of the main strengths of the Hybrid Learning 2.0 Framework is the emphasis on equity and accessibility. Owing to the revelation made by the study that digital divide is among the greatest obstacles of successful implementation of hybrid learning in Jordan, the issue at hand can be considered a key issue. Within the given framework, students are expected to be given the top priority in terms of providing them with access to technology and digital resources when it comes to universities. Students from low-income families or those that come from rural areas should be prioritized in this aspect. By addressing such disparities through specificity of policies, subsidies, and collaboration with private firms, the universities can secure that hybrid learning does not widen the violation of equity, yet it helps come closer to scoring the educational milestone.

Equally important is the emphasis on faculty development and support. The shift to hybrid learning requires instructors to adapt their pedagogical practices and become proficient in the use of new technologies. This needs continuous professional development programs that will not only ensure that faculty knows how to use the hybrid learning tools, but will also equip them with the ability to produce and deliver interesting, interactive and inclusive learning experiences. The framework identifies the readiness and motivation of faculty as key factors that determine the success of hybrid learning. Therefore, universities must continuously provide ongoing support in terms of training, adequate resources, and collaborative milieu of teaching.

The technological infrastructure of universities is another pillar of the proposed framework. Although a number of academic institutions in Jordan have taken great strides in embracing the digital tools and platforms, others, especially some of the public universities, are still grappling with challenges related to old generation technology, poor connectivity as well as inefficient support systems. For Hybrid learning to be successful, universities need to invest in improving their digital infrastructure, deploy strong Learning Management Systems (LMS), and make online learning platform accessible seamlessly. The involvement of national stakeholders, including the Ministry of Higher Education, is important in creating policies that promote the development of digital infrastructures across the country.

The diversity of learning experiences offered by the student-centered design of Hybrid Learning 2.0 is made possible to the varying needs and preferences of students. By utilizing flexible learning alternatives including asynchronous courses, individually designed learning approaches and combination assessment models, the framework acknowledges that students have varied learning styles and routines. This flexibility is especially important for non-traditional students, working professionals, who are committed to family or other duties that may be detrimental to their ability to attend synchronous classes. Hybrid Learning 2.0 allows students to become the masters of their own learning with the help of making education more available and personal; however, higher institutions are expected to continue providing future support.

The institutional and policy support for hybrid learning is essential to the long-term success and sustainability of the model. Universities have the mandate of creating comprehensive strategic plans that incorporate hybrid learning into their academic objectives, accreditation needs, and institutional mission.

Moreover, national policies should be revised in order to follow the increasing importance of digital education and to provide incentives for universities that invest in hybrid learning infrastructure. The responsibility of policymakers in facilitating a friendly environment for hybrid learning cannot be extravagant; it needs a collaborative effort between government, universities, and the private sector to ensure that all stakeholders are ready to support this shift.

In conclusion, the Hybrid Learning 2.0 Framework is a dynamic and holistic model that focuses on how Jordanian universities can cope with unique challenges they are expected to face post-pandemic situation. It provides a guideline for institutions to come up with a more inclusive, flexible and resilient learning ground that can prepare students for the future. The application of this model will contribute positively to the quality of education in Jordan. It will also pave the way for a paradigm shift in the operation of higher learning institutions in the Arab world towards learning and teaching in the country. If Jordan selects a thoughtful and strategic approach in terms of hybrid learning, it is possible to emerge as a leading country in the region and ensure students are prepared to become the leaders in the rapidly changing global society.

Ultimately, the success of Hybrid Learning 2.0 will rely on the collective commitment of all educational stakeholders. Universities, policymakers, faculty, and students must collaborate to overcome the challenges and seize the opportunities that hybrid learning offers. With the right investments, training, and policies in place, Jordanian higher education cannot only recover from the disruptions caused by the COVID-19 pandemic but arise stronger, more inclusive, and better prepared to meet the demands of the future.

REFERENCES

Abu-Alruz, J., Jabarin, R., & Zaza, H. (2021). Challenges and opportunities of e-learning in Arab higher education: A post-COVID perspective. *International Journal of Educational Research Open*, 2, 100057.

Almahasees, Z., Mohsen, K., & Amin, M. O. (2021). Faculty's and students' perceptions of online learning during COVID-19: A case study of a Jordanian university. *Humanities and Social Sciences Communications*, 8, Article 197.

Al-Shallakh, M. A. I. (2024). Embedding artificial intelligent applications in higher educational institutions to improve students' pronunciation performance. Theory and Practice in Language Studies, 14(6), 1897-1906. https://doi.org/10.17507/tpls.1406.31

Badrakhan, S., Shaban, E. A., Oqilat, O., Taha, J., & Doudeen, H. (2024). Efforts towards achieving sustainable development goals in light of national strategy "Jordan's Vision 2025": Reality and challenges. Journal of Educational and Social Research, 14(4), 468-484. https://doi.org/10.36941/jesr-2024-0116

Boelens, R., De Wever, B., & Voet, M. (2017). Four key challenges to the design of blended learning: A systematic literature review. *Educational Research Review*, 22, 1–18.

Braun, V., & Clarke, V. (2006). Using thematic analysis in psychology. *Qualitative Research in Psychology*, 3(2), 77–101. https://doi.org/10.1191/1478088706qp063oa

Garrison, D. R., & Vaughan, N. D. (2008). *Blended Learning in Higher Education: Framework, Principles, and Guidelines*. Jossey-Bass.

Hodges, C., Moore, S., Lockee, B., Trust, T., & Bond, A. (2020). The difference between emergency remote teaching and online learning. *Educause Review*.

Lincoln, Y. S., & Guba, E. G. (1985). *Naturalistic inquiry*. SAGE Publications.

Mayring, P. (2000). Qualitative content analysis. *Forum: Qualitative Social Research*, 1(2). https://doi.org/10.17169/fqs-1.2.1089

Reeves, T. C. (2006). Design research from a technology perspective. In J. V. van den Akker, K. Gravemeijer, S. McKenney, & N. Nieveen (Eds.), *Educational design research* (pp. 52–66). Routledge.

Sa'di, R., Sharadgah, T., Abdelraziq, A., & Yaseen, M. (2023). Future proposals for e-learning at conventional tertiary institutions as they move on past the COVID experience. *Journal of Educational and Psychological Studies*, 17(3), 520–538. https://doi.org/10.24200/jeps.vol17iss3pp520-538

AN AI-ASSISTED WORKFLOW FOR MULTILINGUAL ACCESSIBLE VIDEO PRODUCTION: THE EDUNEXT CASE STUDY

Alice Pierluigi, Marina Linacero Martín, Annamaria De Santis, Claudia Bellini and Tommaso Minerva
Università degli Studi di Modena e Reggio Emilia
Reggio Emilia, Italy

ABSTRACT

The adoption of online modalities in higher education has not automatically ensured effective learning or inclusivity, particularly in multilingual and multicultural contexts. This paper aims to present a practical framework for enhancing accessibility in digital education through audiovisual translation, with a focus on subtitling and dubbing. Using the EDUNEXT Digital Education Hub - an Italian initiative involving 35 universities and 5 conservatories/academies - as a case study, we describe an AI-assisted workflow for creating multilingual educational video content. The proposed workflow requires minimal infrastructure and supports the creation of accessible, inclusive, and multilingual video content. Nonetheless, it involves software licensing costs and significant human post-editing, despite its cost-effectiveness and AI-assisted. The systematic integration of AI-based accessibility tools into digital learning environments has the potential to significantly reduce linguistic and cognitive barriers, supporting the development of inclusive teaching models. By making multilingual educational materials accessible, this approach advances the broader goal of ensuring equitable access to higher education for diverse learner populations.

KEYWORDS

Online Education, Higher Education, Inclusion, Audiovisual Translation, AI-tools, Accessibility

1. INTRODUCTION

Worldwide integration of online modalities into higher education has become widespread, with several institutions offering blended programs, digital platforms, and e-learning systems (Minerva *et al.,* 2024a). However, the mere availability of digital environments does not guarantee effective learning for students. Designing successful online courses requires careful attention to various aspects: e-content cognitive load (Paas *et al*., 2003), the application of the principles of multimedia learning (Low and Sweller, 2014; Mayer, 2011), and the choice of media format (video, text, or audio) that align with both the learning objectives and the intended outcomes (Piras, Reyes and Trentin, 2020).

Incorporating video into online education further necessitates addressing learner diversity through user-centered design (Piras, Reyes and Trentin, 2020). In actual multilingual and multicultural educational contexts, this involves the integration of high-quality translations, particularly through subtitling and dubbing. These practices enhance accessibility and learner engagement, responding to educational imperatives, as well market-driven demands. The increasing importance of accessibility, inclusivity and equity, supported by initiatives such as the EU Accessibility Act and the Audiovisual Media Services Directive, has amplified the relevance of audiovisual translation in educational contexts (Van Hoecke, Schrijver and Rober, 2022).

Subtitles and dubbing function as linguistic translations and also as cognitive aids, supporting comprehension and retention.

Subtitles in simplified language can support note-taking and reduce cognitive overload during video-based learning (Kruger-Marais and Kruger-Roux, 2023), especially in linguistically diverse classrooms where information density can be high (Kruger, Hefer and Matthew, 2014). Additionally, bilingual subtitles have been shown to promote second language acquisition by facilitating simultaneous listening and reading, aiding vocabulary development and listening skills (García, 2017). Nevertheless, the cognitive effects of subtitles vary, also depending on elements such as speed, accuracy, content complexity, and learner proficiency (Pannatier and Betrancourt, 2024). Subtitle speed is particularly critical, as excessive rates can hinder reading and information processing (Kruger, Wisniewska and Liao, 2022; Chan, Kruger and Doherty, 2019). Automatic subtitles generated through speech recognition offer efficiency but often suffer from errors in grammar, segmentation, and synchronization, especially under noisy conditions or with limited vocabularies (Szarkowska *et al.*, 2024). As a result, post-editing remains necessary to ensure clarity and reduce cognitive load, confirming the continued need for human oversight (Chan, Kruger and Doherty, 2019).

Dubbing, meanwhile, provides an alternative that may better suit learners who benefit from uninterrupted visual attention. In contexts where subtitles may compete with visual elements, dubbing allows learners to focus on audio-visual content without the need to divide their attention (Hughes, Montagud and Pesch, 2019). Its effectiveness increases when paired with active listening strategies that enhance audio processing (Kruger-Roux and Angu, 2020).

In summary, the integration of subtitling and dubbing into educational content plays a crucial role in fostering accessible, inclusive and cognitively supportive learning environments, addressing the challenges of contemporary multilingual education. Technological developments support more adaptive subtitling solutions, allowing customization based on individual learning preferences and needs (Gorman, Crabb and Armstrong, 2021). Such personalization enhances accessibility for students with disabilities and the inclusion of learners engaging with content in other languages (Vitikainen and Koponen, 2021).

Within this framework, the central research questions guiding this investigation are: What are the core components of an AI-assisted workflow for the translation of educational content delivered by universities in online environments? What characteristics define the AI tools that are most suitable for implementation in academic contexts? What is the optimal sequence for deploying these tools and what are the specific stages involved in the workflow?

Thus, our study offers a practical framework for the implementation of inclusive digital educational content, with a focus on accessibility through audiovisual translation. After a description of the research methods including an overview of the EDUNEXT project as case study, we propose an exploration of AI-powered tools employed for transcription, translation, subtitling, dubbing and content distribution; and a detailed account of the workflow stages leading from initial audio transcription to the final publication of accessible video materials on the Moodle platform. The implemented subtitling system includes both intralingual subtitling (Italian audio with Italian subtitles) and interlingual subtitling (Italian audio with subtitles in a second language, or vice versa), thereby addressing a range of linguistic needs.

2. METHODS

Through the analysis of a case study, we provide a proposal to achieve accessibility through audiovisual translation by detailing an AI-assisted workflow and software tools.

Case studies (Ary *et al.*, 2010) concentrate on a single individual/entity/program to give a detailed description and understanding of a phenomenon or situation. Our focus is the translation workflow adopted within the Italian Digital Education Hub EDUNEXT, a project currently in its initial stages (see section 2.1). The research involves benchmarking of AI tools for transcription and translation and collaborative experimentation within the EDUNEXT work team. Thus, the research takes a design-based approach (McKenney and Reeves, 2021) with the aim of improving the described workflow through analysis, application, and definition. This is a descriptive paper based on the initial phase of the project, which currently focuses on the design and implementation of the workflow; as such, empirical data and pilot studies are planned for future stages.

2.1 The EDUNEXT Project

The Digital Education Hub EDUNEXT, established as part of the National Recovery and Resilience Plan (NRRP), represents a large-scale initiative launched in 2024 and aimed at rethinking the delivery of higher education in Italy and beyond. Coordinated by University of Modena and Reggio Emilia, EDUNEXT brings together 35 universities and 5 Higher Education Institutions in Art, Music, and Design (edunext.eu) supported by a panel of 55 external partners with advisory roles. From a collaborative perspective, EDUNEXT re-designs and delivers inclusive, accessible digital learning programs tailored to individual needs, offering blended degree courses, specializations, and online lifelong learning programs such as Massive Open Online Courses (MOOCs).

EDUNEXT operates under a unified didactic model known as ECOBI (Educational Clusters with Open Badges and Blended Intensive Program) that combines competency-based curriculum design, blended intensive formats, micro-design methodologies, personalized tutoring, and open badging to provide a flexible and scalable educational infrastructure. Aligned with European standards such as the ESCO EU Framework, EDUNEXT also aims to implement systems for issuing micro-credentials and managing e-portfolios to certify students' acquired competencies in a transparent and portable manner (Minerva *et.al.*, 2024b).

3. RESULTS

The following subsections describe the two main areas of investigation of the case study: AI tools for multilingual subtitling, audio and transcript creation, and the workflow recently developed for publishing educational content to be delivered in EDUNEXT degree courses starting in September 2025.

3.1 AI-Powered Tools for Audiovisual Translation

The tools selected (with no financial interests from the authors) in EDUNEXT for the workflow outlined in this section are: Happy Scribe, Narakeet, Vimeo, and ChatGPT. Softwares were chosen based on their distinguishing characteristics and also on the configurations necessary within EDUNEXT, such as the use of shared licenses, centralized access, and equitable distribution of resources, to ensure the sustainable and coordinated use of these technologies in all the universities of the network.

Happy Scribe (happyscribe.com) is an online platform that leverages artificial intelligence to facilitate the transcription, subtitling, translation, and dubbing of audiovisual content in over 120 languages. Designed for media professionals, academic institutions, and organizations, it aims to enhance the accessibility and global dissemination of educational and communicative materials. Among its core features are high-precision transcription (both automated and human), synchronized and customizable subtitle generation, multilingual translation, synthetic voice dubbing, an interactive editor that allows users to modify text while listening to the corresponding audio, and API integration for workflow automation.

The second platform to be used in the workflow presented is Narakeet (narakeet.com), an online platform that harnesses artificial intelligence to convert written texts into narrated audio and video with a highly realistic delivery. It supports over 800 synthetic voices across more than 100 languages and accents, thereby offering a versatile solution for the creation of multilingual content. The platform provides a wide selection of high-quality synthetic voices, distinguished by their natural and credible vocal performance. Thanks to the ability to customize elements such as intonation, pacing, and emphasis, Narakeet is well-suited to a variety of content types and proves particularly effective for multilingual projects aimed at international audiences. In addition, it allows for the import of Word documents, PDFs, EPUB files, and subtitle files (in SRT format), facilitating the conversion of diverse formats into audio (MP3) or video (MP4) outputs, ready for distribution across various platforms.

For publication purposes, we have chosen Vimeo (vimeo.com) as an online video hosting and distribution platform designed to offer a high-quality viewing and content creation experience. Among its advanced features, Vimeo provides artificial intelligence-based tools for automatic translation, subtitling, voice cloning, dubbing, and content publishing. The platform supports the translation of both subtitles and audio into over 30 languages, enabling rapid content localization and enhancing accessibility. In addition, it offers automatic subtitle generation, which can be easily edited and synchronized with the video. Regarding dubbing, Vimeo employs voice cloning technology to reproduce the original voice in multiple languages with a natural and authentic outcome, significantly reducing the cost and time associated with traditional dubbing methods. Also, it allows for the direct upload and distribution of video content, which can be easily embedded into educational portals or shared via link.

Together with Vimeo, PeerTube (joinpeertube.org) was considered as one of the potential tools by the EDUNEXT to serve as an open publishing platform for Content Library since it offers an open-source alternative for educational content distribution, with external tools for subtitles and translations, providing a privacy-respecting solution for academic video publishing.

To conclude, we selected ChatGPT (chatgpt.com) because it is particularly distinguished by its ability to translate texts with a high degree of accuracy and to generate clear and coherent transcriptions from audio recordings or preliminary drafts. Furthermore, it can be employed as an interactive booklet, enabling users to organize information, synthesize complex content, produce didactic explanations, and structure documents in a clear and systematic manner.

Upstream of the choices made, other possibilities and tools were explored through careful benchmarking. Among tools which represent a comprehensive AI-driven ecosystem capable of transforming how educational video content is produced, localized, and shared, we find: HeyGen (HeyGen.com) and Synthesia (Synthesia.io), both use realistic avatars capable of delivering multilingual narration, facilitating the automatic generation of dubbed content, offering high-quality voice synthesis and lip synchronization, thus enabling scalable localization of learning materials. Also, Miraflow (Miraflow.ai), although less widely known, provides streamlined AI transcription and translation workflows, enabling educators to convert spoken language into written text and translate it into multiple languages with high accuracy. Another platform which was designed specifically for education and enterprise training was Panopto (Panopto.com). It includes automatic speech recognition (ASR) for converting speech to text, enables video search through keywords, and offers multilingual captioning to enhance both accessibility and content discovery. Even YouTube (youtube.com), within an educational context, can offer automatic subtitling with the possibility of manual editing to improve accuracy.

3.2 EDUNEXT AI-Assisted Workflow for Translation and Transcription

Figure 1 shows the EDUNEXT AI-assisted workflow, which facilitates multilingual, inclusive access to educational video content through automated transcription, translation, subtitling, and dubbing. This process is conceived as a cost-effective and easily integrable solution for universities and educational institutions, requiring minimal infrastructure and technical overhead. The workflow presupposes that EDUNEXT video lectures, realised with video recording of the lecturer in person plus synchronous visual support content, are recorded with care and with high-quality audio, which is critical for achieving accurate transcription and optimal results throughout the process. All contents are made available in Italian (L1) and, following this workflow, in English and Spanish (L2), both as subtitles and dubbed audio. Integrated accessibility tools such as transcriptions further support diverse user needs. English and Spanish have been selected as second languages to maximize international outreach: English, as the dominant language of academic communication, and Spanish widely spoken across Spain and Latin America. Various combinations of audio and subtitles in L1 and L2 can be enabled; alternative L2 languages can be selected depending on specific linguistic and cultural needs. This flexibility ensures that the framework can be adapted to diverse educational contexts, supporting multilingualism and inclusion on a broader scale.

The process employs three core platforms: *Happy Scribe* for automated transcription and subtitle generation; *Narakeet* for text-to-speech dubbing; and *Vimeo* for the integration and publication of multilingual versions. Final distribution occurs through *Moodle*.

3.2.1 Phase 1: Transcription, Subtitling, and Translation via Happy Scribe

The Happy Scribe workflow is organized into two sequential phases: production in the original language (L1) followed by production in L2. Once the video is uploaded to the Happy Scribe platform, an initial transcription in the source language is automatically generated. Subsequently, this transcription serves as the basis for producing translated versions in English and Spanish. To generate the transcription and subtitle files in the original language, the workflow consists of six sequential steps to be carried out within the Happy Scribe platform. The process begins with a video file (the recorded lecture) and concludes with the export of two files: the transcription (in TXT or DOCX format) and the subtitles (in SRT format). The following steps must be completed in chronological order:

1. upload the video file (MP4 format) to Happy Scribe.

2. generate the automatic transcription in the original language using Happy Scribe's integrated AI-powered video-to-text functionality.
3. verify the transcription for grammatical, syntactic, and semantic accuracy. Any errors must be corrected manually within the Happy Scribe interface.
4. generate subtitles, including automated segmentation and synchronization with the video, using the dedicated Happy Scribe function available in the transcription panel.
5. review the subtitle text to ensure alignment between audio and text. Any necessary adjustments should be made manually within Happy Scribe.
6. export the two final files from Happy Scribe: (i) the transcription in TXT or DOCX format and (ii) the subtitles in SRT format.

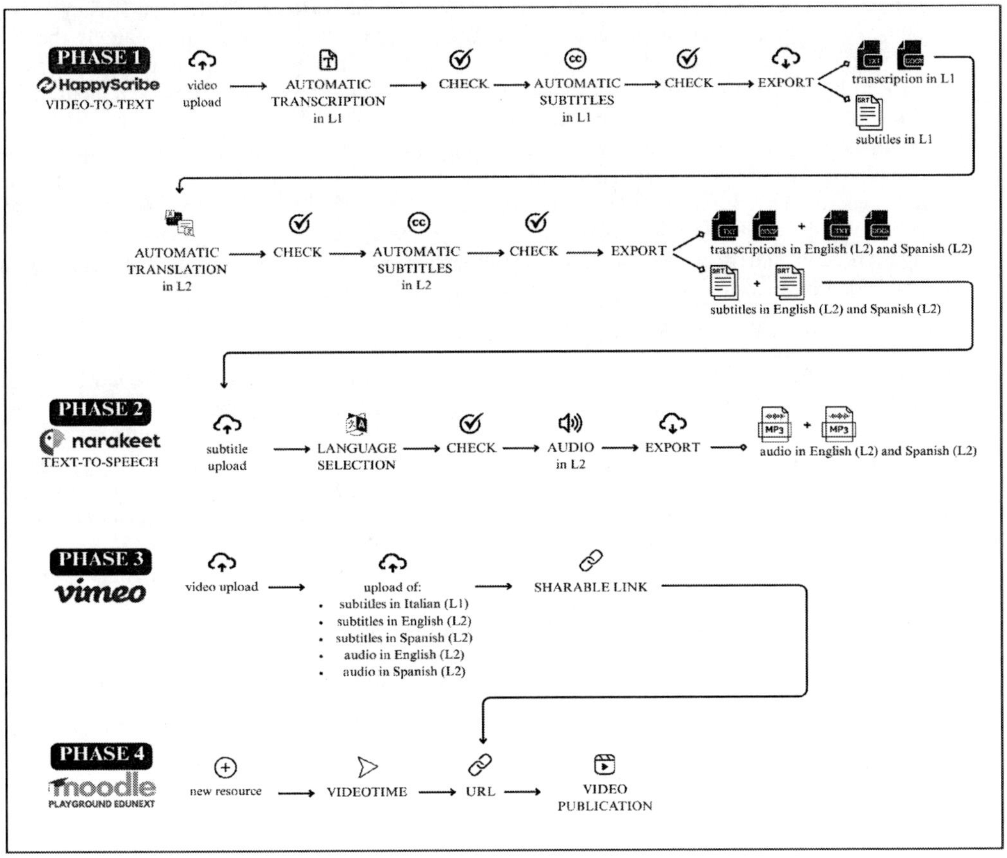

Figure 1. EDUNEXT's workflow model. L1 refers to the native language (Italian), which is used as the original language of content production and instruction, while L2 refers to second languages such as English and Spanish

Once the six steps related to the original-language transcription and subtitling have been completed, the translation process into L2 can begin. As previously stated, the workflow must be executed twice: once for English and once for Spanish. The procedures for English L2 and Spanish L2 are identical and independent: the workflow can begin with either language, with the same method subsequently applied to the other. The starting point is the approved and verified transcription in the original language, already available on Happy Scribe. The final output consists of the translated transcription (TXT or DOCX) and the translated subtitles (SRT). The workflow for producing translated transcription and subtitles comprises five steps:

1. within the original transcription file in Happy Scribe, select the target language (English/Spanish) to automatically generate the translated version.

2. conduct a human review to ensure the translated transcription is accurate, consistent, and faithful to the original. Edits are made directly in Happy Scribe.
3. generate subtitles, including automated segmentation and synchronization with the video, using the dedicated Happy Scribe function available in the transcription panel.
4. verify the subtitle text for alignment with the audio. Manual corrections should be made in Happy Scribe as needed.
5. export the two final files from Happy Scribe: (i) the translated transcription in TXT or DOCX format and (ii) the translated subtitles in SRT format.

The following aspects of the process require particular attention:

- each transcription must be thoroughly reviewed to ensure grammatical, orthographic, syntactic, and lexical accuracy, as well as overall consistency with the video content. This review is a pivotal phase of the process and is primarily the responsibility of the content owner, namely the instructor who designed and recorded the video lecture. To enhance the quality of both original and translated transcriptions, AI-based tools such as ChatGPT may be employed. Nevertheless, human intervention remains essential to ensure the final output's accuracy and reliability.
- subtitles must not be generated before a comprehensive revision of the transcription has been completed. In particular, the generation of subtitles in L2 must begin from the reviewed and approved translated transcription. Starting from the transcription rather than directly translating the subtitles reduces the margin of error and the time required for subsequent corrections.
- upon review and approval, subtitles are exported in SRT format as closed captions and not embedded in the video, allowing users to activate or deactivate them as needed.

3.2.2 Phase 2: Voice-over Production via Narakeet

Following the completion of the transcription and subtitling phase, the next step in the workflow involves generating dubbed audio (voice-over) based on the translated subtitles. This phase applies exclusively to L2 (English and Spanish), as the original audio in L1 is already present in the video. Audio generation is performed using Narakeet which, as said, provides Text-to-Speech (TTS) functionality. The process must be executed separately for each target language. The audio dubbing process in Narakeet requires selecting the target language, including a regional variant (e.g., British or American English), as well as choosing the preferred voice type (e.g. male or female). The input files for this phase are the translated subtitle files in SRT format, while the output consists of dubbed audio files in the respective foreign language (MP3 format), exported from Narakeet. The process is composed of six sequential steps:

1. upload the translated subtitle file (SRT format) to Narakeet.
2. choose the target language, including the appropriate geographical variant.
3. select the preferred voice (male or female) from the available options.
4. use the Preview feature to verify pronunciation, intonation, and pauses by selecting relevant portions of the text. This step is critical to ensure clarity and natural speech.
5. generate the final dubbed audio file using the Create Audio option.
6. export the finalized audio file.

3.2.3 Phase 3: Integration and Publication on Vimeo

At the conclusion of the second phase of the workflow, in addition to the original video in Italian, users should have the following files: (i) Transcription in Italian, the original language; (ii) Subtitles in the original language (Italian), (iii) Transcription in English; (iv) Subtitles in English; (v) Audio dubbed in English; (vi) Transcription in Spanish; (vii) Subtitles in Spanish; and (viii) Audio dubbed in Spanish. Vimeo enables the integration of these elements into a single video lecture in Italian, with the option to activate subtitles in Italian, English, or Spanish, and to switch the audio track to English or Spanish voice-over.

To correctly configure the content on Vimeo, the following eight steps must be complete:

1. upload the original video file to Vimeo.
2. add the subtitle file in Italian via the *Languages* section.
3. add the English subtitle file via the *Languages* section.
4. upload the English dubbed audio file in the *Languages* section.
5. add the Spanish subtitle file via the *Languages* section.
6. upload the Spanish dubbed audio file via the *Languages* section.

7. creation of Shareable Link of unlisted video. This step is essential for content integration into the Moodle platform, where Vimeo will function as the embedded video player.
8. copy the Link.

3.2.4 Phase 4: Publication on Moodle

In the final stage of the workflow, after accessing the course page on the Moodle platform, a new activity must be added to insert the video with subtitles and transcriptions. EDUNEXT proposes a Moodle plugin used in educational settings called Video Time (moodle.org/plugins/mod_videotime), which offers the possibility to display captions hosted on Vimeo and interactive transcripts. In the Settings page of the activity, various metadata can be entered, including the title, description, and any additional notes relevant to the video resource. Here the core element is the insertion of the Vimeo URL previously generated. This URL enables the seamless embedding of the video into the Moodle platform, ensuring that users can view the lesson with the configured subtitles and dubbed audio options directly within the learning environment.

4. CONCLUSIONS

Integrating AI-assisted audiovisual translation and online modalities into higher education presents significant benefits and challenges. Advantages include enhanced accessibility, in compliance with legal and educational standards, cognitive support through subtitles and dubbing, and indirect promotion of language acquisition. Technological advancements enable adaptive, personalized subtitle and audio options that improve inclusivity, particularly for non-native speakers and students with disabilities. AI tools streamline transcription, translation, subtitling, and dubbing, facilitating scalable production of multilingual educational content. Case studies like EDUNEXT's aim to experiment with workflows that are easily integrated and aligned with European standards, supporting flexible credentialing, faculty development and shared resource libraries. However, our experience shows how effective implementation demands high-quality audio recordings, rigorous human quality control, and mindful management of cognitive load. Other limitations include inaccuracies in automatic subtitles, potential distraction from visual elements by students, and ongoing development of AI dubbing technologies. The costs and time involved in implementing the workflow on large productions should also not be neglected. Furthermore, instructors may express reservations regarding the dubbing of their likeness with synthetic voices that do not represent their own, raising ethical considerations that warrant careful examination. These concerns include issues such as consent, voice ownership, and the potential for misrepresentation, especially when AI-generated voices are used without clear attribution or approval. Future research aims to overcome these initial limitations identified during this first phase of production.

As a concrete application of the study, the proposed workflow may serve as a valuable reference for large-scale digital content production. Specifically, it is our intention to demonstrate how emerging technologies can be harnessed to overcome linguistic and cognitive barriers, promote the creation of alternative learning formats, and strengthen technical and intercultural competencies.

ACKNOWLEDGEMENT

EDUNEXT (Project Name DEH2023-00003; SingleProject Code - CUP: E83C23003480007) is financed by National Recovery and Resilience Plan (NRRP), Mission 4 "Education and research" - Component 1 "Strengthening the provision of education services: from creches to universities", Investment 3.4"Advanced university teaching and skills", Sub-investment 3 "Digital Education Hub", funded by the European Union – NextGenerationEU, referred to Directorial Decree No. 2100 of Dec. 15, 2023.
The described workflow for translation and transcription is developed by the EDUNEXT technical team and is the result of shared insights and ideas of the network on innovation in higher education.

AUTHORS' CONTRIBUTIONS

According to CRediT System, Alice Pierluigi and Marina Linacero Martín: Writing - Original Draft, Writing - Review & Editing, Visualization, Resources, Investigation; Annamaria De Santis: Methodology, Writing - Review & Editing, Supervision; Claudia Bellini: Supervision, Writing - Review & Editing; Tommaso Minerva: Conceptualization, Project administration, Supervision, Resources.

REFERENCES

Ary, D. *et al.* (2010) Introduction to Research in Education (8th ed.). Belmont (CA): Wadsworth Cengage Learning.

Chan, W.S., Kruger, J.L. and Doherty, S. (2019) Comparing the impact of automatically generated and corrected subtitles on cognitive load and learning in a first- and second-language educational context. *Linguistica Antverpiensia, New Series: Themes in Translation Studies*, 18, pp.237–272.

García, B. (2017) Bilingual subtitles for second-language acquisition and application to engineering education as learning pills. *Computer Applications in Engineering Education*, 25(3), pp.468–479. https://doi.org/10.1002/cae.21814

Gorman, B., Crabb, M. and Armstrong, M. (2021) Adaptive subtitles: preferences and trade-offs in real-time media adaption. *Proceedings of the ACM Conference*, pp.1–11. https://doi.org/10.1145/3411764.3445509

Hughes, C. Montagud, M. and Pesch, P. (2019) Disruptive approaches for subtitling in immersive environments. *Proceedings of the ACM Conference*, pp.216–229. https://doi.org/10.1145/3317697.3325123

Kruger, J., Hefer, E. and Matthew, G. (2014) Attention distribution and cognitive load in a subtitled academic lecture: L1 vs. L2. *Journal of Eye Movement Research*, 7(5). https://doi.org/10.16910/jemr.7.5.4

Kruger, J., Wisniewska, N. and Liao, S. (2022) Why subtitle speed matters: Evidence from word skipping and rereading. *Applied Psycholinguistics*, 43(1), pp.211–236.

Kruger-Marais, E. and Kruger-Roux, H. (2023) Academic, keyword, and plain English subtitles for natural sciences students: intralingual views. *Transformation in Higher Education*, 15(1). https://doi.org/10.4102/the.v8i0.275

Kruger-Roux, H. and Angu, P. (2020), Bimodal and unimodal comprehension and reception of discipline-specific audiovisual texts among students in a faculty of health sciences. *The International Journal of Learning in Higher Education*, 27(2), pp.63–77. https://doi.org/10.18848/2327-7955/cgp/v27i02/63-77

Low, R. and Sweller, J. (2014) The modality principle in multimedia learning. In: R.E. Mayer, ed. *The Cambridge Handbook of Multimedia Learning*. 2nd ed. Cambridge, England: Cambridge University Press, pp.227–246. https://doi.org/10.1017/CBO9781139547369.012

Mayer, R.E. (2011) Applying the science of learning to multimedia instruction. In: J.P. Mestre and B.H. Ross, eds. *The Psychology of Learning and Motivation: Cognition in Education*, pp.77–108. Elsevier Academic Press. https://doi.org/10.1016/B978-0-12-387691-1.00003-X

McKenney, S., and Reeves, T. C. (2021) Educational design research: Portraying, conducting, and enhancing productive scholarship. *Medical education*, 55(1), 82-92.

Minerva, T. *et al.* (2024a) Students in Italian online universities: enrollments time series analysis from 2005 to 2021. *Journal of E-Learning and Knowledge Society*, 19(4), 50-75. https://doi.org/10.20368/1971-8829/1135872

Minerva, T. *et al.* (2024b) Integration of Educational Clusters with Open Badges and Blended Intensive Program (ECOBI): a comprehensive approach to future university education. *Journal of e-Learning and Knowledge Society*, 20(3), pp.112–125. https://doi.org/10.20368/1971-8829/1136068

Paas, F. *et al.* (2003) Cognitive Load Measurement as a Means to Advance Cognitive Load Theory. *Educational Psychologist*, 38(1), pp.63–71.

Pannatier, M. and Betrancourt, M. (2024) Learning from academic video with subtitles: When foreign language proficiency matters. *Learning and Instruction*, 90, 101863.

Piras, V., Reyes, M.C. and Trentin, G. (2020) *Come disegnare un corso online. Criteri di progettazione didattica e della comunicazione*. Franco Angeli, Milano, Italy.

Szarkowska, A., Ragni, V., Szkriba, S., Black, S., Orrego-Carmona, D., & Kruger, J. L. (2024). Watching subtitled videos with the sound off affects viewers' comprehension, cognitive load, immersion, enjoyment, and gaze patterns: A mixed-methods eye-tracking study. *PLoS One*, *19*(10), e0306251.

Van Hoecke, S.M., Schrijver, I. and Robert, I. S. (2022) Methodological Preparation of a Within-Subject Audiovisual Cognition, Reception and Perception Study. *Journal of Audiovisual Translation*, 5(1), pp.94–128. https://doi.org/10.47476/jat.v5i1.2022.163

Vitikainen, K. and Koponen, M. (2021) Automation in the intralingual subtitling process. *Journal of Audiovisual Translation*, 4(3). https://doi.org/10.47476/jat.v4i3.2021.197

INTEGRATING SLA AND GENAI IN A PRACTICE-ORIENTED INTRODUCTION TO INFORMATION SYSTEMS COURSE: A MIXED METHODS STUDY

Cherifa Boudia[1], Amina Houari[2] and Kamal Benhaoua[1]
[1]Computer Science Department, Technological Laboratory in Artificial Intelligence and Food Security (LABTEC-IA), University Mustapha Stambouli of Mascara, PB 305 Rouad of Mamounia, 29000 Mascara, Algeria
[2]LISYS Lab, University of Mascara, PB 305 Rouad of Mamounia, 29000 Mascara, Algeria

ABSTRACT

Bridging theory and practice in introductory Information Systems education remains a key challenge, particularly when students must master structured modeling methods such as Merise. This mixed-methods study examines the combined effect of Social Learning Analytics dashboards and generative AI (ChatGPT) on student performance and collaboration in a 13-week IS course. Ninety second-year engineering students completed six Merise modeling exercises manually and then with AI support; their Moodle forum interactions were visualized. AI assistance improved logical data model accuracy from 50 % to 78 %, while SLA dashboards increased forum density by 20 %. Thematic analysis of 540 AI-student transcripts and reflective journals highlights the importance of iterative prompt refinement and critical vetting of AI outputs. We conclude with evidence-based recommendations for embedding SLA-triggered interventions and AI-mediated scaffolding in practice-oriented IS curricula.

KEYWORDS

Social Learning Analytics, Generative AI, Merise Modeling, Mixed-Methods, Information Systems Education, ChatGPT

1. INTRODUCTION

The effective teaching of Information Systems (IS) fundamentals presents a persistent pedagogical challenge, particularly when bridging the gap between abstract theoretical concepts and their practical application in complex, structured tasks. Among these tasks, database modeling using formal methodologies like Merise is notoriously difficult for introductory students (Nguyen et al., 2024). Merise requires learners to navigate multiple stages, from identifying entities and relationships in a real-world scenario to constructing a Conceptual Data Model (MCD), and subsequently transforming this into a normalized Logical Data Model (MLD) suitable for relational database implementation. This process demands not only comprehension of core concepts like entities, attributes, relationships, cardinalities, and normalization rules but also the ability to apply these rules systematically and accurately, often under the constraints of typical coursework (Wu, 2024).Traditional pedagogical approaches, often centered around lectures and textbook examples followed by practical exercises, frequently fall short in providing the individualized, real-time feedback necessary for students to overcome common hurdles in Merise modeling. Errors in cardinality specification, primary/foreign key designation, or normalization steps can easily become ingrained if not addressed promptly, hindering the development of robust modeling skills. This challenge is amplified in larger class settings where personalized instructor attention is limited. Theoretical frameworks like Self-Regulated Learning (SRL) highlight the importance of learners actively engaging in cyclical processes of forethought (planning, goal setting), performance (monitoring, strategy use), and self-reflection (evaluation, adaptation) (Zimmerman, 2002). However, fostering effective SRL in complex domains like database modeling requires appropriate support structures. Students need timely feedback to monitor their understanding and adjust their strategies when they encounter difficulties. This is where emerging educational technologies offer significant potential. Generative Artificial Intelligence (GenAI), exemplified by large language models such as ChatGPT, has emerged as a powerful tool capable of providing interactive, dialogic support for complex cognitive tasks (Boudia & Bengueddach, 2024a, Khosravi et al., 2025). GenAI can act as a virtual tutor, offering explanations, generating examples, providing step-by-step guidance, and giving feedback on student work, potentially supporting all phases of SRL. Simultaneously, SLA provides methods to analyze and visualize student interaction data from online learning environments (Moodle), offering insights into collaborative processes, identifying students needing support, and fostering metacognitive awareness about participation (Cohn et al., 2024). SLA dashboards can make learning processes visible, potentially enhancing the social and cognitive presence aspects central to frameworks like the Community of Inquiry (Garrison et al., 2001). While the potential benefits of GenAI and SLA have been explored individually, there is a notable gap in empirical research examining their combined effects within a single pedagogical intervention, particularly in the context of practice-oriented IS education focused on structured modeling. Could SLA insights into collaboration patterns inform the deployment of AI scaffolding? Can AI assistance for individual tasks complement the social learning fostered through online discussions visualized by SLA? This study aims to address this gap by investigating the synergistic potential of integrating SLA dashboards and GenAI conversational support (ChatGPT) in an introductory IS course focused on Merise modeling.

Specifically, this mixed-methods investigation seeks to answer the following research questions:

RQ1: To what extent does AI conversational support (ChatGPT) improve the accuracy and quality of students' logical data models derived from conceptual models using the Merise methodology?

RQ2: How does the introduction of SLA dashboards visualize forum interaction metrics influence students' collaborative behaviors and participation levels during Merise modeling exercises?

RQ3: What interaction strategies do students employ when using generative AI for Merise modeling tasks, and how do they perceive the usefulness and challenges of the combined SLA dashboard and AI support intervention?

This study aims to provide robust, evidence-based insights. The findings are intended to inform the design of more effective technology-enhanced learning environments for teaching complex, structured methodologies in IS and related fields, offering practical recommendations for educators seeking to leverage the complementary strengths of social analytics and artificial intelligence.

The remainder of this paper is organized as follows. Section 2 reviews related work on generative AI as cognitive scaffolding, SLA for metacognition, and their integration in technology-enhanced learning. Section 3 describes our mixed-methods methodology, including course context, intervention design, scoring rubric, data collection, and analysis procedures. Section 4 presents the quantitative and qualitative results addressing RQ1–RQ3, with statistical tests, SLA engagement metrics, survey findings, and illustrative transcript excerpts. Section 5 discusses these findings in light of self-regulated learning and Community of Inquiry frameworks, contrasts them with prior studies, and outlines practical implications. Finally, Section 6 concludes with a summary of contributions, acknowledges limitations, and offers evidence-based recommendations for instructors, students, and AI developers, as well as directions for future research.

2. RELATED WORK

This study builds upon existing research in three key areas: the use of Generative AI (GenAI) as a cognitive tool in education, the application of SLA to understand and support collaborative learning, and the theoretical underpinnings for potentially synergistic integration of these technologies, particularly within the context of practice-oriented IS education.

2.1 GenAI as Cognitive Scaffold in Complex Domains

The advent of powerful GenAI models like ChatGPT has opened new avenues for supporting learners tackling complex cognitive tasks. Unlike static resources, GenAI offers interactive, dialogic feedback tailored to individual user inputs (Boudia & Bengueddach, 2024b, Wu, 2024; Nguyen et al., 2024). This capability is particularly relevant for procedural domains like programming or database modeling, where students must learn not just concepts but also the steps to apply them correctly. GenAI can function as a cognitive scaffold by: a) providing explanations of abstract concepts on demand, b) generating concrete examples (e.g., sample code, model snippets), c) offering step-by-step guidance through a process, and d) providing formative feedback on student-generated artifacts (Khosravi et al., 2025). Research in engineering education, for instance, has shown that integrating GenAI tools into problem-based learning (PBL) modules can significantly enhance students' analytical reasoning and design skills, often yielding large effect sizes (Dehkhoda et al., 2024; Kim, 2025). Similarly, studies in business contexts demonstrate AI's utility in simulating real-world scenarios and providing rapid feedback, accelerating skill acquisition (Wu, 2024; Allgood & Musgrave, 2024). The potential for GenAI to reduce cognitive load during complex tasks, allowing students to focus on understanding rather than getting stuck on syntax or procedural steps, is a key theoretical benefit (Nguyen et al., 2024).

2.2 SLA for Metacognition and Collaboration Awareness

While GenAI often focuses on individual cognitive support, learning rarely occurs in isolation, especially in higher education. SLA focuses on leveraging the digital traces left by student interactions in online environments (e.g., discussion forums, collaborative platforms) to gain insights into the social dynamics of learning (Glick et al., 2019). By analyzing metrics such as post frequency, reply patterns, network centrality (identifying influential or isolated students), and content analysis of discussions, SLA can provide a window into collaborative processes (Cohn et al., 2024). When these insights are visualized through dashboards and presented back to students and instructors, they can serve as powerful tools for metacognitive reflection. Students may become more aware of their own participation patterns relative to their peers, identify opportunities to contribute, or recognize when the group discussion is stalling (Boudia & Krismadinata, 2024). Instructors can use SLA dashboards to monitor group health, identify students needing support, and make timely pedagogical interventions (Khosravi et al., 2025). The underlying principle is that making the social learning process visible can empower both learners and educators to regulate it more effectively.

2.3 Theoretical Rationale for Synergistic Integration of SLA and AI

While GenAI supports individual cognition and SLA enhances social awareness, their true potential in educational settings might lie in their synergistic integration. Theoretical frameworks like the Community of Inquiry (CoI) model emphasize the interplay between cognitive presence (constructing meaning through discourse), social presence (projecting oneself socially and emotionally), and teaching presence (designing and facilitating the learning experience) (Garrison et al., 2001). SLA primarily addresses social presence by visualizing interactions and fostering awareness, while GenAI primarily supports cognitive presence by scaffolding individual understanding and task execution. An integrated approach could leverage SLA insights to inform the deployment of AI support. For example, if SLA metrics indicate a student group is struggling with a specific concept in forum discussions (low cognitive presence), AI prompts could be automatically triggered or recommended to provide targeted explanations or examples. Conversely, if an individual student relies heavily on AI without engaging with peers (low social presence), SLA data could flag this for instructor intervention or prompt the student to share their AI-assisted findings with the group. This aligns with Self-Regulated Learning (SRL) principles (Zimmerman, 2002), where external feedback (from both SLA and AI) informs the monitoring and regulation phases of learning, potentially leading to more adaptive and effective learning strategies. Despite this strong theoretical rationale, few empirical studies have operationalized and evaluated such an integrated SLA-AI approach within a single course design, particularly in demanding technical domains like IS modeling. This study seeks to fill this gap by exploring the practical implementation and effects of combining these technologies.

3. METHODOLOGY

This study employed a mixed-methods research design, integrating quantitative and qualitative data collection and analysis techniques (Creswell & Plano Clark, 2011) to provide a comprehensive understanding of the impact of combining SLA dashboards and GenAI (ChatGPT) support in an introductory IS course focused on Merise modeling.

3.1 Course Context and Participants

3.1.1 Setting and Curriculum

The study was conducted within a mandatory introductory IS course for second-year undergraduate computer engineering students at a large public university in Algeria during their fourth semester. The course met weekly for 13 weeks, totaling approximately 45 contact hours (typically structured as 1.5 hours of lecture followed by 1.5 hours of practical work). The course curriculum was structured around three core modules delivered sequentially: **1) Weeks 1-3:** Enterprise Systems Concepts (Chapter 1): Covered foundational topics such as organizations as systems, types of information systems, and their role within the enterprise. **2) Weeks 4-6:** Introduction to Software Engineering (Chapter 2): Introduced the Software Development Life Cycle (SDLC), various process models (Waterfall, V-Model, etc.), and concepts of information flow. **3) Weeks 7-13:** Merise Modeling Methodology (Chapter 3): Focused intensively on the Merise method, covering Data Flow Diagrams, MCD, and the rules for transforming MCD into MLD. It involved significant hands-on modeling practice.

3.1.2 Participants

Ninety students enrolled in the course participated in the study. They were already organized into three distinct groups within the Moodle LMS (approximately 30 students per group) for administrative and collaborative purposes. All participants had received the standard in-class instruction on Merise modeling principles before the intervention phases began. Prior experience with GenAI for academic purposes was assessed via a pre-survey. Of the 90 second-year computer engineering students, 37 identified as female and 53 as male. None had prior experience with the Merise methodology. No prior academic performance (e.g., GPA) data were collected.

3.2 Intervention Design and Procedure

The study unfolded over several weeks, incorporating distinct phases and technological components:

1. Baseline (Manual) Phase (Approx. Weeks 7-9): After initial instruction on Merise, students were required to individually solve a series of six practical modeling exercises. These exercises covered diverse domains (hospital management, e- commerce, library system, course registration, social media platform, university enrollment) requiring the creation of both MCD and MLD. During this phase, students relied solely on standard course materials (lecture slides, notes) and instructor-provided solutions for feedback after submission. **2. AI-Assisted Phase (Approx. Weeks 10-12):** Students revisited the same six modeling exercises, but this time they were explicitly permitted and encouraged to use ChatGPT (freely available version) as a support tool. They were instructed to use the AI to ask questions, seek clarification, request feedback on their models, or get help with specific transformation steps. A critical requirement was that students submit the full, unedited transcripts of their interactions with AI for each exercise. **3. SLA Dashboard Deployment (Weeks 6-13):** Concurrently with the modeling exercises, starting from Week 6 (before the AI-assisted phase began), weekly SLA dashboards were deployed within each Moodle group. These dashboards visualized key forum interaction metrics calculated from the previous week's activity: network density (ρ, indicating the overall connectedness of the discussion), average posts per student, and individual network centrality scores (e.g., degree centrality, indicating direct interaction levels). Instructors explicitly prompted students at the beginning of each practical session (Weeks 7-13) to review their group's dashboard and reflect on the interaction patterns before starting the exercises. The SLA dashboard was first released at the start of Week 7 (when Merise lectures began) and updated daily through Week 13. AI-scaffolding sessions using the free GPT-3.5 ChatGPT were triggered by the instructor immediately after students completed each manual exercise.

3.3 Data Collection Instruments

Multiple data sources were used to capture different facets of the learning experience: **Modeling Exercise Submissions:** Students submitted their MCD and MLD solutions for all six exercises in both the Manual and AI-Assisted phases. **AI Conversation Transcripts:** Approximately 540 complete transcripts (90 students * 6 exercises) of student interactions with ChatGPT during the AI-Assisted phase were collected. **Reflective Journals:** After completing the AI-Assisted phase, students submitted brief written reflections on their experience using AI and the SLA dashboards, focusing on perceived usefulness, challenges, and strategies employed.**Survey Instrument:** A custom survey was administered at the end of the study. It included: (Binary items assessing prior use of AI for Merise modeling or related tasks. Ten 5-point Likert-scale items adapted from established scales (e.g., Technology Acceptance Model constructs) measuring perceived usefulness of AI, trust in AI suggestions, confidence in using AI, and intention for future use. The scale demonstrated good internal consistency (Cronbach's $\alpha = 0.92$)). **SLA Metrics:** Weekly forum interaction data (posts, replies) were extracted from Moodle logs. Network density and centrality metrics were calculated for each group before (Weeks 1-5 average) and after (Weeks 6-13 average) the dashboard deployment. **Performance Metrics:** In addition to exercise scores, students' overall continuous assessment (Contrôle Continu - CC) marks for the course module were collected as a general performance indicator. At the end of Week 13, immediately following the final practical session, all students (100 % response rate) completed a survey comprising 10 Likert-scale items on trust, confidence, and intention to use AI.

3.4 Scoring and Reliability

To ensure consistent evaluation of the modeling exercises, a detailed 10-point analytic rubric was developed. The rubric assessed the correctness and completeness of both the MCD and MLD components, allocating 2 points each for: 1) accurate identification of entities, 2) correct definition of attributes, 3) appropriate specification of relationships and cardinalities (MCD), 4) correct designation of primary and foreign keys (MLD), and 5) adherence to normalization rules (MLD). To establish inter-rater reliability, two independent raters (e.g., the course instructor and a teaching assistant, both familiar with Merise) scored a randomly selected subset of 20% of the submissions (across both phases). Cohen's kappa coefficient was calculated, yielding a value of $\kappa = 0.83$, indicating substantial agreement between the raters. No formal IRB approval was obtained. Students were informed at the start of the course that anonymized chat transcripts and forum data would be used for research, and participation was voluntary.

3.5 Analysis Procedures

Quantitative Analysis: **Performance Comparison (RQ1):** Paired-samples t-tests were used to compare the mean MLD scores obtained in the Manual phase versus the AI-Assisted phase. Effect size was calculated using Cohen's d for paired samples, and 95% confidence intervals for the mean difference were computed. **SLA Impact (RQ2):** Paired-samples t-tests compared pre-dashboard (average of Weeks 1-5) and post-dashboard (average of Weeks 6-13) forum network density (ρ) for each group. Effect size (Cohen's d) was calculated. **Survey Analysis (RQ3):** Descriptive statistics (frequencies, means, standard deviations) were calculated for survey items. **Correlations:** Pearson correlations were explored between AI usage patterns (derived from transcripts), survey responses (e.g., perceived usefulness), and performance metrics (MLD gain scores, CC marks). **Statistical Rigor:** A Bonferroni correction was applied to adjust the alpha level for multiple comparisons ($\alpha' = 0.05$ / number of primary tests, e.g., 0.025 if focusing on the two main t-tests), although p-values are reported alongside effect sizes and confidence intervals for full transparency. **Qualitative Analysis (RQ3):** Thematic Analysis: An inductive thematic analysis approach (Braun & Clarke, 2019) was applied to the 540 AI conversation transcripts and 90 reflective journals. This involved iterative coding to identify recurring patterns in the types of questions asked to the AI, the strategies used for prompting (e.g., iterative refinement vs. direct solution requests), evidence of critical evaluation of AI responses, and student perceptions of the SLA dashboards and the overall intervention. Codes were grouped into broader themes related to AI interaction strategies, perceived benefits/challenges, and dashboard interpretation. **Trustworthiness:** Qualitative rigor was enhanced through triangulation (comparing findings from transcripts, reflections, and survey data) and potentially member checking (sharing preliminary themes with a subset of students for feedback, if feasible). **Qualitative Coding Procedure:** All 540 chat transcripts and 90 reflective journals were manually coded by the instructor in a spreadsheet. Three themes were identified: Iterative Refinement, Critical Vetting, and Dashboard Literacy Gap.

4. RESULTS

4.1 RQ1: Impact of AI Scaffolding on MLD Accuracy

To better understand students' baseline modeling skills, they were asked to complete exercises manually before using AI tools. These initial diagrams served as a reference point for evaluating how AI influenced their understanding and precision. Figure 1 presents one such hand-drawn Merise MCD created during the early phase of the exercise, prior to AI consultation.

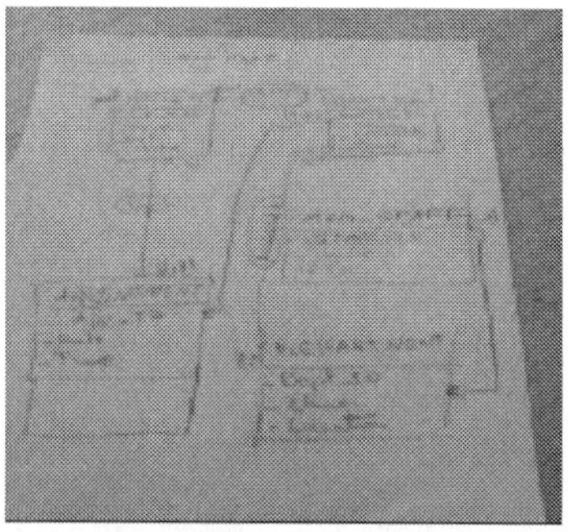

Figure 1. Sample manually drawn MCD diagram by a student for the Hospital Management System task

To assess the extent to which AI assistance improved students' ability to correctly transform Merise (MCDs) into Logical Data Models (MLDs), we compared performance on the six modeling exercises across the two phases (Manual vs. AI-Assisted). A paired-samples t-test revealed a statistically significant and large improvement in MLD accuracy scores (evaluated using the 10-point rubric) when students used ChatGPT for assistance. The mean accuracy increased from 50% (SD = 12.3) in the manual phase to 78% (SD = 10.1) in the AI-assisted phase. This represents a substantial mean increase of 28 percentage points. The difference was statistically significant, $t(89) = 18.7$, $p < .001$ (Bonferroni-corrected $\alpha' = 0.025$). The effect size, calculated using Cohen's d for paired samples, was $d = 1.9$, indicating a very large effect according to conventional benchmarks. The 95% confidence interval for the mean difference was [24%, 32%], further underscoring the robustness of the improvement. A paired-samples t-test revealed a significant improvement in students' MLD design accuracy after using AI assistance (M = 28.1 pp, SD = 14.5), compared to their manual performance, $t(89) = 18.7$, $p < .001$, $d = 1.90$, 95% CI [24.0, 32.0] pp. This qualitative data from AI transcripts illustrated how this improvement occurred (see Figure 1). Students used AI to verify specific transformation rules or get feedback, as exemplified by one student's interaction regarding a many-to- many relationship:

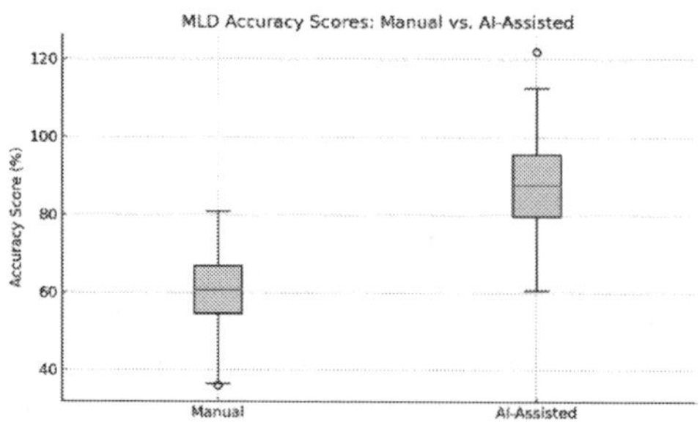

Figure 2. Boxplot comparing MLD accuracy scores between manual attempts and AI-assisted modeling

Table 1. Comparison of Mean MLD Accuracy Scores (N=90)

Phase	Mean MLD Score (%)	Standard Deviation (SD)	95% Confidence Interval (CI)
Manual	50	12.3	[46, 54]
AI-Assisted	78	10.1	[75, 81]

> *"After the AI suggestion to add an 'Enroll' junction table with composite keys based on StudentID and CourseCode, my logical model finally passed validation and made sense."* (Student G03, Transcript Excerpt)

4.2 RQ2: Influence of SLA Dashboards on Forum Engagement

To evaluate the impact of introducing weekly SLA dashboards on collaborative behavior, we compared forum interaction metrics before (Weeks 1-5 average) and after (Weeks 6-13 average) the dashboard deployment. Network density (ρ), representing the proportion of actual connections relative to all possible connections in the forum interaction network, was used as a primary indicator of overall group engagement. A paired-samples t-test (comparing average pre- vs. post-dashboard density across the three Moodle groups, though analysis ideally uses individual data if possible or appropriate MLM) indicated a statistically significant increase in forum density. The average density rose from $\rho = 0.30$ (SD = 0.05) before dashboard implementation to $\rho = 0.36$ (SD = 0.06) afterwards. This increase was statistically significant. The effect size was Cohen's d = 0.44, suggesting a medium effect of the dashboards on increasing interaction density. A comparison of MLD accuracy between students with prior exposure to AI tools and those without showed a significant difference, $t(89) = 4.2$, $p < .001$, $d = 0.44$, with higher accuracy among students who had prior experience. (See Figure 2).

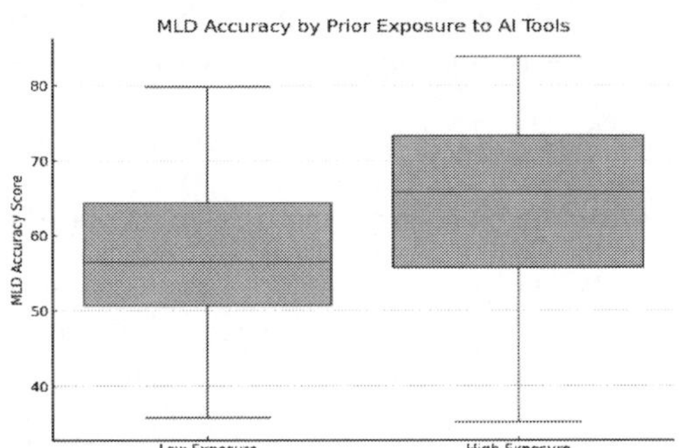

Figure 3. Comparison of MLD accuracy scores between students with and without prior AI exposure

Reflective journals provided context for this quantitative finding. Some students explicitly mentioned the dashboards prompting them to engage more:

> *"The dashboard showed my group's low centrality score compared to others. It made me realize I wasn't contributing much, so I posted extra explanations for the library exercise to help out." (Student G02, Reflection)*

4.3 RQ3: Student Perceptions, Strategies, and AI Interaction Patterns

Survey data provided insights into students' overall perceptions of the AI tool: A large majority (82%) agreed or strongly agreed that using ChatGPT helped clarify Merise rules (Mean = 4.1 on a 5-point scale, SD = 0.7). Trust in the AI's suggestions for conceptual data modeling was relatively high, with 70% agreeing or strongly agreeing (M = 3.8, SD = 0.9). Intention to use AI tools for similar tasks in the future was also positive, with 65% agreeing or strongly agreeing (M = 3.6, SD = 1.0).

Analysis of the 540 AI conversation transcripts revealed dominant interaction patterns and common query types: Common Query Types: The most frequent type of support sought was Conceptual Clarification (42% of interactions coded), where students asked for explanations of Merise concepts (e.g., "What is the rule for a 1:1 relationship?"). Model Transformation assistance (35%) was also common (e.g., "Convert this MCD to MLD"), followed by requests related to Syntax and Representation (23%) (e.g., "Show me this MLD in PlantUML format"). Typical Workflow: A common pattern observed in transcripts involved students working through the modeling steps sequentially with AI validation: 1) Identifying entities/attributes, 2) Defining relationships and cardinalities (often asking AI to verify), 3) Requesting the MCD-to-MLD conversion, and 4) Asking for the output in a specific format like PlantUML or SQL table definitions.

Thematic Analysis of Transcripts and Reflections: Qualitative analysis identified key themes regarding interaction strategies and perceptions:

1. Iterative Prompt Refinement: Successful interactions often involved students asking follow-up questions, providing corrections to the AI, or refining their prompts based on initial AI responses. This iterative dialogue appeared crucial for resolving complex issues.

> Student (G03): "I see a N:N between Student and Course. Should I create a separate table?"

> ChatGPT: "Yes---use ENROLL(StudentID, CourseCode) as a composite PK/FK table."

> Student (G03): "Okay, but what about the enrollment date attribute? Where does that go?"

> ChatGPT: "Good point. The enrollment date should be an attribute within the ENROLL junction table."

2. Critical Vetting of AI Output: While students generally trusted the AI, many transcripts showed evidence of critical evaluation, where students questioned AI suggestions that seemed incorrect or inconsistent with lecture material.

> Student (G01): "Why did you put UNIQUE constraint only on CourseCode in the Enroll table? Shouldn't the combination of StudentID and CourseCode be unique?"

> ChatGPT: "You are correct. My apologies. The primary key for the ENROLL table should indeed be a composite key comprising both StudentID and CourseCode to ensure each student can enroll in a specific course only once."

> Student (G01): "Understood, that makes more sense based on the rules."

3. Dashboard Literacy Gap and Interpretation: Reflections indicated that while the SLA dashboards spurred engagement for some, their initial interpretation was not always intuitive.

> Reflection (Student G01): "At first, I didn't know what the density score $\rho=0.36$ meant or why my centrality was low. After the instructor explained how to read the dashboard in the class session, I understood better and tried to engage more with peers' questions on the forum."

AI Response Evaluation (Based on Researcher Analysis of Transcripts): Our analysis of ChatGPT's responses within the transcripts revealed consistent strengths and weaknesses: **Strengths:** The AI demonstrated high accuracy (estimated >95% correctness in sampled transcripts) in designating primary and foreign keys during MLD transformation and correctly identifying the need for junction tables for many-to- many relationships. It also handled requests in multiple languages (French and English observed). **Weaknesses:** Common issues included inconsistencies in output formatting, occasional failure to enforce specific constraints (like UNIQUE constraints for identifying relationships derived from 1:1 MCD relationships), and sometimes omitting relationship verbs or details crucial for full semantic understanding of the Merise model.

5. DISCUSSION

This study investigated the combined impact of SLA dashboards and GenAI (ChatGPT) assistance on student learning and collaboration within a practice-oriented introductory IS course focused on Merise database modeling. The findings offer valuable insights into the potential and challenges of integrating these technologies, interpreted here in relation to our research questions and the broader literature.

5.1 RQ1: The Potent Effect of AI Scaffolding on Modeling Accuracy

The most striking finding was the dramatic improvement in MLD accuracy when students utilized ChatGPT assistance, evidenced by a 28-percentage point increase and a very large effect size (d=1.9). This confirms the potent capability of current GenAI models to act as effective cognitive scaffolds for complex, rule-based procedural tasks like Merise modeling (Wu, 2024; Nguyen et al., 2024). The AI likely achieved this by offloading significant cognitive load associated with recalling specific transformation rules, identifying correct primary/foreign key relationships, and applying normalization steps, allowing students to focus on understanding the process rather than getting bogged down in syntactic or minor rule errors. The qualitative data supports this, showing students actively using the AI for just-in-time conceptual clarification and step-by-step validation. This aligns with theories suggesting that reducing extraneous cognitive load facilitates deeper learning and schema acquisition (Paas et al., 2003). The high accuracy in PK/FK designation (>95%) observed in our analysis of AI responses further underscores the tool's reliability for core transformation tasks, making it a powerful practice aid.

5.2 RQ2: SLA Dashboards as Catalysts for Engagement, Mediated by Literacy

The introduction of SLA dashboards resulted in a statistically significant, medium-sized increase (d=0.44) in forum interaction density. This suggests that making collaborative activity visible can indeed act as a catalyst, prompting increased engagement within the learning community, consistent with SLA literature promoting metacognitive reflection on participation (Cohn et al., 2024; Khosravi et al., 2025). However, the effect size, while significant, was moderate compared to the AI's impact on individual performance. Furthermore, the qualitative findings revealed a crucial mediator: dashboard literacy. Students only translated the visualized metrics (like density or centrality) into behavioral change (posting more, helping peers) after receiving explicit instruction on how to interpret the dashboards. This highlights that simply providing analytics is insufficient; effective pedagogical integration requires dedicated efforts to develop students' data literacy skills so they can understand and act upon the insights provided. Without this, dashboards risk being perceived as interesting but not actionable.

5.3 RQ3: Navigating AI Use: Strategies, Perceptions, and the Synergy Gap

Students generally perceived the AI positively, particularly valuing its ability to clarify rules (82% agreement) and expressing relatively high trust (70% agreement). The thematic analysis revealed sophisticated interaction strategies beyond simple question- answering. The prevalence of iterative prompt refinement and critical vetting of AI outputs suggests that many students engaged with the AI not as a passive oracle but as an interactive partner, aligning with effective learning practices highlighted in recent GenAI research (Dehkhoda et al., 2024; Khosravi et al., 2025). This active engagement is crucial, given the AI's identified weaknesses in areas like constraint enforcement and formatting consistency. Students who critically evaluated and questioned the AI were better positioned to leverage its strengths while mitigating its weaknesses. Despite the individual benefits of AI and the moderate engagement boost from SLA, the synergy between the two technologies was not strongly perceived by students, nor was it explicitly designed into the intervention beyond concurrent deployment. Students largely viewed the AI as a tool for individual task completion and the dashboards as a separate indicator of group activity. The theoretical potential for SLA insights to trigger or guide AI use (e.g., suggesting AI help for groups struggling with a concept identified in forums) was not realized in this implementation. This points to a need for more sophisticated integration designs in future work, potentially involving system-level connections or more structured pedagogical strategies that explicitly link collaborative awareness (from SLA) with targeted cognitive support (from AI).

5.4 Implications for Practice-Oriented IS Education

These findings have significant implications for educators teaching structured modeling or similar technical skills: Leverage AI as a Practice Tool: GenAI is a powerful tool for providing scalable, individualized practice and feedback on rule-based tasks. Educators should embrace it but structure its use. Teach AI Interaction Skills: Explicit instruction and practice in iterative prompting and critical evaluation of AI outputs are essential to move students beyond passive solution seeking towards active co-construction of knowledge. Integrate SLA with Purpose: SLA dashboards can foster collaboration but require dedicated instruction on data interpretation. Their impact may be amplified if linked to specific collaborative activities or pedagogical interventions. Design for Synergy: Future interventions should explore explicit links between SLA and AI, e.g., using forum analysis to recommend AI prompts, or prompting students to share and critique AI-generated solutions within their groups. Ultimately, the effective integration of these technologies requires a pedagogical approach that guides students in using them thoughtfully as tools to enhance, not replace, their own critical thinking and collaborative skills.

6. CONCLUSION

This study provides empirical evidence on the integration of SLA dashboards and GenAI (ChatGPT) within a practice-oriented introductory Information Systems course focused on the challenging task of Merise database modeling. Our mixed-methods investigation demonstrates that the thoughtful integration of these technologies can yield significant educational benefits. GenAI proved to be a highly effective cognitive scaffold, substantially improving students' accuracy in transforming conceptual to logical data models. It facilitated learning by providing immediate, interactive support for rule application and procedural steps. Concurrently, the deployment of SLA dashboards led to a moderate but significant increase in collaborative engagement within course forums, suggesting that visualizing interaction data can foster greater participation and awareness. However, the study also underscores critical nuances: the effectiveness of AI is mediated by students' ability to engage in iterative prompting and critical evaluation, while the impact of SLA dashboards is contingent upon students' data literacy and explicit pedagogical guidance on interpreting and acting upon the visualized metrics. The anticipated synergy between SLA and AI requires deliberate instructional design to bridge the gap between social awareness and individual cognitive support. Several limitations should be acknowledged when considering the implications of these findings. First, the study was conducted with a specific cohort of second-year computer engineering students at a single Algerian university; the results may not directly generalize to different student populations, academic disciplines, or cultural contexts. Second, the AI tool used was the freely available version of ChatGPT prevalent during the study period; future iterations or different AI models might yield different results regarding accuracy or interaction patterns. Third, the comparison between 'Manual' and 'AI-assisted' phases on the same exercises might introduce practice effects, although the large effect size for AI assistance suggests a substantial tool-specific impact. Fourth, the measurement of MLD accuracy relied on a specific rubric, and while inter-rater reliability was high, evaluating complex models inherently involves some subjectivity. Similarly, survey data relies on self-reporting, which can be subject to biases. Fifth, tracking the precise cognitive processes and the depth of understanding gained through AI interaction solely via transcripts is inherently limited. Finally, while we attempted to control the context, factors like instructor variations across the three Moodle groups or individual student motivation levels were not fully isolated. Based on our findings, we offer the following recommendations: **For Educators:** Embrace GenAI as a powerful tool for guided practice in structured domains like Merise modeling, but integrate it purposefully. Crucially, dedicate class time to developing students' AI literacy, including effective prompt engineering strategies and techniques for critically evaluating AI-generated outputs. When using SLA dashboards, provide explicit training on interpreting the metrics and facilitate discussions on how insights can inform collaborative strategies. Consider designing activities that explicitly link SLA insights (e.g., identifying common errors in forums) with targeted AI-based exercises or prompts. **For Students:** Approach GenAI as an interactive assistant, not an infallible oracle. Engage in iterative prompting, ask clarifying questions, and always critically verify AI suggestions against course materials and established rules. Utilize SLA dashboards, when available and understood, to reflect on personal and group participation, seeking opportunities to contribute and collaborate effectively. **For AI Developers and Learning Technologists:** Enhance educational

AI tools by incorporating domain-specific knowledge and validation checks (built-in Merise schema validation). Explore features that promote critical thinking rather than just providing answers. Investigate ways to integrate SLA insights more seamlessly with AI tools, potentially developing platforms where alerts about collaboration patterns or conceptual misunderstandings (identified via SLA) could trigger automated, context-aware AI interventions or hints. Future research should explore these integrated approaches longitudinally to assess impacts on knowledge retention and transfer to novel problems. Comparative studies examining different SLA dashboard designs, AI models, and integration strategies are needed. Investigating the development of students' critical AI evaluation skills over time and exploring the evolving role of the instructor in AI-augmented learning environments are also crucial next steps.

REFERENCES

Allgood, M. and Musgrave, P., (2024), Leveraging ChatGPT in public sector human resource management education. *Journal of Public Affairs Education*, 30, 494–510.

Boudia, C. and Bengueddach, A., (2024a), ChatGPT in Higher Education: Advantages, Limitations, and Perspectives.

Boudia, C. and Bengueddach, A., (2024b), Innovating Higher Education: The Rise of ChatGPT in Collaborative Learning - A Short Review. In *Proceedings of ICRES 2024 - International Conference on Research in Education and Science*, Antalya, Turkiye:, pp. 1347–1370.

Boudia, C. and Krismadinata, (2024), Early insights into SLA with chatGPT: Navigating CS teachers and student perspectives in an opinion-based exploration. *Edelweiss Applied Science and Technology*.

Braun, V. and Clarke, V., (2019), Reflecting on reflexive thematic analysis. *Qualitative Research in Sport, Exercise and Health*, 11, 589–597.

Cohn, C., Snyder, C., Fonteles, J.H., T. S., A., Montenegro, J., and Biswas, G., (2025), A multimodal approach to support teacher, researcher and AI collaboration in STEM +C learning environments. *British Journal of Educational Technology*, 56, 595–620.

Creswell, J.W., Shope, R., Plano Clark, V.L., and Green, D.O., (2006), How interpretive qualitative research extends mixed methods research. *Research in the Schools*, 13, 1–11.

Dehkhoda, A.M., Kipp, D., and Chow, C.F., (2024), Transformative Pedagogy: Leveraging Generative AI Tools for Enhanced Learning Experiences. *Proceedings of the Canadian Engineering Education Association (CEEA)*.

Glick, D., Cohen, A., Festinger, E., Xu, D., Li, Q., and Warschauer, M., (2019), Predicting Success, Preventing Failure: Using Learning Analytics to Examine the Strongest Predictors of Persistence and Performance in an Online English Language Course. In D. Ifenthaler, D.-K. Mah, and J.Y.-K. Yau (eds), *Utilizing Learning Analytics to Support Study Success* ((Cham: Springer International Publishing)), pp. 249–273.

Khosravi, H., Shibani, A., Jovanovic, J., Pardos, Z.A., and Yan, L., (2025), Generative AI and Learning Analytics: Pushing Boundaries, Preserving Principles. *Journal of Learning Analytics*, 12, 1–11.

Kim, H.A., (2025), A Case Study on SAL-Based Teaching and Learning in a Functional Clothing and Materials Course Using Generative AI. *IEEE Access*, 13, 73430–73439.

Nguyen, A., Ilesanmi, F., Dang, B., Vuorenmaa, E., and Järvelä, S., (2024), Hybrid intelligence in academic writing: examining self-regulated learning patterns in an AI-assisted writing task. In *HHAI 2024: Hybrid Human AI Systems for the Social Good*, pp. 241–254.

Paas, F., Renkl, A. and Sweller, J., (2003). Cognitive load theory and instructional design: Recent developments. *Educational Psychologist*, 38.

Wu, S., (2024), Artificial intelligence-enhanced learning: A new paradigm in the "business data analysis and application" course. *Journal of Contemporary Educational Research*, 8, 164–175.

Wu, T.-T., Lee, H.-Y., Li, P.-H., Huang, C.-N., and Huang, Y.-M., (2024), Promoting Self-Regulation Progress and Knowledge Construction in Blended Learning via ChatGPT-Based Learning Aid. *Journal of Educational Computing Research*, 61, 3–31.

Zimmerman, B.J., (2002), Becoming a Self-Regulated Learner: An Overview. *Theory Into Practice*, 41, 64–70.

PROBLEM SOLVING AND ARTIFICIAL INTELLIGENCE: A BLENDED INTENSIVE PROGRAMME EXPERIENCE

Vittoria Bonanzinga
Dipartimento di Ingegneria dell'Informazione, delle Infrastrutture e dell'Energia Sostenibile, Università Mediterranea di Reggio Calabria, via Graziella (Feo di Vito), 89100, Reggio Calabria, Italy

ABSTRACT

This study presents the outcomes of a mixed-methods evaluation of the "Problem Solving and A.I." Blended Intensive Programme (BIP), an Erasmus+ initiative conducted at the DIIES Department of the Mediterranea University of Reggio Calabria. Designed to foster transversal skills and AI literacy through experiential and collaborative learning, the program engaged international students in interdisciplinary activities combining theoretical seminars, practical labs, and intercultural teamwork. Grounded in experiential learning and social constructivist theories, the study employed a structured post-course questionnaire (available in Appendix), (N=20) and final test performance (N=36) to assess teaching effectiveness, skill acquisition, and student engagement. Quantitative data revealed high satisfaction (average score: 4.3/5) and strong knowledge acquisition (average test score: 90.65%). Qualitative feedback, analyzed through thematic coding, highlighted the program's success in fostering transversal skills (e.g., critical thinking, teamwork) and intercultural engagement, particularly in problem-solving and intercultural communication. Challenges related to time management and content density were noted. The integration of theoretical AI foundations (e.g., neural networks, optimization algorithms) with hands-on applications (e.g., cryptography exercises, predictive maintenance models) proved particularly effective. Recommendations for future editions include optimizing session duration and offering modular content pathways. The study contributes to the growing literature on BIPs by offering insights into the pedagogical design and impact of AI-focused international education initiatives.

KEYWORDS

Blended Intensive Programme, Erasmus+, Artificial Intelligence, Problem Solving, Digital Didactics, International Collaboration

1. INTRODUCTION

The landscape of higher education is constantly evolving, driven by the need to train professionals capable of addressing complex and rapidly changing challenges—particularly in fields like Artificial Intelligence (AI). The rapid integration of AI across industries has created an urgent demand for educational models that develop both technical proficiency and human-centric problem-solving skills (Russell & Norvig, 2021). However, traditional computer science curricula often emphasize algorithmic knowledge without sufficient connection to real-world applications (Ng et al., 2020). This gap highlights the importance of situated learning experiences that mirror the interdisciplinary and ethical challenges of AI in practice (Wenger, 2019; Floridi et al., 2018). In this context, Blended Intensive Programmes (BIPs) represent an innovative teaching modality, introduced within the Erasmus+ program, which combines the scalability of digital learning with the rich interaction of short-term physical mobility (European Commission, 2023). Grounded in Kolb's (2014) experiential learning theory and Vygotsky's (1978) social constructivism, BIPs aim to bridge critical gaps in contemporary education by: Linking algorithmic theory with practical implementation in real-world contexts (Ng et al., 2020); Fostering intercultural collaboration in technical disciplines, which is often limited in traditional programs (Gögele & Kletzenbauer, 2023); Developing critical thinking about AI's ethical and societal dimensions (Floridi et al., 2018). The primary objective of BIPs is to promote the internationalization of curricula, a multidisciplinary approach, and the development of transversal skills (soft skills) through intensive, collaborative activities. This model builds on the success of earlier initiatives like the Erasmus Intensive Programme (IP), which demonstrated the value of international cooperation and hands-on learning in industrial settings (Martin Mellado et al., 2024). Recent studies further highlight BIPs' role in advancing higher education, from enhancing internationalization (Gögele & Kletzenbauer, 2023) to enriching discipline-specific learning in fields like physics (González-Pavón et al., 2024). The "Problem Solving and A.I." BIP was designed to address these needs, offering a dynamic, multicultural environment where students engage with both the technical foundations and broad applications of AI across industries. By integrating problem-solving methodologies—a critical 21st-century competency—with AI education, the programme equips learners to analyze complex challenges, structure solutions, and innovate responsibly. This approach not only aligns with industry demands but also prepares students to navigate the ethical and intercultural dimensions of AI's expanding role in society.

2. THE "PROBLEM SOLVING AND A.I." COURSE

2.1 Structure and Participants

The BIP involved 36 students from four European universities, selected based on academic background and interest.
The hybrid format included:

- **Asynchronous online module**: Pre-course theoretical content, including the use of MATLAB.
- **In-person intensive week (March 31–April 4, 2025)**: Lectures, labs, and group work.

2.2 Methodologies and Content

The course structure integrated a variety of teaching methodologies:

- Theoretical Seminars: Frontal lectures on the fundamental principles of Artificial Intelligence, including core AI sub-fields such as Expert Systems, Robotics, Deep Learning, Computer Vision, Machine Learning, Natural Language Processing, Neural Networks, and Fuzzy Logic. In particular, the topic of Deep Learning was explored with reference to the comprehensive work of Goodfellow et al. (2016), which is widely recognized as a foundational text in the field. These sessions also covered the AI training process, the biological inspiration behind neural networks, and the mathematical underpinnings of problem-solving and optimization algorithms. A significant focus was placed on the AI training process, detailing how AI systems learn through input data, processing functions, training with past data, and iterative error correction, mirroring human brain learning. Furthermore, the seminars explored the biological inspiration behind AI, notably discussing the McCulloch-Pitts experiment (1943) which introduced the first mathematical model of an artificial neuron, and the Hubel & Wiesel experiment (1962), which revealed how the brain processes visual information and inspired the development of artificial neural networks and deep learning architectures like Convolutional Neural Networks (CNNs). The core components of AI—algorithms, data, and computing power—were also thoroughly examined. These sessions also covered mathematical analysis for complex problem modelling and problem-solving methodologies, including a detailed exploration of Problem Solving as a cognitive process involving six key phases: Problem Identification, Problem Analysis, Solution Generation, Evaluation, Implementation, and Results Evaluation. Various strategies for problem solving were discussed, such as Divide and Conquer, Lateral Thinking, Analogy and Reduction, and the Scientific Method. Common barriers in problem solving, including Confirmation Bias, Mental Set & Functional Fixedness, Unnecessary Constraints, and Irrelevant Information, were also addressed. Specific algorithms for optimization and graph theory were introduced, including Prim's Algorithm, Kruskal's Algorithm, Dijkstra's Algorithm, and the Ford-Fulkerson Algorithm, with their algorithmic overviews and pseudocode. The key areas of AI, such as Computer Vision, Natural Language Processing, Robotics, and their applications in the Automotive Industry and Medicine, were also introduced.
- Practical Laboratories: Hands-on sessions where students applied theoretical concepts through programming exercises and the use of specific software tools for AI and data analysis. These labs allowed students to engage directly with AI applications, such as those in industrial automation,

medicine, finance, transportation, customer support, and entertainment. A notable example included Digit Recognition, where students learned to classify images of handwritten digits (e.g., '1' and '7'), demonstrating how CNNs can automate visual tasks like scanning forms or processing checks. Another practical application explored was Machine Learning for Predictive Maintenance, highlighting benefits such as proactive maintenance, cost savings, improved equipment performance, increased equipment lifespan, data-driven insights, and enhanced safety through optimization of design, data-driven decision making, process optimization, and quality control. The progress of model training, showing accuracy convergence and loss reduction, was also a key practical outcome, with examples demonstrating high validation accuracy (e.g., 99.33%) achieved over a few epochs. Examples of applications in autonomous vehicles were also presented, showing how sensors and protocols interact in an in-vehicle system for navigation and safety. Furthermore, students explored real-world applications of optimization algorithms, such as using Dijkstra's algorithm for Traffic Management (shortest path calculation and dynamic traffic signal optimization), Kruskal's/Prim's algorithms for Electricity Networks (optimal grid design, demand prediction, and renewable energy integration), and path-finding algorithms for Social Networks (optimizing connectivity, personalized content, and influential user identification). Practical examples of personal optimization were also discussed, including travel planning (finding optimal routes considering real-time data), study and work schedule optimization (applying scheduling algorithms to maximize efficiency and balance life), and meal planning and grocery shopping optimization (creating cost-effective and nutritionally balanced menus). Students also engaged in exercises directly related to cryptography and cipher analysis, which served as practical demonstrations of problem-solving and algorithmic thinking. These included:

  - Caesar Cipher Decryption: Students tackled the problem of decrypting a message encrypted with a Caesar cipher (a fixed shift of 7 positions), successfully restoring the original text: "Charlie is going to intensify support to the grey zone, please spread disinformation by exposing presumed pollution effects of its industrial activity.". This exercise highlighted the simplicity and vulnerability of such methods, emphasizing that simple substitution ciphers do not provide high levels of security.
  - Analysis of a "Doubling" Affine Cipher: Students investigated a scheme reminiscent of an affine cipher where each letter was "doubled" (e.g., A -> A, B -> C, C -> E, corresponding to $E(x)=2x (\mathrm{mod}\,26)$). They discovered that attempts at decryption lead to ambiguity because the function is not invertible (due to 2 and 26 not being coprime), meaning each ciphertext letter can originate from two different plaintext letters. For instance, a ciphertext 'A' (value 0) could come from plaintext 'A' (value 0) or 'N' (value 13). This demonstrated the crucial concept of non-reversibility in encryption and the importance of coprime multipliers in modular arithmetic for unique decryption. Furthermore, they explored how many different 5-letter plaintext words could result in the same ciphertext, reinforcing the understanding of mapping ambiguities in such ciphers.
  - Vernam Cipher (One-Time Pad) Security Analysis: Students examined the Vernam cipher, or one-time pad (OTP), understanding its theoretical advantage of absolute secrecy when the key and message have the same length and the key is truly random and used only once. They were challenged to consider if there could be a "nontrivial key" that would compromise this absolute secrecy, prompting critical thought on the practical implementation aspects of even theoretically secure ciphers.

- Interdisciplinary Group Work: The central component of the program, where students, organized into multicultural teams, tackled real or simulated problem-solving challenges, using AI-based approaches and applying the problem-solving methodologies and optimization algorithms learned. These projects fostered the exchange of ideas, negotiation, and task division, demonstrating how AI mimics and extends human problem-solving abilities. The cryptographic exercises mentioned above also contributed to this, as students likely collaborated to analyse and solve these problems, applying their understanding of algorithms and mathematical concepts in a team setting.
- Critical Thinking Sessions: Interactive sessions to delve deeper into specific topics and facilitate comparison between different cultural and academic perspectives. These discussions also touched upon the Role of Critical Thinking in AI Development, emphasizing its objective analysis for better judgment and its phases (analyse statements, subdivide reasons, check for contradictions, weigh

evidence, evaluate opinions independently). The analysis of cipher vulnerabilities and the discussion on Vernam cipher security were prime examples of how critical thinking was applied to evaluate complex systems.

The main educational objective was to provide students with a solid theoretical and practical foundation in problem-solving techniques and Artificial Intelligence, while simultaneously promoting the development of transversal skills such as critical thinking, collaboration, intercultural communication, and adaptability.

3. METHODOLOGY

To evaluate the effectiveness of the "Problem Solving and A.I." Blended Intensive Programme, a post-course survey was conducted. The adopted methodology, inspired by Creswell's (2014) framework for mixed methods research, involved the administration of a final questionnaire designed to collect detailed feedback on the training experience from the participants.

The data collection instrument was a structured questionnaire, composed of both closed questions (multiple choice and 5-point Likert scales, where 1 = "Not at all satisfied/agree" and 5 = "Very satisfied/agree") and open questions, to allow participants to freely express opinions and suggestions. The areas investigated included:

- Teaching Effectiveness: Evaluation of the quality of lectures, the preparation of lecturers, and the clarity of content.
- Course Organization: Logistics, time management, and program coherence.
- Active Participation and Engagement: Perceived opportunity to contribute, interaction with peers and lecturers.
- Skill Development: Self-assessment on the acquisition of new knowledge and the improvement of transversal skills (e.g., problem solving, teamwork, intercultural communication).
- Critical Issues and Suggestions: Identification of any difficulties encountered and proposals for future improvement.

The questionnaire was administered to all 36 students participating in the BIP at the end of the in-person activities, ensuring anonymity to encourage sincere responses. A total of 20 students completed the questionnaire, corresponding to a response rate of approximately 56%. Quantitative data were analysed using descriptive statistics (means, percentages), while responses to open questions were subjected to thematic analysis to identify patterns and recurring themes.

4. RESULTS

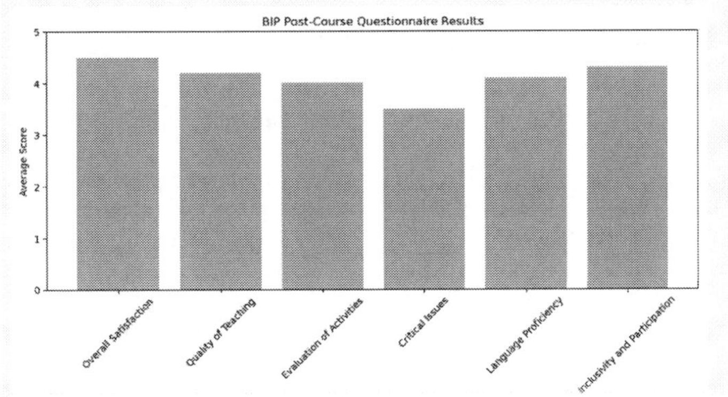

Figure 1. BIP Post-Course Questionnaire Results. This figure illustrates the average scores for various aspects of the "Problem Solving and A.I." Blended Intensive Programme, showing high satisfaction and positive evaluations across most categories, with "Critical Issues" being the lowest scored area

The analysis of the questionnaires revealed an overall positive picture of the "Problem Solving and A.I." BIP experience, while highlighting some areas for improvement.

- Overall Satisfaction: Over 80% of students expressed high satisfaction with the knowledge acquired and the overall program experience, with an average score of 4.3 out of 5. This suggests that the course effectively met participants' expectations in terms of learning.
- Quality of Teaching: Lecturers were rated very positively, with an average score between 4 and 5 out of 5 for their preparation, presentation skills, and availability for interaction. Students particularly appreciated the clarity of explanations and the lecturers' ability to stimulate debate and active participation.
- Evaluation of Proposed Activities: Practical activities and group work were considered extremely stimulating and relevant. Ninety percent of students emphasized the importance of teamwork and the practical application of acquired skills as key elements for effective learning and the development of transversal skills.
- Emerging Critical Issues: Among the open comments, a significant percentage of students (approximately 35%) reported the perception of prolonged times for some lecture sessions and, in some cases, an excessive density of content. Some suggested the need to dedicate more time to in-depth study of specific topics or to reduce the number of concepts introduced in a single session.
- Language Proficiency: The majority of students (approximately 95%) reported no significant difficulties with English, the working language of the course. This indicates a good level of language preparation among participants and the effectiveness of communication by the lecturers.
- Inclusivity and Participation: All participants stated that they actively participated in the activities, feeling an integral part of the working group. This result highlights the program's success in creating an inclusive and collaborative environment, fostering interaction among students of different nationalities and academic backgrounds.

  The outcomes of the final assessment administered at the end of the in-person phase further reinforce the positive feedback gathered through the questionnaires. All participants successfully completed the test, which evaluated their understanding of key concepts in artificial intelligence, optimization algorithms, and critical thinking applied to real-world problems. The average score was remarkably high, with most students achieving between 13 and 15 out of 15 points. This result demonstrates not only the high level of engagement and preparation among the students but also the effectiveness of the instructional design and the pedagogical approach adopted in the BIP. The alignment between theoretical content, practical applications, and interdisciplinary problem-solving appears to have strongly supported knowledge retention and skill acquisition.

5. DISCUSSION

The results of this survey confirm the effectiveness of Blended Intensive Programmes as innovative tools to promote educational innovation and interdisciplinarity in higher education. The "Problem Solving and A.I." BIP experience demonstrated how the combination of virtual learning and physical mobility can enhance the exchange between students from different academic cultures, mutually enriching perspectives and problem-solving methodologies.

Artificial Intelligence, as the central theme of the program, proved particularly suited to the "learning by doing" approach. The practical application of AI concepts to real-world problem solving, informed by an understanding of its fundamental components and biological inspirations (as highlighted by the McCulloch-Pitts and Hubel & Wiesel experiments), made the training experience more engaging and immediately relevant for the participants. This facilitated the comprehension of complex concepts and their translation into applied contexts, significantly contributing to the development of students' computational and analytical thinking. The program also contributed to the development of computational thinking, a key 21st-century skill that supports problem-solving and analytical reasoning in digital contexts. As highlighted by Shute et al. (2017), computational thinking involves practices such as decomposition, pattern recognition, abstraction, and algorithmic thinking, all of which were actively fostered through the course's AI-based activities and interdisciplinary challenges. The inclusion of concrete examples such as digit recognition and the benefits of machine learning for predictive maintenance, alongside discussions on broader AI applications in areas like robotics, computer vision, and autonomous vehicles, further solidified this practical understanding and showcased the diverse impact of AI. The deep dive into specific optimization algorithms (Prim's, Kruskal's, Dijkstra's, Ford-Fulkerson) and their real-world relevance in traffic management, electricity networks, and social networks, as well as their utility in personal optimization scenarios, provided a comprehensive view of how AI enhances problem-solving capabilities. The student exercises in deciphering messages and analysing cryptographic schemes further reinforced these practical applications, demonstrating how even seemingly simple ciphers involve fundamental problem-solving principles and a rigorous analytical approach.

The program also implicitly addressed the broad applications of AI, showcasing its transformative impact across various industries, including those highlighted in the student presentations such as industrial automation, medicine, finance, transportation, customer support, and entertainment. This practical exposure, coupled with theoretical insights, provided a holistic understanding of AI's role in modern society. The analysis of cipher vulnerabilities by the students provided a tangible example of understanding system weaknesses and the importance of robust algorithmic design, a critical aspect in AI development.

The critical issues that emerged, in particular those related to time management and content density, are consistent with the challenges typical of intensive programs. The compressed nature of BIPs requires a careful balance between the depth and breadth of the topics covered. The perception of "too much content in too little time" can potentially limit deep assimilation and critical reflection, despite the high overall satisfaction. This suggests that, while maintaining intensity, greater attention to students' "learning curve" and the distribution of workloads could further optimize teaching effectiveness.

The high level of participation and inclusivity is a strength of the program, highlighting its success in creating a stimulating and supportive learning environment. The cultural and academic diversity of the participants fostered a rich environment for insights, where the resolution of problems was approached from multiple angles, strengthening intercultural communication and teamwork skills. The emphasis on critical thinking as a tool for objective analysis and judgment, particularly in the context of AI development, was a crucial aspect of the program, exemplified by the students' work in evaluating the security of different ciphers.

Furthermore, the program's content implicitly raised important ethical considerations related to AI, such as data privacy, algorithmic bias, job automation, and accountability for AI decisions. While not explicitly detailed in the survey results, these aspects are crucial for a comprehensive understanding of AI and are likely to have been part of the broader discussions and awareness fostered during the program. The student presentations further underscored the need for engineers to develop both technical and ethical awareness in AI, and the benefits of AI + Human Collaboration, where human creativity and judgment combine with AI's data processing and optimization capabilities.

6. CONCLUSIONS

The "Problem Solving and A.I." Blended Intensive Programme represented a successful training experience, demonstrating how the strategic combination of hybrid activities, specialized content, and international collaboration can significantly contribute to the development of students' transversal and professional skills. The program effectively promoted active learning and engagement, providing participants with essential tools and methodologies to face the challenges of the 21st century, particularly those related to Artificial Intelligence, including its foundational principles, various sub-fields, practical applications, optimization algorithms, and emerging ethical considerations. The student exercises in analysing and deciphering various cryptographic schemes provided concrete examples of applied problem-solving and critical thinking within the AI context.

For future editions of the BIP, the following optimizations are suggested:

- Optimization of Session Duration: Review lesson planning to ensure more frequent breaks or slightly shorter sessions, in order to maintain high attention and facilitate content assimilation.
- Balance of Theory and Practice: Ensure an even more refined balance between theoretical exposition and practical activities, perhaps by introducing more time for autonomous application or for "deep dive" sessions on specific topics of greater interest to students.

- Content Flexibility: Consider the possibility of offering slightly customizable pathways or optional modules to allow students to delve deeper into areas of greater interest, mitigating the perception of excessive density.

In summary, the "Problem Solving and A.I." BIP is configured as a promising model for higher education, capable of training students who are not only technically competent but also equipped with the soft skills essential for navigating an increasingly complex and interconnected world.

APPENDIX

https://forms.office.com/e/6p8KhQUMyw?origin=lprLink

ACKNOWLEDGEMENT

Vittoria Bonanzinga is a member of the Gruppo Nazionale per le Strutture Algebriche, Geometriche e le loro Applicazioni, Istituto Nazionale di Alta Matematica, (GNSAGA- INdAM), Italy. The author is grateful to the anonymous referees for their insightful remarks and constructive feedback, which significantly contributed to the improvement of this work.

REFERENCES

Creswell, J. W., (2014). *Research design: Qualitative, quantitative, and mixed methods approaches*. Sage.

European Commission, (2023).*Erasmus+ Programme Guide* (latest edition).

Floridi, L. et al., (2018). AI4People—An ethical framework for a good AI society. *Nature Human Behaviour*, 2(12), 1052-1064.

Gögele, S. and Kletzenbauer, P., (2023). Blended Intensive Programmes: Promoting Internationalization in Higher Education. *Education and New Developments*, Vol.2, pp 381–383.

González-Pavón, C. et al. (2024). Bridging Learning and International Practice for Enhancing Physics Education. *European Journal of Education and Pedagogy*, Vol. 5, No. 4, pp 1–12.

Goodfellow, I. et al, (2016). *Deep Learning*. MIT Press, Cambridge, Massachusetts.

Gregurec, I. et al. (2024). Mastering the Maze of Blended Intensive Programmes Procedure - Case of Croatia. *EDULEARN24 Proceedings*. IATED, Palma, Spain, pp. 5341–5346.

Hubel, D. H. and Wiesel, T. N., (1962). Receptive fields, binocular interaction and functional architecture in the cat's visual cortex. *The Journal of Physiology*, Vol. 160, No. 1, pp 106-154.

Kolb, D. A. (2014). *Experiential learning: Experience as the source of learning and development*. FT Press.

Martin Mellado, E. et al., (2024). An International and Multidisciplinary Teaching Experience with Real Industrial Team Project Development. *arXiv*: 2403.15398.

McCulloch, W. S. and Pitts, W., (1943). A logical calculus of the ideas immanent in nervous activity. *The Bulletin of Mathematical Biophysics*, Vol. 5, No. 4, pp 115-133.

Ng, D. T. K.et al., (2020). AI literacy: Definition, teaching, evaluation and ethical issues. *Proceedings of the HKAIS International Symposium*.

Russell, S. J. and Norvig, P., (2021). *Artificial Intelligence: A Modern Approach* (4th ed.). Pearson.

Shute, V. J. et al., (2017). Demystifying computational thinking. *Educational Research Review*, 22, 142-158.

Vygotskij, L. S., (1978). *Mind in Society: The Development of Higher Psychological Processes*. Edited by M. Cole, V. John-Steiner, S. Scribner e E. Souberman. Cambridge, MA: Harvard University Press.

Wenger, E., (2019). *Communities of practice: Learning, meaning, and identity*. Cambridge University Press.

ADVANCING INTERNATIONAL ACCOUNTING EDUCATION WITH ARTIFICIAL INTELLIGENCE

Marja-Liisa Tenhunen
Dimitrie Cantemir Christian University, Bucharest, Romania

ABSTRACT

In today's rapidly evolving digital landscape, artificial intelligence (AI) has emerged as a transformative force across various disciplines, including education. This article explores the role of AI in advancing international accounting education at the university level. By integrating AI-driven tools and technologies, universities can enhance the quality and accessibility of accounting education, fostering a global perspective that meets the demands of an interconnected world. The study investigates the potential of AI to personalize learning experiences, automate administrative tasks, and provide real-time feedback, thereby increasing student engagement and improving outcomes. Furthermore, the article highlights the challenges and ethical considerations associated with adopting AI in academia, such as data privacy, algorithmic bias, and the digital divide. Through a review of current practices, case studies, and expert interviews, the paper presents actionable recommendations for effectively incorporating AI into accounting curricula. Emphasis is placed on preparing students not only with technical accounting knowledge but also with critical thinking and adaptability skills to thrive in the AI-driven workplace. Ultimately, this research underscores the importance of a collaborative effort among educators, policymakers, and technology providers to harness the potential of AI responsibly and equitably. By doing so, universities can cultivate a new generation of accountants who are equipped to navigate the complexities of the global economy and drive innovation within the field.

KEYWORDS

Artificial Intelligence, Accounting Education, University Learning, International Perspectives, Digital Transformation

1. INTRODUCTION

In an increasingly globalized world, the field of international accounting has gained prominence due to the harmonization of financial reporting standards, the mobility of capital, and the emergence of multinational enterprises. University-level education in international accounting plays a pivotal role in preparing future professionals to navigate the complexities of cross-border transactions, diverse regulatory frameworks, and varying ethical standards. With the adoption of International Financial Reporting Standards (IFRS) in over 140 jurisdictions, the demand for accounting graduates with a deep understanding of global accounting norms has significantly grown (Nobes & Parker, 2020, p. 24). However, many universities continue to rely on traditional pedagogical approaches that may not fully equip students for the dynamic challenges of international accounting practice. These methods often emphasize theoretical knowledge over analytical or adaptive skills. As such, there is a pressing need for higher education institutions to modernize their teaching strategies and curricula to foster not only technical competency but also technological agility and critical thinking (Jackling, de Lange & Natoli, 2013, p. 242).

Artificial Intelligence (AI) is transforming the landscape of higher education, offering powerful tools to enhance learning personalization, automate administrative processes, and support data-driven decision-making. In educational settings, AI applications such as intelligent tutoring systems, learning analytics, natural language processing, and automated feedback mechanisms are increasingly deployed to improve both teaching efficiency and student outcomes (Luckin et al., 2016, p. 12). AI can also support adaptive learning platforms that respond to individual student needs and performance in real time. These platforms utilize machine learning algorithms to analyze student data and recommend tailored learning paths, thereby improving retention, engagement, and mastery of complex topics (Baker & Inventado, 2014, p. 61).

Furthermore, AI-driven tools facilitate formative assessments, allowing instructors to identify learning gaps early and intervene more effectively.

Integrating AI into the teaching of international accounting holds transformative potential. It can enable educators to simulate real-world accounting environments, automate routine assessments, and offer immersive experiences through AI-powered case studies and interactive dashboards. This integration aligns educational outcomes with the evolving expectations of the accounting profession, which increasingly values digital literacy, data interpretation skills, and the ability to work with intelligent systems (IFAC, 2021, p. 6). By embedding AI into international accounting education, universities can cultivate a more inclusive and responsive learning environment. AI tools can assist non-native English speakers through real-time translation or transcription, accommodate learners with disabilities, and support diverse learning styles (Holmes, Bialik & Fadel, 2019, p. 28). For educators, AI enables data-informed instructional design and continuous improvement in teaching strategies. Given the convergence of globalization and digitalization, integrating AI into university-level international accounting education is not merely beneficial - it is imperative. Universities that embrace this integration will not only enhance their educational quality but also contribute to shaping a future-ready accounting workforce equipped to thrive in the digital economy.

2. LITERATURE REVIEW

The integration of artificial intelligence (AI) into higher education has garnered increasing attention as universities seek to enhance student engagement, improve learning outcomes, and personalize instruction. AI technologies such as intelligent tutoring systems, natural language processing, and learning analytics are transforming pedagogical approaches across disciplines (Luckin et al., 2016, p. 3). Holmes et al. (2019) argue that AI in education goes beyond automation by enabling real-time feedback, adaptive content delivery, and predictive analytics to identify at-risk students (p. 12). Their work categorizes AI applications into content creation, personalization, and assessment, highlighting the importance of ethical considerations and institutional preparedness (p. 14). A systematic review by Zawacki-Richter et al. (2019) analyzed 146 peer-reviewed articles on AI in higher education, revealing a concentration on learning analytics and automated assessment, with limited application in specialized fields such as accounting (p. 19). The review also emphasizes the need for disciplinary-specific AI strategies to avoid a one-size-fits-all implementation (p. 22).

Teaching international accounting at the university level traditionally relies on lectures, case studies, and comparative analyses of International Financial Reporting Standards (IFRS) and local standards (Gray & Needles, 1997, p. 88). This content-heavy approach often emphasizes compliance and reporting rather than judgment and strategic decision-making. Alon and Dwyer (2016) examined pedagogical strategies in international accounting and found that simulation and experiential learning - though effective - are underutilized in many curricula (p. 106). These strategies enable students to apply IFRS in real-world scenarios, promoting deeper learning and cross-border financial literacy. More recent approaches incorporate blended learning and flipped classrooms, particularly in response to the COVID-19 pandemic (Bryant et al., 2020, p. 238). However, despite digital delivery, the core curriculum has largely remained unchanged, focusing on static knowledge rather than dynamic problem-solving skills essential for future professionals in a global economy (p. 241).

Despite growing interest in EdTech, international accounting education still lacks integration with AI tools that foster personalized and contextual learning. One major gap is the absence of intelligent tutoring systems tailored to the complexities of IFRS and International Auditing Frameworks (IAF). As AI tools increasingly assist professionals in real-time decision-making (Deloitte, 2023, p. 17), universities lag behind in equipping students with similar tools for experiential learning. Another limitation is the minimal focus on ethical and cross-cultural challenges associated with AI-assisted accounting practices. While curriculum reforms often highlight international standards, they rarely address the implications of automated compliance systems or AI-driven audit technologies (Richins et al., 2017, p. 47). This creates a disconnection between academic training and professional realities. Moreover, faculty development in AI-enhanced pedagogy remains insufficient. According to a survey by PwC (2022), only 24% of accounting educators feel confident using AI-based tools in classroom settings (p. 9). The lack of institutional support, technical infrastructure, and interdisciplinary collaboration contributes to the slow adoption of AI in international accounting programs.

3. THE ROLE OF ARTIFICIAL INTELLIGENCE IN EDUCATION

Artificial Intelligence (AI) is rapidly transforming the educational landscape, enabling novel teaching methodologies, personalized learning experiences, and enhanced academic support systems. In the context of higher education, particularly international accounting education, AI technologies are opening new pathways to improve student engagement, instructional efficiency, and curriculum relevance in a globally connected world.

AI exhibits a wide array of capabilities that are reshaping how teaching and learning occur at the university level. These capabilities include personalized learning pathways, intelligent tutoring systems, automated grading, predictive analytics, and the facilitation of collaborative learning through virtual assistants and chat bots. One of the most significant contributions of AI in education is **personalized learning**. Through machine learning algorithms, AI can adapt instructional content to meet the needs of individual learners based on their pace, performance, and preferences. Platforms like *Carnegie Learning* and *Squirrel AI* have demonstrated how adaptive technologies can tailor educational content, particularly in complex subjects like mathematics and accounting, to improve learning outcomes (Woolf, 2020).

Intelligent tutoring systems (ITS) provide another transformative application of AI. These systems mimic human tutors by analyzing student responses and providing immediate, context-sensitive feedback. In accounting education, ITS can guide students through accounting exercises, offering hints and correcting misconceptions in real-time. AI also enables **automated assessment and feedback**, relieving instructors from repetitive grading tasks and allowing more time for curriculum development and student mentoring. For example, AI-powered platforms can assess student submissions for accounting journal entries, identify common errors, and provide standardized feedback across diverse student cohorts (Luckin et al., 2016). Furthermore, **predictive analytics** can identify students at risk of underperforming, allowing for early interventions. By analyzing patterns in student behavior, attendance, and assessment results, universities can tailor support services to improve retention and performance, especially among international students who may face linguistic and cultural barriers. Finally, **AI-powered chat bots** and virtual assistants can support administrative functions and student advising, ensuring 24/7 assistance for queries related to course registration, deadlines, or accounting terminology, thus enhancing the overall student experience.

4. AI APPLICATIONS IN INTERNATIONAL ACCOUNTING TEACHING

Artificial Intelligence (AI) is reshaping the landscape of international accounting education by introducing adaptive, data-driven solutions that enhance both teaching effectiveness and student learning experiences. This chapter explores four key areas where AI can be effectively applied: enhancing engagement, automating assessments, personalizing learning, and increasing accessibility.

AI-powered tools such as intelligent tutoring systems, chat bots, and virtual teaching assistants foster greater interactivity in international accounting classrooms. By simulating real-time feedback and conversation, these tools help students practice complex concepts like IFRS interpretations or international taxation scenarios in a dynamic, low-risk environment. For example, platforms integrating natural language processing (NLP) allow students to engage in dialogic learning, which has been shown to improve critical thinking and comprehension in accounting topics (Zawacki-Richter et al., 2019, p. 17). Gamified AI platforms also increase student motivation by transforming passive learning into interactive experiences.

AI facilitates the automation of objective assessments such as multiple-choice questions, numerical problems, and case analysis quizzes. Machine learning algorithms can also evaluate more complex submissions such as written reflections or audit simulations by analyzing patterns in student responses. This automation significantly reduces faculty workload while ensuring consistent and timely feedback. Studies show that automated grading systems, when carefully designed, can reach up to 90% agreement with human markers in accounting and business education (Holmes et al., 2021, p. 55).

Adaptive learning technologies tailor instructional content to individual student needs based on real-time performance data. In international accounting, where students often have varying levels of prior exposure to standards like Generally Accepted Accounting Principles (GAAP) or IFRS, AI can identify gaps and adjust learning paths accordingly. For instance, platforms like *Coursera* and *Edmodo* have integrated AI algorithms that adapt quizzes and video materials to individual learning speeds and styles (Lu et al., 2022, p. 103). This ensures that students remain engaged and appropriately challenged, improving mastery and retention.

AI enhances accessibility in accounting education for students with disabilities, those studying in a non-native language, or learners from different cultural backgrounds. Real-time translation, speech-to-text transcription, and text simplification engines help remove language and cognitive barriers. Moreover, AI-supported accessibility tools can convert technical accounting terminology into layman's terms, making complex concepts more digestible (Krouska et al., 2021, p. 88). This aligns with the broader inclusion goals of internationalized education frameworks.

In conclusion, the integration of AI into international accounting education offers transformative potential by fostering interactive, equitable, and efficient learning environments. As institutions globally seek to modernize curricula and improve outcomes, AI stands out as a strategic enabler of innovation in accounting pedagogy.

5. CASE STUDIES AND EXAMPLES OF AI IN UNIVERSTY TEACHING

Artificial Intelligence (AI) is transforming the way accounting is taught at universities, fostering more interactive, efficient, and industry-relevant education. Several universities have integrated AI tools into their teaching environments, offering valuable insights into best practices and challenges. The following case studies showcase how leading institutions have successfully integrated AI technologies into their accounting curricula, improving both teaching outcomes and student engagement.

Brigham Young University, USA, case *AI-Driven Data Analytics in Accounting.* Brigham Young University has taken a pioneering approach to incorporating AI and data analytics into its accounting programs. Through the use of AI-powered tools such as *Tableau, Power BI*, and *Alteryx*, students engage in hands-on learning to analyze large datasets in real-world accounting scenarios. The university also launched a dedicated Accounting Data Analytics course that emphasizes AI-based decision support and pattern recognition. Students work on case simulations using anonymized data from accounting firms, allowing them to practice AI-supported auditing and fraud detection techniques. According to Hansen et al. (2021, p. 134), BYU's model has enhanced students' understanding of the intersection between technology and accounting principles, preparing them for modern audit environments.

Monash University, Australia, case: *AI-Enhanced Simulation Platforms for Forensic Accounting*

Monash University has embraced immersive AI tools through its Forensic Accounting Simulation Platform (FASP). This platform mimics real-world fraud investigations, allowing students to act as forensic accountants who use AI tools to detect anomalies in financial transactions. It incorporates machine learning models that simulate fraudulent behaviors for students to identify and explain. This approach has received positive reviews from both students and professional bodies. According to Davis et al. (2020, p. 201), students reported greater engagement and practical skill development, aligning academic learning with workplace demands.

A notable example is **University of Georgia's,** USA *AI CPA Review System*, which uses machine learning to adapt review materials for CPA exam candidates. The system analyzes user performance and tailors the review path to focus on weak areas, leading to improved pass rates and greater student confidence (Davenport & Ronanki, 2018).

In Asia, Shanghai **Jiao Tong University** has implemented *AI-driven platforms for accounting courses* that enable students to simulate financial transactions and receive real-time feedback. This hands-on approach allows students to engage deeply with practical accounting tasks in a controlled, error-tolerant environment.

One case is **Coventry University** in the UK, which deployed the *AI Course Companion*, a chat bot developed in collaboration with IBM Watson, to support accounting students. The chat bot answers questions about course content, explains accounting terms, and even quizzes students on IFRS standards, contributing to higher engagement and exam preparedness (Holmes et al., 2021).

Arizona State University, USA has utilized *adaptive learning platforms* in business and accounting programs, which resulted in increased student retention rates and satisfaction. The adaptive systems continuously assess student knowledge levels and recommend personalized study plans and resources (Siemens et al., 2013).

At **University of Oulu**, the Oulu Business School Finland conducted a case study to explore *the integration of artificial intelligence (AI) in management accounting education.* The study involved interviews with faculty and second-year master's students, revealing that while there is an awareness of AI's role in accounting, there is a lack of comprehensive textbooks combining AI with management accounting. Despite this, students expressed a willingness to adopt AI skills, and some had already taken AI-related courses during their undergraduate studies. The study emphasizes the need for continuous improvement in AI knowledge and skills to remain competitive in the global job market (Kwafo, 2019).

The Chinese **University of Hong Kong** (CUHK): case, *Chat bots and Natural Language Processing (NLP) in Accounting Courses.* CUHK has developed a conversational AI assistant for their accounting students, using Natural Language Processing (NLP) to respond to students' questions about course content, homework, and accounting standards. The chat bot - integrated with the university's Learning Management System (LMS) - has shown success in enhancing out-of-classroom learning. In a recent pilot, more than 70% of students used the chat bot regularly, and 85% reported improved understanding of complex accounting principles. According to Li and Wong (2022, p. 77), the system reduced the teaching staff's workload and enabled a more scalable support model for large classes.

These case studies demonstrate that AI integration in higher education is not merely theoretical but is being actively implemented across continents. Importantly, they highlight the potential of AI to support international students by bridging language gaps, offering flexible learning environments, and aligning instructional delivery with diverse learning styles.

6. IMPLEMENTATION CHALLENGES AND CONSIDERATIONS

The integration of Artificial Intelligence (AI) into international accounting education at the university level presents immense potential. However, effective implementation is not without its challenges. This chapter explores the principal barriers and considerations - technological, ethical, and pedagogical - that institutions must address to harness AI meaningfully and responsibly.

6.1 Technological Barriers

One of the primary challenges in implementing AI tools in accounting education is the **technological infrastructure required** to support these systems. Many universities, particularly in developing countries, lack the necessary hardware, stable internet connectivity, or cloud-based services to run advanced AI applications effectively (Bond et al., 2018, p. 110). Incompatibility between existing Learning Management Systems (LMS) and AI-enhanced platforms can further hinder smooth integration (Luckin et al., 2016, p. 45). Moreover, the cost of deploying and maintaining AI systems - including machine learning models for grading, personalized learning analytics, or natural language processing tools - and be prohibitive. Institutions must consider not only the initial investment but also long-term sustainability, regular updates, cyber security needs, and data storage capabilities (Selwyn, 2019, p. 78).

6.2 Ethical Concerns with AI in Education

Ethical issues surrounding the use of AI in education are multifaceted and particularly sensitive in international accounting programs, which emphasize professional responsibility and trust. One major concern is **data privacy**. AI systems rely on large datasets, often involving personal and academic records of students. Inadequate data protection may lead to breaches of student confidentiality, violating laws such as the EU's General Data Protection Regulation (GDPR) (Williamson & Piattoeva, 2022, p. 133). Another issue is **algorithmic bias**. If the datasets used to train AI models are skewed or unrepresentative, the AI tools may inadvertently perpetuate discrimination or inequality in learning outcomes (Eubanks, 2018, p. 74). For example, AI-powered grading systems may favor students with specific linguistic or cultural backgrounds, undermining fairness in international classrooms. Furthermore, the **opacity of AI decision-making** - often referred to as the "black box" problem - a challenge the principles of transparency and accountability. In disciplines like accounting, where transparency is a core ethical value, the inability to explain how an AI system arrived at a decision can be deeply problematic (Binns et al., 2018, p. 5).

6.3 Faculty Training and Adaptation

Probably the most underestimated challenge is the human factor - specifically, faculty readiness to adopt and integrate AI technologies into their pedagogical practice. Instructors may lack confidence or understanding in using AI-based systems, especially if they perceive these technologies as a threat to their professional autonomy (Zawacki-Richter et al., 2019, p. 18). Successful AI adoption requires significant **faculty training**, both in terms of technical literacy and pedagogical strategies. Educators need to understand how AI works, how to interpret its outputs, and how to adjust teaching approaches accordingly. Without adequate support, there is a risk that AI tools may be misused or underused, resulting in minimal educational gains (Holmes et al., 2021, p. 92). Moreover, **interdisciplinary collaboration** between educators, data scientists, and IT professionals is crucial. Such partnerships can ensure that AI applications are aligned with accounting learning goals and ethical standards while remaining technically sound (Roll & Wylie, 2016, p. 56). While AI offers significant opportunities for enhancing international accounting education, implementation must be approached with caution, planning, and inclusivity. Addressing technological barriers, confronting ethical dilemmas, and supporting faculty adaptation are all critical steps. Universities must develop strategic frameworks that prioritize not just innovation but also equity, transparency, and pedagogical soundness.

7. FUTURE DIRECTIONS

As artificial intelligence (AI) technologies continue to evolve, their integration into accounting education at the university level opens new avenues for pedagogical innovation, global collaboration, and curriculum advancement. This chapter outlines key potential developments in AI tools for accounting education and highlights opportunities for strategic partnerships between academia and AI developers.

7.1 Potential Developments in AI Tools for Accounting Education

AI is expected to play an increasingly transformative role in the way accounting is taught and learned. One future direction is the development of **intelligent tutoring systems (ITS)** tailored specifically to accounting topics such as financial reporting, auditing, and international taxation. These systems can provide real-time feedback, simulate accounting scenarios, and adapt content based on a learner's pace and performance (Woolf, 2010, p. 22). Moreover, AI-driven platforms will likely become more adept at **natural language processing (NLP)**, enabling sophisticated automated grading and feedback for narrative responses in exams or assignments - especially useful for auditing case analyses or ethics essays. The use of **generative AI** in creating custom learning materials, quizzes, and case studies based on the latest international accounting standards (IFRS) can significantly enhance teaching efficiency and personalization (Pan, Zhang & Wang, 2023, p. 84).

Another promising area is **virtual reality (VR) integrated with AI**, where students can engage in simulated audit environments or real-time financial decision-making under varying international conditions. When coupled with AI, VR could enable assessment of not just knowledge, but also judgment, communication, and decision-making skills (Johnson et al., 2016, p. 118). In addition, predictive analytics will be embedded into **learning analytics systems**, providing early warnings about student disengagement or academic risk. These systems can guide interventions, making education more inclusive and responsive to diverse learner needs (Ifenthaler & Yau, 2020, p. 46).

7.2 Collaboration Opportunities between Academia and AI Developers

To fully realize the transformative potential of AI in accounting education, robust collaboration between **universities and AI developers** is essential. Such partnerships can take several forms:

Co-development of Educational AI Tools: Universities can partner with EdTech companies or AI startups to design custom tools aligned with accounting curricula. Joint innovation labs and incubators within business schools can serve as test beds for such collaboration (Brynjolfsson & McAfee, 2014, p. 202).

Industry-Academic Research Projects: Interdisciplinary research teams involving accounting educators, AI engineers, and cognitive scientists can work on projects that explore how AI affects learning outcomes, ethics in AI-based assessment, or cross-cultural implications in international accounting education (Kaplan & Haenlein, 2019, p. 19).

Internships and Real-World Case Integration: AI companies can offer students' opportunities to intern on projects involving automation of accounting processes, thus blending classroom learning with industry practice. Universities can also integrate AI use cases from the industry directly into course modules (Davenport & Ronanki, 2018, p. 110).

Standardization and Policy Influence: Academic institutions, in collaboration with international accounting bodies and AI developers, can shape emerging policies and standards related to AI in education and financial services, ensuring that future applications align with ethical and educational values (OECD, 2021, p. 67). By establishing continuous dialogue between technologists and educators, the design and deployment of AI in accounting education can be made more context-aware, learner-centered, and globally relevant.

8. CONCLUSION

This study has explored the transformative potential of Artificial Intelligence (AI) in the context of international accounting education at the university level. Through a multidisciplinary lens, it has examined how AI-driven tools, platforms, and methodologies are redefining pedagogical practices, enhancing curriculum content, and reshaping the competencies required of future accounting professionals.

Key findings indicate that AI can significantly enhance the learning experience by offering personalized learning paths, automating routine tasks, and providing real-time feedback. Adaptive learning technologies and intelligent tutoring systems have demonstrated the capacity to accommodate diverse learner profiles and promote mastery of complex international accounting standards and principles. Furthermore, AI-based analytics allow educators to monitor student performance in detail, thereby enabling timely pedagogical interventions.

In addition, the integration of AI into accounting curricula encourages the development of critical skills such as data literacy, analytical thinking, and ethical reasoning, which are essential for navigating the globalized, technology-driven landscape of contemporary accounting practice. Case studies and empirical evidence reviewed in this article suggest that institutions embracing AI-enhanced instruction are better positioned to prepare students for emerging roles in international financial reporting, auditing, and compliance.

The infusion of Artificial Intelligence into international accounting education represents a pivotal advancement in bridging the gap between academic instruction and real-world practice. AI not only modernizes the delivery of accounting education but also aligns it with the demands of a rapidly evolving global economy. By leveraging AI, universities can foster a more dynamic, inclusive, and future-oriented learning environment. This technological integration promotes lifelong learning, cultivates adaptability, and ensures that graduates are equipped with both the technical expertise and cognitive ability to thrive in diverse international accounting roles.

As educators, researchers, and institutions continue to explore AI's full potential, a deliberate and ethically grounded approach is essential to ensure that the benefits of AI are accessible, equitable, and aligned with the overarching goals of higher education. With sustained investment, collaboration, and innovation, AI has the power to redefine international accounting education - not just as a field of study, but as a cornerstone of global professional excellence in the 21st century.

REFERENCES

Alon, A., & Dwyer, P. D. (2016). *Globalization and the Future of International Accounting Education.* Journal of Accounting Education, 36, pp. 102–117.

Baker, R. S., &Inventado, P. S. (2014). Educational data mining and learning analytics. In J. A. Larusson & B. White (Eds.), *Learning Analytics: From Research to Practice,* Springer. pp. 61–75.

Binns, R., Veale, M., Van Kleek, M., & Shadbolt, N. (2018). *'It's Reducing a Human Being to a Percentage': Perceptions of Justice in Algorithmic Decisions.* Proceedings of the 2018 CHI Conference on Human Factors in Computing Systems, p. 5.

Bond, M., Zawacki-Richter, O., & Nichols, M. (2018). *Digital Transformation in Higher Education: A Survey of Strategic Approaches*. International Journal of Educational Technology in HigherEducation, 15(1), p. 110.

Bryant, S. M., Stone, D. N., & Wier, B. (2020). *Technology and pedagogy in accounting education during COVID-19*. The Accounting Educators' Journal, 30(2), pp. 233–244.

Brynjolfsson, E., & McAfee, A. (2014). *The Second Machine Age: Work, Progress, and Prosperity in a Time of Brilliant Technologies*. New York: W. W. Norton & Company, p. 202.

Davenport, T. H., & Ronanki, R. (2018). Artificial intelligence for the real world. *Harvard Business Review*, 96(1), 108–116.

Davis, R., Jackson, A., & McCarthy, I. (2020). *"Simulated Learning in Forensic Accounting: The Impact of AI-Based Tools."* Journal of Forensic & Investigative Accounting, 12(1), pp. 195–210.

Deloitte. (2023). *AI in Audit and Accounting: Redefining the Profession*. Deloitte Insights, pp. 15–20.

Eubanks, V. (2018). *Automating Inequality: How High-Tech Tools Profile, Police, and Punish the Poor*. St. Martin's Press, p. 74.

Gray, S. J., & Needles, B. E. (1997). *Issues in International Accounting Education*. Issues in Accounting Education, 12(1), pp. 83–94.

Holmes, W., Bialik, M., & Fadel, C. (2021). *Artificial Intelligence in Education: Promises and Implications for Teaching and Learning*. Center for Curriculum Redesign.

Ifenthaler, D., & Yau, J. Y. K. (2020). *Utilising Learning Analytics to Support Study Success*. Springer. p. 46.

International Federation of Accountants (IFAC). (2021). *The Accountancy Profession—Playing a Positive Role in the Reporting Ecosystem*, p. 6.

Jackling, B., de Lange, P., & Natoli, R. (2013). Transitioning to IFRS in Australian classrooms: Impact on teaching approaches. *Accounting Education*, 22(3), pp. 240–263.

Johnson, L., Adams Becker, S., Cummins, M., Estrada, V., Freeman, A., & Hall, C. (2016). *NMC Horizon Report: 2016 Higher Education Edition*. Austin, Texas: The New Media Consortium. p. 118.

Kaplan, A., & Haenlein, M. (2019). Siri, Siri, in my hand: Who's the fairest in the land? On the interpretations, illustrations, and implications of artificial intelligence. *Business Horizons*, 62(1), 15–25. p. 19.

Krouska, A., Troussas, C., &Virvou, M. (2021). "AI-powered personalization in educational systems: A review of the state of the art." *Applied Sciences*, 11(4), 1676. p. 88.

Kwafo, D. (2019). *The impacts of artificial intelligence on management accounting students: A case study at Oulu Business Schoo*. Publications of, University of Oulu.

Lu, J., Yang, J., & Recker, M. (2022). "The role of AI in enhancing learning analytics in MOOCs." *Educational Technology Research and Development*, 70(1), 98–110. p. 103.

Luckin, R., Holmes, W., Griffiths, M., & Forcier, L. B. (2016). *Intelligence Unleashed: An Argument for AI in Education*. Pearson Education.

Nobes, C., & Parker, R. (2020). *Comparative International Accounting* (14th ed., p. 24). Pearson Education.

OECD. (2021). *Artificial Intelligence in Education: Challenges and Opportunities for Sustainable Development*. OECD Publishing. p. 67.

Pan, Y., Zhang, T., & Wang, H. (2023). Leveraging Generative AI for Personalized Learning in Business Education. *Journal of Educational Technology Development and Exchange*, 16(2), 75–90. p. 84.

PwC. (2022). *Digital Readiness of Accounting Educators*. PwC Future of Learning Survey, 6–10.

Richins, G., Stapleton, R., Stratopoulos, T., & Wong, C. (2017). *Big Data Analytics: Opportunity or Threat for the Accounting Profession?* Journal of Information Systems, 31(3), pp. 63–79.

Roll, I., & Wylie, R. (2016). *Evolution and Revolution in Artificial Intelligence in Education*. International Journal of Artificial Intelligence in Education, 26(2), p. 56.

Selwyn, N. (2019). *Should Robots Replace Teachers? AI and the Future of Education*. Polity Press, p. 78.

Siemens, G., Gasevic, D., Haythornthwaite, C., Dawson, S., & Shum, S. B. (2013). *Open Learning Analytics: An Integrated & Modularized Platform*. Society for Learning Analytics Research.

Van der Meijden, A., &Veenman, M. (2020). *"Adaptive Learning Technologies in Accounting Education."* Innovations in Education and Teaching International, 57(1), 90–103.

Williamson, B., & Piattoeva, N. (2022). *Objectivity as Standardization in Data-Driven Education*. Learning, Media and Technology, 47(2), p. 133.

Woolf, B. P. (2020). *Building Intelligent Interactive Tutors: Student-Centered Strategies for Revolutionizing E-Learning*. Morgan Kaufmann.

Zawacki-Richter, O., Marín, V. I., Bond, M., & Gouverneur, F. (2019). *Systematic Review of Research on Artificial Intelligence Applications in Higher Education – Where are the Educators?* International Journal of Educational Technology in Higher Education, 16(1), pp.1–39.

ENHANCING GEOMETRIC CONSTRUCTIONS IN ELEMENTARY EDUCATION USING GEOGEBRA: A CONSTRUCTIVIST APPROACH

Lilla Korenova[1], Angelika Schmid[2] and Zuzana Gemrot[2]
[1]Comenius University Bratislava, Faculty of Education
Racianska 59, Bratislava, Slovakia
[2]University of Ostrava,Faculty of Education
Frani Sramka 3, Ostrava, Czech Republic

ABSTRACT

This paper investigates the integration of GeoGebra software into lower secondary school geometry lessons using an action research approach. The aim was to explore how the use of GeoGebra affects students' experiences, attitudes, and motivation when solving triangle construction tasks. The study was carried out in two ninth-grade classes in a Czech primary school, drawing on the principles of the constructivist Hejny method. A mixed-methods design was employed, combining classroom observations with open-ended questionnaires and a semantic differential to collect qualitative feedback. The findings indicate that many students found GeoGebra to be a helpful and engaging tool, particularly in terms of visual clarity and flexibility. While some preferred traditional paper-based methods, the digital environment was generally well received. The study provides insights into the practical implementation of GeoGebra in geometry teaching and offers recommendations for supporting students with varying levels of digital competence.

KEYWORDS

GeoGebra, Constructivist Learning, Digital Mathematics Education, Geometric Constructions, Primary Education

1. INTRODUCTION

In recent years, digital technologies have become an integral part of modern education, transforming traditional teaching methods across various subjects, including mathematics (Prodromou & Lavicza, 2017). GeoGebra, in particular, has emerged as a widely used tool in geometry instruction due to its ability to visualise abstract concepts, promote interactivity, and enhance student engagement (Hohenwarter & Lavicza, 2010; Kaya & Öçal, 2018). Its dynamic environment offers students the opportunity to explore geometric relationships and carry out constructions with precision and clarity (Shadaan & Leong, 2013).

Alongside technological advancements, constructivist pedagogical approaches have gained growing attention. One such approach is the Hejny Method, a scheme-oriented and problem-based teaching strategy widely implemented in Czech primary schools. It encourages students to actively discover mathematical principles through structured tasks, collaboration, and cognitive autonomy (Hejný, 2012).

This study combines the affordances of GeoGebra with the constructivist framework of the Hejny Method in the context of geometry education. Conducted as an action research project, it was implemented in two ninth-grade classes at a Czech lower secondary school. The primary aim was to investigate students' experiences and attitudes toward using GeoGebra when solving triangle construction problems. Rather than measuring academic performance, the research focused on students' motivation, engagement, and practical feedback concerning the use of digital tools in geometry lessons.

The study draws on a mixed-methods approach, integrating classroom observations with open-ended questionnaires and semantic differential scales. The findings are used to identify key factors that influence students' receptiveness to digital geometry tools and to provide evidence-based recommendations for teachers seeking to support diverse learners in a digital classroom environment.

2. THEORETICAL BACKGROUND

Digital technologies are increasingly shaping mathematics education by introducing new opportunities for visualisation, interaction, and personalised learning. In geometry instruction, dynamic digital tools can support students' conceptual understanding and foster motivation, particularly when used in combination with learner-centred pedagogies (Hohenwarter & Lavicza, 2010; Prodromou & Lavicza, 2017).

2.1 The Role of Digital Tools in Geometry Education

Tools such as GeoGebra offer students the ability to experiment with geometric constructions, receive instant feedback, and visualise relationships dynamically (Shadaan & Leong, 2013). Earlier research on digital platforms for Euclidean geometry has highlighted the potential of e-learning environments to support visual and interactive learning (Santos & Quaresma, 2008). Digital environments also contribute to the development of essential 21st-century competencies, including digital literacy and collaborative problem-solving (PISA, 2021). However, successful integration depends on appropriate infrastructure, teacher preparation, and student readiness (Kostolányová & Šarmanová, 2014). Challenges include unreliable internet connections, unequal access to devices, and differences in students' digital competences.

2.2 Constructivist Learning and the Hejny Method

Constructivism posits that learners build knowledge actively through experience and reflection. In Czech educational practice, the Hejny Method embodies this philosophy by encouraging students to discover mathematical ideas through problem-solving and collaboration (Hejný, 2012). It is based on a set of twelve principles, including building on students' prior knowledge and promoting reasoning in social contexts (Žilková et al., 2018).

The teacher's role in this method is to guide exploration rather than transmit procedures. Students are encouraged to develop and test their own solution strategies, leading to deeper conceptual understanding and increased autonomy. This approach is particularly suitable for introducing geometric constructions in a way that is meaningful and engaging for learners.

2.3 GeoGebra as a Tool for Constructivist Teaching

GeoGebra, as a free and open-source software, integrates geometry, algebra, and measurement tools into a single interactive platform (Hohenwarter & Lavicza, 2010). Its dynamic features support discovery-based learning, as students can manipulate geometric objects and test their hypotheses in real time (Shadaan & Leong, 2013). Moreover, GeoGebra facilitates independent and collaborative learning, allowing students to work at their own pace or engage with peers in solving visual tasks (Arbain & Shukor, 2015; Kaya & Öçal, 2018). Previous studies have also confirmed the positive influence of GeoGebra on students' conceptual understanding, especially when applied in carefully structured learning environments (Gurmu et al., 2024).

In the context of this study, GeoGebra was used in the design of a custom-made GeoGebra Book developed by the teacher–researcher and applied during geometry lessons. The digital materials were aligned with tasks from Czech textbooks and adapted to support construction-based learning through the Hejny framework (Gemrot, 2024). Similar approaches have been explored in previous research, where digital applets were developed to support primary students' understanding of geometric concepts such as symmetry (Žilková, 2020).

2.4 Conceptual Framing of the Study

The conceptual framing of this research is built around the integration of three core dimensions:

- **Digital tools**: their accessibility, accuracy, and visual feedback;
- **Constructivist pedagogy**: guided discovery, peer collaboration, and student agency;
- **Student experience**: affective engagement and usability perceptions.

Rather than predicting learning gains, this study aims to explore how the combination of GeoGebra and the Hejny Method is perceived by learners and what conditions may support or hinder the integration of dynamic geometry in real classrooms. These principles informed both the design of the teaching intervention and the structure of the data collection instruments.

3. METHODOLOGY

This study employed a mixed-methods action research design to explore students' experiences with the use of GeoGebra during geometry lessons based on the Hejny Method. Mixed-methods research, as defined by Creswell and Clark (2007), combines qualitative and quantitative approaches to provide a more comprehensive understanding of educational phenomena. The research was carried out by a pre-service teacher during her teaching practicum at a Czech lower secondary school. The aim was to examine students' attitudes, challenges, and feedback related to working with a GeoGebra Book focused on triangle constructions.

3.1 Research Design

The research was framed as a proactive action research project with a positivist orientation, complemented by qualitative data collection tools (Hendl, 2012). The structure of the research followed the iterative cycle of planning, action, and reflection, as commonly defined in action research literature (Phillips & Carr, 2014), allowing the teacher–researcher to respond to classroom realities and adjust instruction accordingly. The researcher implemented four lessons in the subject *Mathematics Exercises* in two ninth-grade classes (9.A and 9.B) at Primary School Příbor (Základní škola Příbor), Jičínská 486, Czech Republic. Each class was taught twice, with the researcher both teaching and observing. The lessons were based on tasks adapted from Hejny's textbooks and implemented through a custom GeoGebra Book (Gemrot, 2024).

3.2 Research Sample

The sample consisted of two intact classes of ninth-grade students. These students were typically 14 to 15 years old, corresponding to the final year of lower secondary education in the Czech system.

- **Class 9.A**: 15 students (no previous experience with GeoGebra)
- **Class 9.B**: 19 students (prior experience with GeoGebra during COVID-19 remote learning)

In total, 34 students participated in the action research. The school was well-equipped with iPads and internet access. Each student worked individually with a device during the lessons (Figure 1).

Figure 1. Teacher is shown drawing on teacher's PC, students are drawing on iPads. Source: Own Work

3.3 Data Collection Tools

Two main tools were used for data collection:

- **Observation protocol**: The researcher noted classroom dynamics, student interaction, technical challenges, and reactions to digital tasks. Informal reflections and photographs were taken with informed consent.
- **Student questionnaire**: Administered online via Survio after the final lesson. It consisted of:
  - Two filter questions (previous GeoGebra use and class identification)
  - Four open-ended questions (motivation, perceived benefits, likes/dislikes, suggestions)
  - One semantic differential scale (5 bipolar pairs, e.g. "easy–difficult", "fun–boring") rated on a 5-point Likert scale (–2 to +2).

3.4 Digital Materials: GeoGebra Book

The teaching material was a digital collection entitled *Triangles with the Hejny Method*, designed by the researcher in the GeoGebra platform.

- The book was shared via QR code and Teams.
- Each lesson focused on triangle construction using tools such as perpendicular bisectors, altitudes, and Thales' circle.
- Students worked at their own pace with semi-guided interactive applets.

Technical challenges (e.g. login issues, device connectivity) were noted and addressed between lessons.

3.5 Ethical Considerations

The study was conducted with the approval of the host school and in collaboration with the class teacher. No personal data were collected. Students were informed about the purpose of the activity and participation in the post-lesson questionnaire was voluntary.

3.6 Data Analysis

The qualitative data from the questionnaire were analysed thematically (Hendl, 2012), focusing on:

- Students' attitudes toward GeoGebra
- Reported benefits and challenges
- Perceptions of motivation and usability

The semantic differential responses were summarised using frequency tables. The analysis did not aim to generalise, but to understand learners' individual experiences and perspectives in context.

4. RESULTS AND DISCUSSION

This section presents the findings of the action research conducted in two ninth-grade classes at a Czech lower secondary school. The results are based on qualitative and quantitative data collected through post-lesson student questionnaires and classroom observations.

Analysis focuses on two main areas:

(1) **students' perceptions of the GeoGebra learning experience**, and

(2) **challenges and suggestions for improving the implementation of digital tools in geometry education**.

4.1 Student Perceptions: Semantic Differential Analysis

A semantic differential scale was used in the post-intervention questionnaire to evaluate students' subjective experiences with GeoGebra. Students rated five pairs of contrasting adjectives on a five-point scale ranging from −2 (negative) to +2 (positive). The categories included: Easy–Difficult, Fun–Boring, Clear–Unclear, Useful–Useless, and Motivating–Demotivating.

Table 1. below summarises the distribution of responses and weighted mean scores for each category.

Table 1. Student Responses to Semantic Differential Scale (N = 32). Source: Own Work

Category	-2	-1	0	+1	+2	Mean Score
Easy–Difficult	1	2	6	15	8	**+0.9**
Fun–Boring	0	2	3	13	14	**+1.2**
Clear–Unclear	1	2	6	14	9	**+1.0**
Useful–Useless	0	2	2	15	13	**+1.2**
Motivating–Demotivating	2	1	3	15	11	**+1.1**

The results indicate an overall positive reception of GeoGebra across all categories. Students found the environment generally useful, fun, and motivating, and they rarely described the experience as boring or useless. The highest positive score was observed for *Fun–Boring* (+1.2), suggesting a high level of engagement.

4.2 Qualitative Feedback from Open-Ended Questions

Open-ended questionnaire items provided further insight into how students perceived GeoGebra during geometry lessons. The following key themes were identified through thematic analysis:

- **Ease of Use and Visual Feedback**:

"GeoGebra helped me because I could easily see if my construction was correct without having to erase and start again."

- **Increased Efficiency**:

"It was faster and more convenient than using a ruler and compass."

- **Learning Curve Challenges**:

"I didn't know how to log in at first and spent time trying to access the book."

"The tools were sometimes confusing, especially when I had to go back a step."

- **Requests for Improvement**:

"It would help if the tasks were explained more clearly at the beginning."

"A colour-coding of task difficulty would make it easier to choose what to solve."

Although the majority of students expressed positive attitudes, several responses noted the importance of technical preparedness and interface clarity—particularly for students unfamiliar with GeoGebra on tablets.

4.3 Interpretation and Pedagogical Implications

The results suggest that GeoGebra, when integrated with structured pedagogical methods like the Hejny framework, can enhance student engagement, usability perception, and autonomy in solving geometric construction problems. Despite occasional technical difficulties (login issues, tool limitations), students appreciated the interactive and exploratory nature of the software.

These findings align with previous research emphasising the motivational and cognitive benefits of dynamic geometry environments (Arbain & Shukor, 2015; Kaya & Öçal, 2018; Shadaan & Leong, 2013). Similar findings have been observed in other geometry domains, such as stereometry, where action research studies report increased student engagement and deeper understanding when GeoGebra is used in combination with constructivist strategies (Korenova et al., 2024).

To support successful implementation, future lessons should:

- Include time for onboarding and orientation,
- Provide printed QR codes and simplified login procedures,
- Differentiate tasks by complexity and student digital readiness,
- Encourage reflection on both process and mathematical reasoning.

4.4 Comparison Between Classes (9.A vs. 9.B)

Although both classes followed the same instructional sequence and used the same digital materials, a comparison between Class 9.A and Class 9.B revealed notable differences in digital fluency and perception of the learning experience.

- **Prior Experience with GeoGebra**:
  - Class 9.B had used GeoGebra during earlier remote learning sessions (COVID-19 period).
  - Class 9.A had no previous experience with the software.
- **Initial Technical Barriers**:
  - In 9.B, students struggled with login issues, forgotten passwords, and slow connectivity during the first lesson.
  - In 9.A, a printed QR code was introduced by the teacher to simplify access, resulting in significantly fewer delays.
- **In-Class Behaviour and Engagement**: The teacher–researcher observed that students in 9.A demonstrated higher enthusiasm and more frequent spontaneous cooperation, perhaps due to the novelty of the tool and the smoother start.
- **Learning Outcomes and Tool Mastery**: By the second lesson, the difference in digital fluency between the two groups was no longer apparent, indicating a steep learning curve and fast adaptation by 9.A students.

This comparison underlines the importance of instructional scaffolding and technical support, especially when introducing new technologies to inexperienced students. Well-prepared onboarding (such as QR codes or simplified interfaces) can offset initial disadvantages and create more equitable learning conditions.

5. CONCLUSION

This study examined the implementation of GeoGebra software in lower secondary geometry education through a classroom-based action research project. Conducted in two ninth-grade classes, the research focused on student motivation, usability, and engagement while solving triangle construction problems using a custom-designed GeoGebra Book.

The results of the semantic differential scale and open-ended feedback indicate that students generally perceived GeoGebra as useful, engaging, and motivating. Positive responses prevailed in all five evaluated categories—especially in terms of ease of use and task clarity. Students appreciated the software's interactive interface, visual precision, and ability to undo actions, which facilitated a smoother construction process compared to traditional paper-based methods.

Despite these positive perceptions, some technical and pedagogical challenges emerged:

- **Technical difficulties** in the initial sessions (login issues, connectivity delays),
- **Variability in digital readiness**, particularly among students with no prior exposure to the tool,
- **Interface limitations**, such as missing or unclear functions on mobile versions of the app.

The action research also revealed that instructional scaffolding, such as using printed QR codes and shared digital displays, played a crucial role in overcoming entry barriers. Moreover, students without prior experience adapted quickly when provided with structured support—highlighting the importance of inclusive onboarding strategies.

Drawing on these findings and previous studies (e.g., Arbain & Shukor, 2015; Kaya & Öçal, 2018; Prodromou & Lavicza, 2017), we offer the following practical recommendations:

1. **Initial guidance and differentiated support** are essential for students with varying digital fluency. Grouping by experience level may be helpful in early sessions.
2. **Blended instruction** that combines GeoGebra with traditional tools should be maintained, both to reinforce manual construction skills and to accommodate diverse learning preferences.
3. **Teachers need professional development** that goes beyond technical usage—focusing instead on pedagogical integration of digital tools into constructivist teaching frameworks such as the Hejny Method.
4. **Digital lesson materials** (e.g., GeoGebra Books) should be carefully designed to avoid interface obstacles and provide toolbars tailored to the structure of each task.

This case study reaffirms the value of combining dynamic digital tools with inquiry-based pedagogies. While the research was limited to a small sample and focused on a specific topic within geometry, it contributes to a growing body of evidence that well-designed digital integration can support deeper engagement and more accessible mathematics learning for diverse student populations.

For future research, we recommend:

- **Extended implementation** over a longer time frame to examine sustained effects,
- **Replication in diverse educational settings**, including rural schools or schools with limited digital infrastructure,
- **Investigations focused on teacher development**, examining how digital tools shape instructional strategies and professional growth.

Ultimately, GeoGebra—when thoughtfully integrated into constructivist frameworks—has the potential to modernise mathematics education, empower learners, and provide inclusive opportunities for mathematical exploration.

ACKNOWLEDGEMENT

The research has been supported by projects:
- Erasmus+ 2023-1-CZ01-KA220-HED-000160664: "Accelerating STEAM-related Knowledge and Skills via 3D Modelling and 3D Printing"
- SGS06/PdF/2025: " Integration of e-learning, e-testing and GeoGebra software in mathematics education"
- KEGA 077UK-4/2025 Edu-Steam laboratory in primary education teaching

REFERENCES

Arbain, N. and Shukor, N.A., (2015). The effects of GeoGebra on students achievement. *Procedia-Social and Behavioral Sciences*, *172*, pp.208-214.

Creswell, J.W. and Clark, V.P., (2007). Mixed methods research. *Thousand Oaks, CA.*

Gemrot, Z., (2024). GeoGebra Book. https://www.geogebra.org/m/gnjd8pd2

Gurmu, F. et al, (2024). Effects of GeoGebra-assisted instructional methods on students' conceptual understanding of geometry. *Cogent Education*, *11*(1), p.2379745.

Hejný, M., (2012). Exploring the cognitive dimension of teaching mathematics through scheme-oriented approach to education. *Orbis scholae*, *6*(2), pp.41-55.

Hendl, J., (2012). *Kvalitativní výzkum: Základní teorie, metody a aplikace. 3. Vyd. Praha: Portál. 480 s.* ISBN 978-80-262-0219-6.

Hohenwarter, M. and Lavicza, Z., (2010). GeoGebra, its community and future. In *Asian Technology Conference in Mathematics* (pp. 1-10).

Kaya, A. and Öçal, M.F., (2018). A Meta-Analysis for the Effect of GeoGebra on Students' Academic Achievements in Mathematics. *Necatibey Faculty of Education Electronic Journal of Science & Mathematics Education, 12*(2).

Korenova, L. et al, (2024), September. Exploring the Use of GeoGebra Software in Teaching Stereometry at Primary Schools: An Action Research Study. In *International Scientific Conference on Distance Learning in Applied Informatics* (pp. 149-169). Cham: Springer Nature Switzerland.

Kostolányová, K. and Šarmanová, J., (2014). Use of Adaptive Study Material in Education in E-Learning Environment. *Electronic Journal of e-Learning, 12*(2), pp.172-182.

Phillips, D.K. and Carr, K., (2014). *Becoming a teacher through action research: Process, context, and self-study*. Routledge.

PISA, (2021). *21st-century readers: Developing literacy skills in a digital world.* OECD Publishing.

Prodromou, T. and Lavicza, Z., (2017). Integrating technology into mathematics education in an entire educational system–Reaching a critical mass of teachers and schools. *International Journal for Technology in Mathematics Education, 24*(3), pp.129-135.

Santos, V. and Quaresma, P., (2008), July. E-learning course for Euclidean geometry. In *2008 Eighth IEEE International Conference on Advanced Learning Technologies* (pp. 387-388). IEEE.

Shadaan, P. and Leong, K.E., (2013). Effectiveness of Using GeoGebra on Students' Understanding in Learning Circles. *Malaysian Online Journal of Educational Technology, 1*(4), pp.1-11.

Žilková, K. et al, (2018). *Young children's concepts of geometric shapes*. Harlow: Pearson.

Žilková, K., (2020). The creation and evaluation of mathematical applets on symmetry for primary school pupils. In *EDULEARN20 Proceedings* (pp. 7886-7893). IATED.

PEDAGOGICAL INTENTION AND LEARNER PERCEPTION OF TRANSVERSAL SKILLS

Marc Beutner and Jan-Phillip Lüttke
University Paderborn
Warburger Str. 100, 33098 Paderborn, Germany

ABSTRACT

Transversal skills, such as critical thinking, resilience, digital literacy, and intercultural awareness, become more and more important in a VUCA world which is characterized by many changes and uncertainties. Therefore, the aim of the presented study is to explore the gap between pedagogical intention and learner perception in the promotion of transversal competencies in education. With regard to the increasing complexity in a VUCA world and its effects on education, transversal skills are considered essential for future-related learning. Based on UNESCO's six-domain framework, this cross-national study conducted a quantitative online survey among 146 educators from school and higher education contexts in 16 European countries. The questionnaire assessed both teacher self-evaluations and their perceptions of student competencies across 24 items, using descriptive statistics to compare perceived competence levels across transversal domains. The results show that while transversal competencies are widely addressed in instructional practice, significant perceptual gaps persist - particularly in intrapersonal development, critical thinking, and media and information literacy. These findings point to a misalignment between curricular intentions and actual learner engagement, and underscore the need for more explicit, reflective, and sustainable integration strategies.

KEYWORDS

Transversal Skills, European Education, VUCA, Digital Literacy, Intercultural Awareness

1. INTRODUCTION

The contemporary society is characterized by profound transformations in ecological, technological, and social domains. These developments are causing fundamental changes in the world of work, education, and societal structures. This transformation is described as VUCA-world, an acronym for a world of volatility, uncertainty, complexity, and ambiguity (see Schütte et al. 2022; Bergmann & Garrecht 2021). Educational institutions are facing an imperative to prepare learners for a reality characterized by disruption, digital transformation, and global crises. In response, the focus of modern education systems is increasingly shifting from the acquisition of technical knowledge to a targeted promotion of interdisciplinary, cross-contextual skills. Transversal competencies encompass a range of skills, including the capacity for critical and creative thinking, self-regulation, empathic and cooperative collaboration, thoughtful engagement with digital information, and proactive promotion of a health-conscious, responsible lifestyle. Such abilities are regarded as pivotal for successfully navigating intricate challenges within education, employment, and society at large. The competencies are associated with lifelong learning, which emphasizes the continuous acquisition of knowledge and the capacity to adapt to new contexts (see UNESCO 2015; OECD 2018). Key competencies have been anchored in education policy since the 1990s (see EU Commission 2006), while the concept of transversal competencies aims a flexible, cross-contextual, and sustainable application of knowledge and skills. Despite the international discourse on transversal skills, there is a notable lack of empirical studies comparing teachers' self-perception with their assessment of student skills, particularly across countries and sectors. In light of societal transformations driven by the VUCA- world, digitalization, and global sustainability challenges, this international study – covering 16 countries and six UNESCO-based skill domains – fills a methodological gap and offers comparative insights into teacher practice across education levels. It focuses specifically on the six dimensions of transversal competencies as defined by UNESCO: (1) critical and innovative thinking, (2) interpersonal skills, (3) intrapersonal skills, (4) global citizenship, (5) media and information literacy, and

(6) physical and mental well-being. This study is guided by the following two research questions: 1. T*o what extent do teachers actively foster transversal competencies in their classrooms?* *2. How do they assess the manifestation of these competencies among their students?*

2. TRENDS AND SKILLS IN THE VUCA WORLD

2.1 Education in the VUCA World and Megatrends as Drivers of Change

The current educational landscape is characterized by a profound transformation fueled by technological, ecological, political, and social upheavals. The VUCA world – an acronym for volatility, uncertainty, complexity, and ambiguity – describes the conditions under which individuals learn, live, and work today (see Schütte et al. 2022; Bauer et al. 2010). These four dimensions pose fundamental challenges to established teaching and learning formats: They require not only adaptability, but also a new understanding of education as a dynamic, reflective, and context-sensitive process. In this context, the promotion of transversal skills is becoming increasingly important. They are key elements in meeting the demands of an increasingly complex world – especially because they enable people to cope with complex situations independently, creatively, and collaboratively. Megatrends such as digitalization, individualization, globalization, demographic change, and neo-ecology are having far-reaching effects on the structure and function of education systems (see Zukunftsinstitut 2022; Harwardt 2022). They not only affect content and methods, but also require a fundamental reorientation of education toward skills development rather than the mere transfer of knowledge. The ability to solve problems individually, engage in lifelong learning, and participate in society is becoming a prerequisite for personal and professional resilience (see Ameln & Wimmer 2016). Educational institutions are therefore called upon not only to respond to these structural changes, but also to play an active role in equipping learners to be forward-thinking.

2.2 Future Skills and 21st Century Skills

In the wake of digital and social transformation, a paradigm shift in the focus on skills is emerging. In addition to technical knowledge, interdisciplinary, cross-cutting skills are becoming increasingly relevant. These so-called '21st century skills' include critical thinking, creativity, collaboration, and communication (Voogt & Pareja Roblin 2012; Schleicher 2018). These skills are defined by international organizations such as the OECD, the EU Commission, and the World Economic Forum as cornerstones of a future-oriented education agenda (OECD 2018; EU 2019). They are considered central to innovation, social participation, and employability in the 21st century. The literature also increasingly points to the need to anchor these skills not only selectively but systematically in everyday teaching (see Scharnhorst 2021). While 21st century skills highlight innovation and collaboration for employability (Voogt & Roblin 2012, OECD 2018), transversal competencies as defined by UNESCO broaden this scope to include ethics, media literacy, and well-being – promoting lifelong, cross-contextual learning (UNESCO 2016).

2.3 The Six UNESCO Dimensions of Transversal Competencies

UNESCO offers a particularly differentiated framework with its definition of transversal skills, as formulated by the ERI-Net (Education Research Institutes Network) in 2016. Transversal skills are characterized by their cross-contextual applicability and are therefore not limited to specific subject areas (cf. UNESCO 2016). The six central dimensions are:

1. Critical and innovative thinking→ Problem-solving skills, creative idea development, reflection
2. Interpersonal skills→ Communication, teamwork, empathy, conflict resolution
3. Intrapersonal skills→ Self-awareness, self-discipline, emotion regulation, resilience
4. Global citizenship→ Ethical thinking, social responsibility, intercultural sensitivity
5. Media and information literacy→ Critical engagement with digital information, use of technologies
6. Physical and mental health→ Stress management, healthy lifestyle, mental resilience

The dimensions form the conceptual backbone of the present study. They were used to systematically record educators' self-assessment of their teaching practice and the assessment of its implementation by learners. The six UNESCO dimensions align closely with international competence frameworks developed by the OECD (2018) and the European Commission (2019), both of which emphasize transversal abilities such as creativity, resilience, self-regulation, and digital literacy as essential skills for the future.

2.4 Transversality and Competence Transfer

Transversal skills are characterized by their applicability across contexts—they are effective not only in a specific subject area, but also in everyday situations, social contexts, and new learning situations (Hart et al. 2021). The question of when skills are actually 'transversal' is closely linked to the concept of transfer. According to Scharnhorst & Kaiser (2018), transversality exists when individuals are able to flexibly transfer existing knowledge or skills to new contexts. It is not the skill itself that is transversal, but rather the ability to shift contexts – in other words, adaptive transfer. However, empirical findings show that transfer does not automatically succeed. Competencies acquired in specific situations often remain tied to the learning context (Weinert 1998). Transfer is particularly successful when learning processes offer a wide range of application possibilities – e.g., through project-based or research-based learning, interdisciplinary tasks, or reflection phases. This study therefore examines not only the presence of individual skills, but also their perceived transferability to different areas of life and learning – a central feature of transversal competence development.

3. METHODOLOGY

This study follows a quantitative, cross-sectional research design to explore how teachers from diverse educational settings across Europe perceive the promotion and implementation of transversal skills in their teaching practice. The investigation is guided by two central research questions: (1) *To what extent do teachers actively foster transversal competencies in their classrooms?* (2) *How do they assess the manifestation of these competencies among their students?* The research focuses on the six domains of transversal skills as defined by the UNESCO ERI-Net framework (UNESCO 2016), encompassing critical and innovative thinking, interpersonal skills, intrapersonal skills, global citizenship, media and information literacy, and physical and mental well-being. Data collection from October to December 2024 was carried out via an online questionnaire. Participation was voluntary, anonymous, and preceded by a detailed information sheet explaining the purpose of the study and the university's data protection policies. To ensure a diverse and representative sample, we recruited participants through various channels including institutional mailing lists from Erasmus+. projects, regional teacher networks, higher education consortia, and professional development associations. Additional outreach was conducted via social media platforms. The questionnaire was available in English to facilitate cross-national participation. 182 responses were received, of which 146 could be included in the analysis. The sample represents teachers from 16 European countries and encompasses all major educational sectors: 45.2% of participants were employed in higher education, 31.5% in secondary schools, 16.4% in vocational training institutions, and 6.9% in primary education.

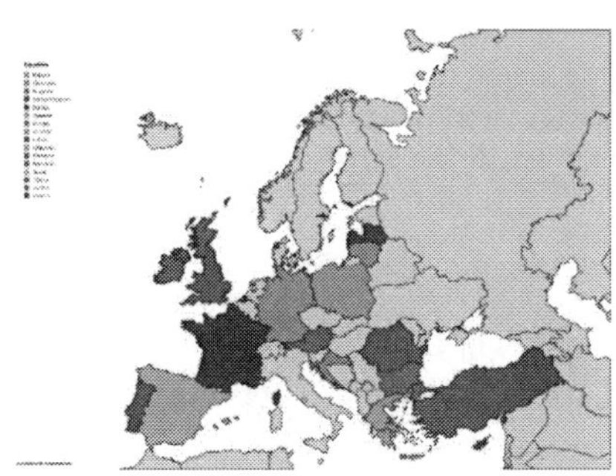

Figure 1. Participating teachers from 16 European countries

Gender distribution was nearly balanced, with 50% identifying as female, 48.6% as male, and 1.4% as non-binary or diverse. The age distribution ranged from under 30 to over 50, ensuring generational diversity in professional perspectives. Teaching experience varied considerably: approximately one-third of respondents had more than 20 years of professional practice, while another third had fewer than ten years. Subject representation was broad, with a concentration in STEM fields (54.8%), followed by language education (19.9%), humanities (15.8%), and other disciplines (9.6%).

The instrument used in this study was a structured questionnaire divided into two parts, both aligned with the six UNESCO domains. Part A focused on teachers' self-assessment regarding their ability to foster transversal skills. Part B asked teachers to evaluate how well their students apply these skills in academic or everyday contexts. Each of the six dimensions was measured using four items per section, for a total of 48 statements. Responses were collected on a five-point Likert scale (1 = strongly disagree to 5 = strongly agree), allowing for nuanced insights into attitudes and perceptions. Sample items from Part A (teacher self-assessment) included statements such as "I actively integrate problem-solving strategies into my teaching" or "I frequently use modern technologies in the design and implementation of my teaching."

Table 1. Sample Items from Part A - Teacher Self-Assessment

Dimension	**Sample Items**
Critical and Innovative Thinking	"I actively integrate problem-solving strategies into my teaching."
Interpersonal Skills	"I establish an open and ongoing culture of communication between everyone involved in my teaching."
Intrapersonal Skills	"I know the strengths and weaknesses of my teaching personality."
Global Citizenship	"In my teaching, I place a high value on ethical thinking in my students."
Media and Information Literacy	"I frequently use modern technologies in the design and implementation of my teaching."
Physical and Mental Health	"I attach great importance to teaching my students a balanced way of life and to setting a good example."

Sample items from Part B (student competence perception) included "Your students communicate effectively with each other" or "Your students reflect on their own thinking." The item construction was theory-driven and operationalized based on descriptors from the UNESCO framework.

Table 2. Sample Items from Part B – Student Competence perception

Dimension	**Sample Items**
Critical and Innovative Thinking	"Your students will reflect on their own thinking."
Interpersonal Skills	"Your students communicate effectively with each other."
Intrapersonal Skills	"Your students can regulate their emotions appropriately."
Global Citizenship	"Your students recognize their responsibility to the community."
Media and Information Litera	"Your students are able to interpret digital information appropriately."
Physical and Mental Health	"Your students take care of their physical well-being."

The data was analyzed using descriptive statistics, including mean values and standard deviations. This analytical approach was chosen due to the heterogeneity of the sample and the exploratory nature of the research. Rather than aiming to generalize to the population through inferential statistics, the analysis focused on identifying patterns, distributions, and potential discrepancies between teachers' self-assessments and their perceptions of student competencies. All analyses were conducted using the statistical software JASP (JASP Team, 2024).

4. RESULTS

For each of the six UNESCO domains of transversal skills, mean values (*M*) and standard deviations *(SD)* were calculated in order to describe central tendencies and variances across the responses. This descriptive approach aligns with the study's objective to capture both the self-assessment of teachers and their perceptions of how learners demonstrate transversal competencies in various educational contexts. Across all six domains, teachers rated their own ability to promote transversal competencies consistently higher than their assessment of the students' actual application of these skills. The largest discrepancies were found in the domains of critical and innovative thinking, intrapersonal skills, and media and information literacy. Across all six domains, teachers rated their own ability to promote transversal competencies consistently higher than their assessment of the students' actual application of these skills. The largest discrepancies were found in the domains of critical and innovative thinking, intrapersonal skills, and media and information literacy. Table 3 summarizes mean scores (*M*) and standard deviations (*SD*) for each item across all six UNESCO dimensions, reporting both teachers' self-assessed ability to promote transversal skills and their evaluation of how well students demonstrate these skills.

Table 3. Descriptive Statistics (*M, SD*) for Teachers' and Students' Ratings by UNESCO Skill Domain

Domain	**Item**	**Teachers *M* (*SD*)**	**Students *M* (*SD*)**
Critical and Innovative Thinking	Critical Reflection	4.01 (0.81)	3.82 (0.76)
	Problem-solving strategies	3.98 (0.88)	3.75 (0.72)
	Creativity	4.05 (0.78)	3.68 (0.81)
	Innovative Thinking	4.00 (0.78)	3.75 (0.80)
Interpersonal Skills	Communication	3.96 (0.87)	3.88 (0.74)
	Teamwork	3.87 (1.00)	3.87 (0.78)
	Conflict resolution	4.03 (0.83)	3.78 (0.85)
	Empathy	4.10 (0.85)	3.86 (0.84)
Intrapersonal skills	Self-discipline	3.82 (0.78)	3.61 (0.75)
	Self-awareness	4.04 (0.84)	3.63 (0.81)
	Emotional regulation	3.95 (0.79)	3.70 (0.80)
	Resilience	3.96 (0.80)	3.68 (0.79)
Global Citizenship	Ethical thinking	3.91 (0.87)	3.69 (0.72)
	Intercultural understanding	4.02 (0.75)	3.68 (0.82)
	Social responsibility	3.99 (0.76)	3.74 (0.78)
	Global perspective	3.72 (0.91)	3.67 (0.81)
Media and Information Literacy	Ethical Use of Digital Media	3.98 (0.83)	3.68 (0.90)
	Critical evaluation of information	4.08 (0.81)	3.86 (0.82)
	Using Media Tools	3.98 (0.96)	3.79 (0.83)
	Awareness of Media Influence	4.12 (0.78)	3.85 (0.84)
Physical and Mental Health	Digital interpretation	3.60 (0.95)	3.39 (0.77)
	Stress management	3.87 (0.82)	3.38 (0.82)
	Physical and Mental Health	3.99 (0.88)	3.52 (0.80)
	Mental health support	3.85 (0.83)	3.56 (0.77)

To visualize these discrepancies, Figure 2 presents the item-level gaps across all domains.

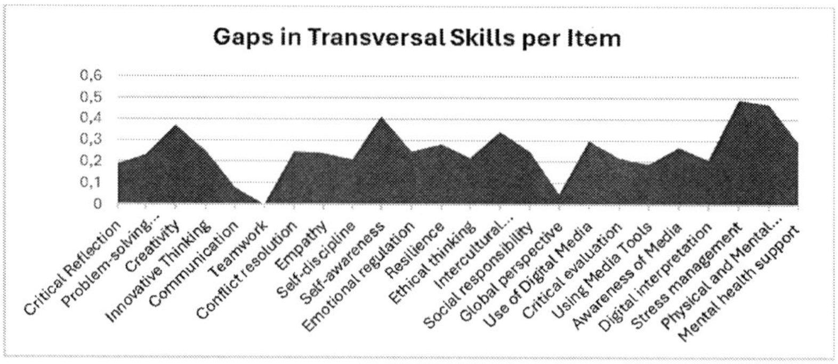

Figure 2. Gaps in Transversal Skills per Item

Figure 3 summarizes the average gap per UNESCO dimension to highlight overarching trends in teacher – student discrepancies.

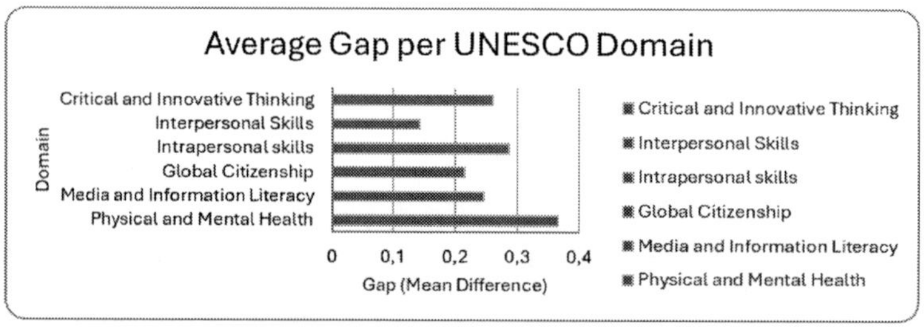

Figure 3. Average Gap per UNESCO Domain

4.1 Critical and Innovative Thinking

This domain encompasses skills such as critical reflection, creativity, problem-solving, and innovative thinking. According to Scriven and Paul (1987), critical thinking involves reasoned judgment and reflective analysis. Teachers reported high engagement in fostering such skills, particularly creativity ($M = 4.05$) and critical reflection ($M = 4.01$). However, students were rated notably lower in these areas (e.g., creativity: $M = 3.68$), indicating a transfer gap. This suggests that while instructional practices are in place, learners may struggle to apply these cognitive skills autonomously, particularly in unfamiliar or complex scenarios.

4.2 Interpersonal Skills and Intrapersonal Skills

Interpersonal skills involve the ability to interact constructively with others through communication, empathy, and collaboration. Here, the gap was comparatively small. Teachers and students provided almost identical ratings for teamwork ($M = 3.87$), and communication ratings were closely aligned. Minor differences in empathy ($\Delta M = 0.24$) and conflict resolution ($\Delta M = 0.25$) suggest that interpersonal skills are not only promoted but also meaningfully perceived by learners in classroom contexts.

This domain includes internal abilities such as self-awareness, resilience, and emotional regulation. Teachers rated themselves highly, especially on self-awareness ($M = 4.04$) and resilience ($M = 3.96$), but perceived student competencies were substantially lower (e.g., resilience: $M = 3.68$). This reveals a pronounced self-regulation gap, reflecting the difficulty of externally assessing and fostering metacognitive and emotional competencies in learners. As such, more explicit support structures like mentoring or reflective practices may be needed. This may be attributable to the internal and less observable nature of such competencies, which renders them challenging to assess externally. Intrapersonal skills such as resilience and self-awareness frequently elude external observation and are often underrepresented in conventional evaluation frameworks (Spitzberg & Cupach, 2011; UNESCO 2016).

4.3 Global Citizenship

Global citizenship is the idea that individuals are not just members of their local or national communities, but also of a broader global community. A global citizen is aware of the wider world, respects and values diversity, and acts responsibly to make the world more just, sustainable, and peaceful (Wintersteiner 2022). Global citizenship encompasses ethical awareness, social responsibility, and intercultural understanding. While teachers perceived themselves as strongly integrating these values (e.g., intercultural understanding: $M = 4.02$), student ratings lagged behind (e.g., global perspective: $M = 3.67$). This engagement gap may stem from a lack of authentic learning opportunities that allow students to experience civic and ethical issues in tangible ways. These findings point to the need for authentic learning environments that allow students to actively engage with real-world ethical, intercultural, and civic challenges.

4.4 Media and Information Literacy

Media and Information Literacy is the ability to create, use, access, analyze, and evaluate information and media in a critical and ethical way. It is designed to empower individuals and especially media creators to make informed decisions as users and creators of media and information (Koltay 2011).

This domain reflects the capacity to access, evaluate, and ethically use media. Despite high teacher ratings for media use (e.g., awareness of media influence: $M = 4.12$), students were rated lower across all items (e.g., ethical use: $M = 3.68$), indicating a digital literacy gap. This points to an implementation challenge commonly noted in international studies, underlining the importance of structured digital education that moves beyond tool use toward critical understanding. This mismatch reflects a wider implementation gap observed in international studies. OECD TALIS (2018), for example, reported that many teachers still feel underprepared for digital instruction, despite high political and curricular emphasis. This underscores the importance of explicitly embedding media literacy in the curriculum.

4.5 Physical and Mental Health

As a transversal skill physical and mental health refers not just to personal well-being, but to the knowledge, attitudes, and behaviors that support a healthy lifestyle and enable individuals to function effectively in all areas of life—education, work, and social settings (Weber et al. 2025). This domain yielded the lowest student ratings overall. While teachers reported solid engagement in modeling healthy behaviors (e.g., $M = 3.87$), student scores for stress management ($M = 3.38$) and mental health support ($M = 3.56$) were significantly lower. This well-being gap underscores the need for stronger institutional support for mental and physical health, including psychosocial resources and preventive education.

5. DISCUSSION AND IMPLICATIONS

This study follows a quantitative, cross-sectional research design to explore how teachers from diverse educational settings across Europe perceive the promotion and implementation of transversal skills in their teaching practice. The investigation is guided by two central research questions: (1) *To what extent do teachers actively foster transversal competencies in their classrooms*? (2) How do they assess the manifestation of these competencies among their students? The research focuses specifically on the six domains of transversal skills as defined by the UNESCO ERI-Net framework (UNESCO 2016), encompassing critical and innovative thinking, interpersonal skills, intrapersonal skills, global citizenship, media and information literacy, and physical and mental well-being. Data collection was carried out via an online questionnaire.

5.1 Interpretation of Key Findings and Implications for Educational Practice

The findings of this study offer valuable insights into how transversal competencies are perceived and fostered in contemporary educational practice. The overall alignment between the self-assessed efforts of teachers and their pedagogical intentions indicates that transversal skills are widely recognized as central to modern teaching. However, the notable discrepancies between teachers' self-perception and their assessment of student competencies across several domains reveal a gap between didactic planning and actual learner integration. This pattern was particularly visible in the domains of critical and innovative thinking, intrapersonal development, and media and information literacy. In all three, teachers rated their own strategies as effective, yet assessed student competencies as comparatively underdeveloped. These gaps suggest that transversal competencies, while addressed conceptually, may not yet be sufficiently internalized or activated by learners. This could be due to a lack of reflective engagement, insufficient practice opportunities, or the absence of explicit scaffolding in these areas. Conversely, the smaller gaps observed in the interpersonal domain (e.g., communication and teamwork) may be attributable to the fact that these skills are inherently practiced in collaborative classroom settings and are more visible in student interactions. These findings align with prior research indicating that visible competencies - such as teamwork or participation - are more easily observed and reinforced by educators than invisible ones, such as self-regulation or ethical reasoning (cf. Scharnhorst, 2021).

The findings of this study yield several significant implications for both pedagogical practice and educational policymaking. As transversal competencies increasingly constitute the foundation of education that is prepared for the future, their role within curricula must be redefined. Rather than remaining marginal or implicit, these competencies require systematic and measurable integration across disciplines. This encompasses the development of modular units that explicitly target skills such as interdisciplinary problem-solving, ethical reasoning, and digital literacy. The integration of transversal competencies in this manner guarantees congruence between the curricular objective and the instructional praxis, while concurrently cultivating coherence across educational strata and contexts.

At the level of professional development, the results indicate a clear need to strengthen teachers' capacity to operationalize transversal learning outcomes. While the participating educators demonstrated high levels of self-reported engagement, the observed discrepancy between teaching intent and perceived student uptake suggested the need for additional support. In particular, evidence-based training formats - emphasizing reflective practice, collaborative design, and context-sensitive adaptation - have the potential to empower educators to translate abstract skill frameworks into pedagogically grounded action. Furthermore, the study underscores the significance of learner activation in the cultivation of transversal competencies. Given the inherent latent or cross-contextual nature of many of these competencies, it is possible for students to engage in transversal learning without being able to recognize or articulate it. In order to address this need, the integration of metacognitive scaffolds is imperative. Such scaffolds include reflective journals, formative peer assessment, and structured feedback mechanisms. These elements not only promote learner agency and self-regulation but also render transversal learning more visible and meaningful.

Institutional support structures are indispensable for sustaining transversal skill development. Educational settings should offer enabling environments that extend beyond subject instruction, including project-based learning ecosystems, mentorship programs, and psychosocial support systems. These structures are designed to facilitate the long-term cultivation of competencies that are not only relevant for academic success but also critical for navigating social, technological, and ethical challenges in an increasingly complex world.

5.2 Limitations and Future Research

The present study is exploratory and based on self-reported data from a non-representative, self-selecting sample of educators, most of whom likely have a strong interest in pedagogical innovation. While the cross-national scope enhances relevance, the exclusive reliance on teacher self-assessments introduces potential bias. Future research should triangulate data sources – e.g., through student self-reports, classroom observations, or performance-based assessments – to increase validity. Longitudinal designs may help capture competence development over time. There remains a need for validated, context-sensitive instruments that allow for reliable and developmentally appropriate assessment of transversal competencies across educational levels. Given the complex and multi – layered nature of these skills, mixed – methods design integrating quantitative scope with qualitative insight are particularly well suited to capture both patterns and contextual nuances in their acquisition and application.

6. CONCLUSION

This study provides a contribution to the expanding corpus of research on transversal competencies. It offers empirical insights into how educators perceive and promote these skills across diverse European educational contexts. While the findings confirm a high level of engagement among teachers, they also reveal a discrepancy between pedagogical intention and perceived student application, particularly in domains such as intrapersonal development, critical thinking, and digital literacy. These discrepancies underscore the necessity for more explicit instructional strategies, enhanced metacognitive scaffolding, and institutional structures that facilitate transversal learning. In the future, the integration of transversal competencies must become a strategic priority across all levels of education. This integration should not only be a response to current societal challenges; it should also serve as a foundation for resilient, reflective, and future-ready learners. To ensure long-term impact, the integration of transversal competencies should be supported by scalable, age. Appropriate and pedagogically sound learning environments. Digital, modular platforms tailored to different age groups and subject areas, featuring adaptive, game-based, or AI-assisted components, offer promising opportunities for skill development. Embedding such environments into national and international curricula can be strengthen the systematic promotion of transversal competencies across education systems. Moreover, comparative implementation strategies, across disciplines, educational levels, and teacher profiles, can inform sustainable and context-sensitive integration. Concurrently, this study may function as the foundation for a comprehensive research agenda that aims to systematically integrate transversal competencies into educational theory, curriculum design, and classroom practice in a sustainable and evidence-based manner. In conclusion, the study of transversal skills highlights their critical role in shaping well-rounded, adaptable, and future-ready individuals. Unlike technical or domain-specific abilities, transversal skills, such as critical thinking, interpersonal and intrapersonal skills, media and information literacy, global citizenship, and physical and mental well-being—are transferable across different fields and life contexts. These skills not only enhance academic and professional performance but also contribute significantly to personal development and social responsibility. In an increasingly complex and interconnected world, cultivating transversal skills equips learners and professionals with the tools to navigate change, solve problems, and work effectively with others. As such, education systems, employers, and policymakers must prioritize the development and integration of these competencies into curricula, workplace training, and lifelong learning initiatives. Ultimately, transversal skills are not just complementary to traditional learning. They are essential for success in the 21st century.

REFERENCES

Ameln, F. v., & Wimmer, R. (2016). Neue Arbeitswelt, Führung und organisationaler Wandel. *Gruppendynamik und Organisationsberatung*, 47 (1), 11-21.

Bauer, M., Knape, J., Koch, P. & Winkler, S. (2010). Dimensionen der Ambiguität. Zeitschrift für Literaturwissenschaft und Linguistik, 40 (158), 7-75.

Bergmann, R. & Garrecht, M. (2021). Organisation und Projektmanagement. In *BA kompakt*. https://doi.org/10.1007/978-3-662-63754-8

European Commission (2000). Memorandum of lifelong learning. Brussel: Office for Official Publications of the European Communities.

European Union (2019). Key competencies for lifelong learning. Publications Office of the European Union.

Hart, J., Noack, M., Plaimauer, C., & Bjørnåvold, J. (2021). Towards a structured and consistent terminology on transversal skills and competences. Brussels.

Harwardt, M. (2022). Management der digitalen Transformation. Springer Fachmedien Wiesbaden.

JASP Team (2024). JASP (Version 0.19.3)[Computer software].

Koltay, T. (2011). The media and the literacies: Media literacy, information literacy, digital literacy. *Media, culture & society*, *33*(2), 211-221.

OECD (2018). The future of education and skills: Education 2030. OECD Publishing.

OECD (2020). *TALIS 2018 Results (Volume II): Teachers and School Leaders as Valued Professionals*, TALIS, OECD Publishing, Paris, https://doi.org/10.1787/19cf08df-en.

Sambaiah, M., & Aneel, Y. (2016). Intrapersonal skills as core of the personality: Some home truths. *Journal of english language and literature*, *3*(1), 225-230.

Scharnhorst, U. (2021). Transversale Kompetenzen- notwendig, erwünscht und schwierig zu erreichen. *BWP*, 1, 18-23.

Scharnhorst, U., & Kaiser, H. (2018). Transversale Kompetenzen für eine ungewisse digitale Zukunft. In J. Schweri, I. Trede, & I. Dauner (Eds.): Digitalisierung und Berufsbildung: Herausforderungen und Wege in die Zukunft (OBS EHB Trendbericht 3, 9-12). OBS EHB.

Schleicher, A. (2018). World Class: How to Build a 21st – Century School System, Strong Performers and Successful Reformers in Education, OECD Publishing, Paris.

Schütte, R., Seufert, S. & Wulfert, T. (2022). IT-Systeme wirtschaftlich verstehen und gestalten: Methoden-Paradoxien- Grundsätze. Springer Gabler.

Scriven, M., & Paul, R. (1987, March). Critical thinking. In *The 8th Annual International Conference on Critical Thinking and Education Reform, CA* (Vol. 7, No. 9).

Spitzberg, B. H., & Cupach, W. R. (2011). Interpersonal skills. *The Sage handbook of interpersonal communication*, *4*, 481-527.

UNESCO (2015) Transversal competencies in education policy and practice. UNESCO Asia-Pacific Regional Bureau for Education.

UNESCO (2016). ERI-Net Asia-Pacific Regional Policy Series 2014: Transversal competencies in education policy and practice. Paris: UNESCO Publishing.

Voogt, J., & Pareja Roblin, N. (2012). A comparative analysis of international frameworks for 21st century competences: Implications for national curriculum policies. *Journal of Curriculum Studies*, 44 (3), 299-321.

Weber, A. M., van Laar, E., Borgonovi, F., Ackerman, P. L., Nixon, N., Graesser, A. C., & Greiff, S. (2025). Psychological Science and the Blind Spot in Education: Learning and Instruction of Transversal Skills in the Twenty-*First Century. Educational Psychology Review, 37(2), 41.*

Weinert, F. E. (1998). Concept of competence: A conceptual clarification. In D. S. Rychen & L. H. Salganik (Eds.), Definition and selection of competencies, 45-66.

Wintersteiner, W. (2022): *Global citizenship education.* In Beutel, W., Gloe, M., Himmelmann, G., Lange, D., Reinhardt, V., Seifert, A. (Hrsg.). Handbuch Demokratiepädagogik. Debus verlag: Frankfurt Main., 347-360.

Zukunftsinstitut (2022). Das Potenzial der Megatrends für Strategie und Management.

THE MEDIATING ROLE OF CREATIVITY IN ACADEMIC PERFORMANCE THROUGH THE USE OF CHATGPT FOR UNIVERSITY STUDENTS

Olger Gutierrez Aguilar, Aleixandre Duche Pérez, Anani Gutierrez Aguilar
and Sandra Chicana Huanca
Universidad Católica de Santa María
Arequipa, Perú

ABSTRACT

The study explores the mediating role of creativity in the academic performance of university students using ChatGPT. It investigates how this artificial intelligence tool improved information acquisition efficiency and language proficiency and how these skills influenced creativity and academic performance. The research adopts a quantitative explanatory approach with a non-experimental cross-sectional design. The sample is made up of 389 university students of Psychology and Advertising and Multimedia at the Universidad Católica Santa María de Arequipa, Peru. A survey is used to measure information acquisition efficiency, creativity, linguistic competence, and academic performance, and it is analysed by structural equation modelling with SmartPLS. The most important result reveals that creativity significantly mediates information acquisition efficiency and linguistic competence with academic performance. The coefficients of determination (R2) indicated that these variables explained 72.6% of the variability in academic performance, demonstrating a strong influence of creativity in this relationship. The study confirms that creativity is a crucial mediator in the impact of ChatGPT use on academic performance. It emphasises the importance of developing educational environments that foster information efficiency and linguistic competence in order to improve students' creativity and academic performance. This research underlines the relevance of integrating pedagogical strategies that enhance these key competencies, preparing students to face academic and professional challenges in an increasingly information-based world.

KEYWORDS

ChatGPT, Academic Performance, Linguistic Competence, Creativity, Efficiency of Information Acquisition

1. INTRODUCTION

ChatGPT emerges as a promising tool in the contemporary educational field, facilitating personalized feedback, query resolution, and content generation that stimulates active student participation, according to (Al Shloul et al., 2024). This participation is crucial, as it is directly linked to intrinsic motivation and commitment to learning, essential aspects for improving academic performance, as indicated by Beltozar-Clemente and Díaz-Vega (2024). A case study in Australia reported by Sandu et al. (2024) highlights that regular use of ChatGPT can increase the perception of academic benefits, suggesting its potential to raise academic performance through structured integration into pedagogical practices.

The implementation of ChatGPT has also been shown to have positive impacts on student motivation and interest by creating interactive environments that favour both collaboration and attention to individual needs, as reported by Alarbi et al. (2024). However, careful didactic planning is crucial to optimize the advantages of artificial intelligence, minimizing technological dependencies and ensuring the preservation of critical skills, as Baltà-Salvador et al. (2025). Sadallah et al. (2025), suggest that ChatGPT is playing a transformative role in academia. It influences the performance of students, teachers, and researchers both directly and indirectly, improving information acquisition and language skills development. This reinforces the hypothesis of its positive impact in the university context.

2. THEORETICAL BACKGROUND

2.1 Efficiency in the Acquisition of Information Through ChatGPT

Speed and accuracy in obtaining educational data determine academic success (Alfarraj & Wardat, 2024). Among the reported benefits are the facilitation of the search for relevant sources, the synthesis of arguments, and the critical exploration of diverse perspectives (Aguilar et al., 2023). These advantages are related to the concept of "efficiency in information acquisition," which refers to the ability of students to collect, filter, and use data optimally to solve tasks. Previous research highlights that efficient use of ChatGPT in university settings increases the likelihood that students achieve an appropriate structuring of the information collected, which translates into a substantial improvement in their performance (Al Shloul et al., 2024; Wang et al., 2024).

On the other hand, the system proposed by Wang et al. (2024) points out that personalized support, adapted to the zone of proximal development, allows students to optimize their reading comprehension. Although the focus is on English reading, the same principle applies to any subject where ChatGPT is used to obtain timely and more accurately organized data. In this way, the efficiency of information acquisition obtained through ChatGPT improves academic performance. Furthermore, such fast and reliable access to diverse information can stimulate the generation of novel ideas and foster creativity.

2.2 Linguistic Competence as a Key Factor in Improving Academic Performance

The development of linguistic competence is linked to students' ability to communicate fluently, clearly, and coherently in academic contexts (Wang & Ren, 2024). In a study conducted in Hong Kong, the integration of generative AI demonstrated the potential to support effective argumentation and the structuring of written information, underpinning the quality of textual productions (Wang & Ren, 2024).

In the university setting, incorporating ChatGPT can foster this competence by correcting grammatical errors and suggesting improvements in writing, as well as by offering different discursive perspectives and a variety of references. Prompt functions and conversational interaction facilitate recurrent practice and experimentation with more complex linguistic structures (Chang & Li, 2025). Furthermore, using ChatGPT for immediate feedback allows students to quickly reflect on their texts, contributing to the iterative improvement of their communication (Chu & Liu, 2024). In this sense, it is inferred that reinforcing linguistic competence through the strategic use of ChatGPT is positively associated with academic performance. In addition, by improving competence in using and appropriating linguistic resources, a creative space is promoted to generate innovative and creative ideas and solutions.

2.3 Creativity as a Mediator Between ChatGPT and Academic Performance

Various studies have investigated how AI tools can foster creativity or, conversely, stagnate it. In a study with engineering students, it was identified that the continuous practice of ChatGPT-assisted ideation generated suspicion about the possible homogenization of results but, at the same time, offered enriching frameworks for the exploration of alternatives (Baltà-Salvador et al., 2025). For their part, Chang and Li (2025) highlight that the methodological guide in using ChatGPT contributes to improving the flexibility and originality of ideas when clear objectives and strategies for creative search are established.

Although there is still debate about the potential restriction of originality imposed by automatically generated responses, the literature indicates that creativity can emerge from the critical and reflective use of these responses (Yan, 2025). In a co-creation process, the user can combine ChatGPT suggestions with his or her own background, producing more elaborate ideas and innovative perspectives (Chu & Liu, 2024). This dynamic coincides with the mediated role of creativity, which postulates that using ChatGPT fosters students' creativity, improving their academic performance. Likewise, mediation hypotheses have been proposed where creativity directly intervenes in the relationship between certain variables and final performance (Aguilar et al., 2023). This mediating role is particularly appreciated in the efficiency in the acquisition of information that enhances creativity, which in turn leads to improvements in performance, and increased linguistic competence promotes creativity, which also translates into better performance. These approaches are consistent with the premise that by facilitating access to broader information and improving language skills, the ground becomes fertile for generating novel solutions and critical approaches, which benefits the quality of learning in the long run. However, attention to the development of fluency, flexibility, originality and elaboration, as key components in creative development, should not be lost sight of (Tomaylla-Quispe et al., 2023). Similarly, pupils' emotional moods and self-confidence play a role in obtaining meaningful learning (Gutierrez-Aguilar et al., 2023).

2.4 Contributions and Tensions of AI in Educational Processes

Although several studies point to an encouraging picture of the benefits of AI in promoting creativity and performance, tensions and ethical challenges are also pointed out (Alfarraj & Wardat, 2024). Plagiarism, academic dishonesty or inappropriate use of resources may arise, which questions the relevance of regulation and training aligned with ethical values (Sandu et al., 2024). At the same time, the importance of active teacher mediation has been emphasized, in which the teacher guides, supervises and guides interactions with ChatGPT so that students do not depend excessively on technology (Baltà-Salvador et al., 2025).

Another aspect of adopting ChatGPT is related to the availability of infrastructure and students' digital skills. Some works highlight the importance of equipping the student population with skills for the critical use of AI and the role of self-efficacy in the motivation to use these tools (Beltozar-Clemente & Díaz-Vega, 2024). If these challenges are addressed, ChatGPT could be consolidated as a transformative pedagogical resource that reinforces information acquisition, linguistic competence, and creativity (Al Shloul et al., 2024).

The body of previous research shows a clear trend: the introduction of ChatGPT in university educational settings is associated with improved academic performance, optimization of information acquisition, and enhancement of creativity, as long as its implementation is carried out in a planned and reflective manner (Aguilar et al., 2023; Wang & Ren, 2024). ChatGPT's contribution to mediating creativity is evidenced by its ability to diversify ideas and assist the user in divergent thinking (Chang & Li, 2025). This reinforces the thesis that postulates creativity as an indispensable link between efficiency in acquiring information or linguistic competence and academic performance.

Likewise, the use of ChatGPT in the university context, managed with appropriate pedagogical strategies and ethical regulations, points to improvements in information acquisition, linguistic fluency and precision, and academic performance. In turn, creativity emerges as one of the core variables that connects the tool's effectiveness with student performance since divergent thinking and the production of original solutions are strengthened when the student combines their skills with the assistance of AI. All of this supports the relevance of examining, in future research, the mechanisms by which creativity mediates the impact of ChatGPT on academic performance, contributing to a comprehensive view of the use of artificial intelligence in higher education. Based on the preceding theoretical considerations, the following hypotheses are proposed:

Direct effects hypothesis:

H1: Using ChatGPT fosters students' creativity (CRE), improving their academic performance (RA).

H2: Information acquisition efficiency (EAI) through ChatGPT improves academic performance (RA).

H3: Information acquisition efficiency (EAI) through ChatGPT fosters students' creativity (CRE).

H4: Improved language competence (CL) through interaction with ChatGPT is associated with higher academic performance (RA).

H5: Language competence (CL) developed through ChatGPT stimulates creativity (CRE).

Indirect or mediation effects hypothesis:

H6: Efficiency of information acquisition (EAI) through ChatGPT enhances creativity (CRE), which in turn improves academic performance (RA).

H7: Linguistic competence (CL) increased by interaction with ChatGPT promotes creativity (CRE), which in turn is reflected in higher academic performance (RA).

3. METHODOLOGY

The research had a quantitative approach of an explanatory nature through a non-experimental sectional design. The study sample comprised 389 university students in the professional careers of psychology, representing 40.9% (159) and advertising and multimedia with 59.1% (230). The distribution by sex was 31% men (N = 121) and 69% women (N = 268), with ages between 16 and 29 years, with an average of 19.5 years (Standard Deviation = 3.33). The research was carried out at the Catholic University of Santa María in Arequipa, Peru.

The data were obtained through a survey conducted between September and October 2024. Before applying the instrument, the participants' authorization was previously obtained through a consent form. The instrument was designed for the study with the following structure of latent variables: Efficiency of information acquisition (EAI = 6 items), Creativity (CRE = 6 items), Linguistic competence (CL = 6 items) and Academic performance (RA= 6 items). To measure the variables, a 5-point Likert scale was used to evaluate the responses: (1) Strongly disagree, (2) Somewhat disagree, (3) Neither agree nor disagree, (4) Somewhat agree (5) Strongly agree. SmartPLS (v: 4.1.0.9) was used for data processing and analysis, applying the CB-SEM module (Structured Equation Modeling based on covariances).

4. RESULTS

An exploratory factor analysis was performed to test the quality of the instrument as a preliminary empirical test with 50 volunteer students. The results to assess the measure of sampling adequacy, using the KMO (Kaiser-Meyer-Olkin), was 0.958; a satisfactory fit was achieved. The results of the Bartlett Sphericity test indicate a Chi-square (χ^2) value of 8550 with 276 degrees of freedom (df) and a p-value significantly lower than 0.001, indicating that the variables are correlated, justifying the use of factor analysis.

The Exploratory Factor Analysis (EFA), performed using the minimum residual extraction method and oblimin rotation, provided a detailed view of the structuring of the key variables in the study (Watkins, 2018). The items associated with each of these constructs showed significant factor loadings, with specific values highlighting the strength of the association within their respective factors. For example, the language competence (LC) items had extremely high loadings, with values ranging from 0.836 to 0.909, pointing to a clear and predominant explanation of the variance of these items by this factor. Similarly, in Factor 2, the creativity items exhibit loadings ranging from 0.636 to 0.896, indicating a strong and consistent association with the creativity construct.

Furthermore, the analysis reveals that item uniqueness, which measures the proportion of variance that is unique to each item and not shared with other items in the analysis, was relatively low for most items in the language competence and creativity factors, with values such as 0.181 for LC1 and 0.210 for CRE5, respectively. This reflects that the identified factors satisfactorily explain most of the variance in these items. In contrast, Factor 4 items related to information acquisition efficiency show higher uniqueness values, such as 0.524 for EAI5, which may indicate additional complexity in fully capturing this construct.

Table 1. Construct reliability and validity

	Cronbach's alpha (standardized)	Cronbach's alpha (unstandardized)	Composite reliability (rho_c)	Average variance extracted (AVE)
Academic performance (RA)	0.933	0.933	0.934	0.701
Creativity (CRE)	0.933	0.933	0.935	0.702
Efficiency of information acquisition (EAI)	0.887	0.885	0.887	0.569
Linguistic competence (CL)	0.961	0.961	0.961	0.803

Table 1 presents a robust assessment of the reliability and validity of the constructs used in the study, reflecting high levels of internal consistency and composite reliability for each of the constructs: academic achievement (RA), creativity (CRE), information acquisition efficiency (EAI), and linguistic competence (CL). Both academic achievement and creativity exhibit Cronbach's alpha and composite reliability values higher than 0.930, with average variance extraction (AVE) exceeding the threshold of 0.700, indicating excellent reliability and convergence of the items measuring these constructs. Although information acquisition efficiency shows a slightly lower value in these metrics (around 0.887 for reliability and 0.569 for AVE), it still suggests adequate reliability. Language competence stands out with exceptionally high reliability (0.961) and an AVE of 0.803, demonstrating the internal cohesion of the items and excellent convergent validity, reinforcing the robustness of the instrument to capture these key constructs within the research context. These results strengthen the use of these measures in future research and validate the application of the constructs to explore educational dynamics in advanced academic studies.

Table 2. Discriminant validity – Heterotrait-monotrait ratio (HTMT)

	(RA)	(CRE)	(EAI)	(CL)
Academic performance (RA)				
Creativity (CRE)	0.717			
Efficiency of information acquisition (EAI)	0.753	0.722		
Linguistic competence (CL)	0.709	0.564	0.487	

Table 2 shows the discriminant validity using the Heterotrait-Monotrait (HTMT) ratio for the studied constructs (Henseler et al., 2015). The values obtained between different constructs are less than 0.90, which is generally considered a good indication of discriminant validity, suggesting that each construct is sufficiently distinct from the others. However, the relationships between EAI and RA (0.753) and EAI and CRE (0.722) are relatively high, indicating a more remarkable similarity between these constructs. Language competence (LC) shows the most significant discrimination from the other constructs, especially with EAI, where the HTMT value is 0.487. These results indicate adequate differentiation between the constructs. However, the closeness between some pairs of constructs suggests that they might share some common attributes or interact in ways that could be explored in future research.

The analysis of the structural model presented in Figure 1, using covariance-based structural equation modelling, highlights the interdependent relationships between information acquisition efficiency (IAE), creativity (CRE), language competence (LC), and academic performance (AP). The model results reveal that creativity has a coefficient of determination (R^2) of 0.558, indicating that approximately 55.8% of the variability in creativity is explained by information acquisition efficiency and language competence. On the other hand, academic performance shows an even higher R^2 of 0.726, suggesting that the combined influences of information acquisition efficiency, creativity, and language competence explain 72.6% of its variability. This high level of explanation underlines the importance of these predictor variables in academic performance.

The standardized path coefficients reinforce the importance of the relationships between the variables. For example, information acquisition efficiency has a direct and significant impact on academic performance (coefficient of 0.408) and creativity (coefficient of 0.554), indicating that increased efficiency in processing and obtaining information can significantly improve academic performance and students' creative ability. Furthermore, language proficiency also strongly and positively influences academic performance (coefficient of 0.411) and contributes significantly to creativity (coefficient of 0.299). These findings highlight how fundamental skills such as information acquisition efficiency and language proficiency foster better academic performance and enhance creativity, evidencing their crucial role in educational and cognitive development.

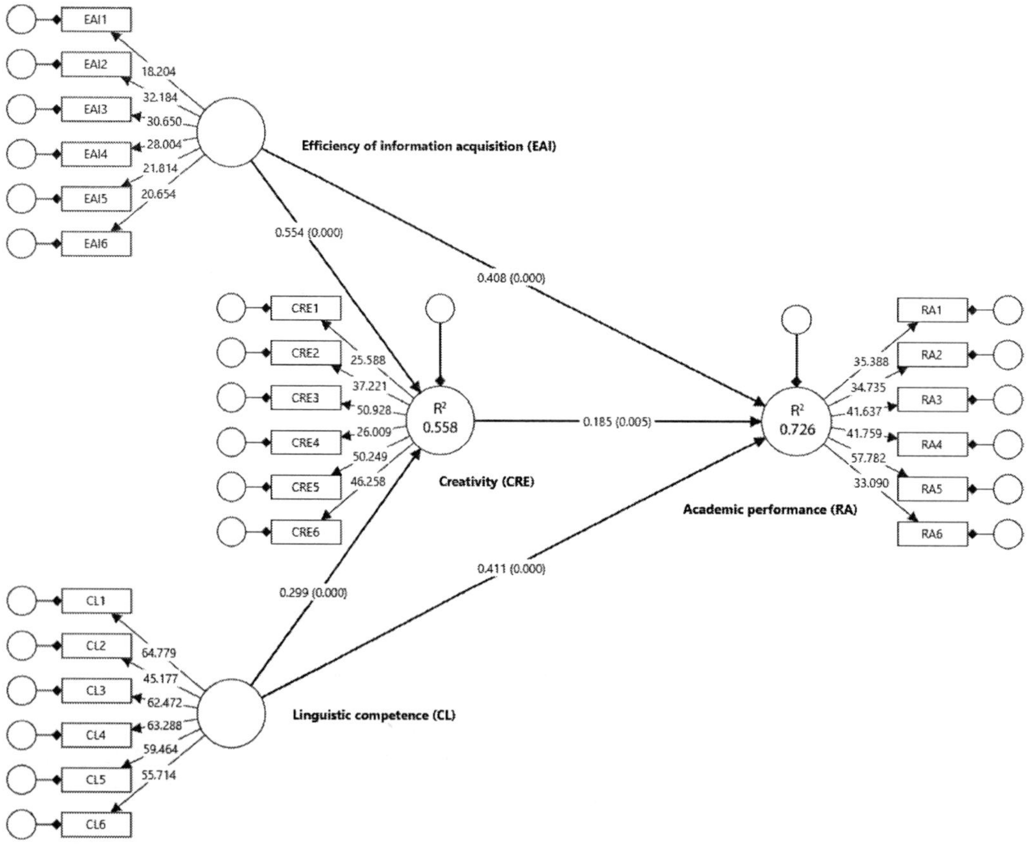

Figure 1. Regression model with standardised coefficients and significance level p-values

Table 3. Path coefficients – Mean, STDEV, T values, p values – Direct effects

	Original sample (O)	Sample mean (M)	Standard deviation (STDEV)	T statistics (|O/STDEV|)	P values
H1: Creativity (CRE) -> Academic performance (RA)	0.208	0.212	0.072	2.876	0.004
H2: Efficiency of information acquisition (EAI) -> Academic performance (RA)	0.587	0.582	0.095	6.150	0.000
H3: Efficiency of information acquisition (EAI) -> Creativity (CRE)	0.709	0.712	0.086	8.224	0.000
H4: Linguistic competence (CL) -> Academic performance (RA)	0.321	0.324	0.044	7.261	0.000
H5: Linguistic competence (CL) -> Creativity (CRE)	0.207	0.206	0.036	5.694	0.000

The evaluation of the hypotheses in the study, based on the results in Table 3, demonstrates a strong validation of the proposed theoretical relationships between the variables, using a significance level of P value less than 0.05. Each of the hypotheses put forward has been statistically significant, evidenced by extremely low P values. Specifically, the relationships between creativity and academic performance and information acquisition efficiency and creativity and academic performance have shown positive and significant influences. This is also reflected in the relationship between linguistic competence and these two dimensions, underlining the importance of these skills in the educational context.

The acceptance of all hypotheses indicates that information acquisition efficiency and linguistic competence capabilities are vital predictors for academic performance and fostering creativity. These results support the theories that link these skills with academic and creative improvements and highlight their practical relevance in educational development. Therefore, the evidence provided by the path coefficients and their corresponding T and P values strengthens the understanding of how these fundamental skills interact, promoting a more effective and enriching academic environment.

Table 4. Specific indirect effects - Mean, STDEV, T values, p values

	Original sample (O)	Sample mean (M)	Standard deviation (STDEV)	T statistics (|O/STDEV|)	P values
H6: Efficiency of information acquisition (EAI) -> Creativity (CRE) -> Academic performance (RA)	0.148	0.151	0.055	2.693	0.007
H7: Linguistic competence (CL) -> Creativity (CRE) -> Academic performance (RA)	0.043	0.044	0.017	2.503	0.013

Table 4 presents the specific indirect effects in the structural equation model, highlighting the mediating role of creativity (CRE) in the relationship between information acquisition efficiency (IAE), linguistic competence (LC), and academic performance (AP). The statistical results underline that creativity effectively acts as a mediating variable in these relationships, which is reflected in the significant T and P values for both mediation hypotheses.

Hypothesis H6, which postulates that information acquisition efficiency impacts academic performance through creativity, shows an indirect effect coefficient of 0.148 with a P value of 0.007. This result validates the mediation hypothesis and confirms that creativity is a significant channel through which information acquisition efficiency influences academic performance. The T value of 2.693 indicates that this effect is statistically significant, exceeding the confidence threshold usually accepted in social and educational studies.

On the other hand, hypothesis H7 assesses the indirect effect of language competence on academic performance through creativity, reporting a coefficient of 0.043 with a P value of 0.013. This result also supports the idea that creativity mediates the relationship between language competence and academic performance, with a T value of 2.503, reinforcing the mediation's statistical significance. These findings highlight the importance of creativity as a fundamental element that enriches the educational process and acts as an essential bridge between various cognitive skills and academic success.

5. CONCLUSIONS

The results show that information efficiency and linguistic competence enhance performance through creativity, as rapid search and synthesis of sources unleash original ideas (Al Shloul et al., 2024; Wang et al., 2024). Creative mediation supports Aguilar et al. (2023) and coincides with the argumentative enhancement found by Wang and Ren (2024). The direct language-performance link confirms the motivation-engagement relationship of Beltozar-Clemente and Díaz-Vega (2024), reaffirming the transformative role of ChatGPT in higher education (Sadallah et al., 2025).

Pedagogically, it is appropriate to link information literacy, rhetorical practice and co-creation with AI, as proposed by Chang and Li (2025). Libraries and writing centres can serve as laboratories for searching, filtering and writing training with ChatGPT, mitigating the dependency described by Baltà-Salvador et al. (2025) and the ethical risks cited by Sandu et al. (2024). Factors such as infrastructure, digital self-efficacy and emotional state can modulate benefits (Gutierrez-Aguilar et al., 2023); there is an urgent need to measure the evolution of perceived usefulness longitudinally (Alarbi et al., 2024). Consolidating informational efficiency and linguistic expression thus emerges as a strategic lever for creative and ethical academic excellence in the age of AI.

Despite its significant findings, this study is not free from limitations that could affect the generalizability of the results. One important limitation is the potential lack of representativeness of the sampling, which might not fully capture the diversity of educational contexts. Furthermore, the study is based on cross-sectional analysis, which limits the ability to infer direct causality between variables. Regarding practical applications, the study's results have valuable implications for curricular design and pedagogical strategies. In particular, integrating programs that promote information acquisition efficiency and linguistic competence can significantly improve students' creativity and academic performance. This approach could be instrumental in formulating innovative teaching methods that prepare students to meet academic and professional challenges in an increasingly information-based world.

REFERENCES

Aguilar, O. G., Delgado-Delgado, F., Meza-Málaga, J., Turpo-Gebera, O., & Apaza, F. T. (2023). Predictors of Academic Performance through the Use Of Chatgpt in University Students [Article]. Human Review. International Humanities Review / Revista Internacional de Humanidades, 21(2), 411-421. https://doi.org/10.37467/revhuman.v21.5077

Al Shloul, T., Mazhar, T., Abbas, Q., Iqbal, M., Ghadi, Y. Y., Shahzad, T., Mallek, F., & Hamam, H. (2024). Role of activity-based learning and ChatGPT on students' performance in education [Review]. Computers and Education: Artificial Intelligence, 6, Article 100219. https://doi.org/10.1016/j.caeai.2024.100219

Alarbi, K., Halaweh, M., Tairab, H., Alsalhi, N. R., Annamalai, N., & Aldarmaki, F. (2024). Making a revolution in physics learning in high schools with ChatGPT: A case study in UAE [Article]. Eurasia Journal of Mathematics, Science and Technology Education, 20(9), Article em2499. https://doi.org/10.29333/ejmste/14983

Alfarraj, Y. F., & Wardat, Y. (2024). Exploring the Impact of ChatGPT on Scientific Research: Assessing Strengths, Weaknesses, Opportunities, and Threats [Article]. Education as Change, 28, Article 16006. https://doi.org/10.25159/1947-9417/16006

Baltà-Salvador, R., El-Madafri, I., Brasó-Vives, E., & Peña, M. (2025). Empowering Engineering Students Through Artificial Intelligence (AI): Blended Human–AI Creative Ideation Processes with ChatGPT [Article]. Computer Applications in Engineering Education, 33(1), Article e22817. https://doi.org/10.1002/cae.22817

Beltozar-Clemente, S., & Díaz-Vega, E. (2024). Physics XP: Integration of ChatGPT and Gamificatioto Improve Academic Performance and Motivation in Physics 1 Course [Article]. International Journal of Engineering Pedagogy, 14(6), 82-92. https://doi.org/10.3991/ijep.v14i6.47127

Chang, H. F., & Li, T. (2025). A framework for collaborating a Large Language Model tool in brainstorming for triggering creative thoughts [Article]. Thinking Skills and Creativity, 56, Article 101755. https://doi.org/10.1016/j.tsc.2025.101755

Chu, H., & Liu, S. (2024). Can AI tell good stories? Narrative transportation and persuasion with ChatGPT [Article]. Journal of Communication, 74(5), 347-358. https://doi.org/10.1093/joc/jqae029

Gutierrez-Aguilar, O., Chicana-Huanca, B., Postigo-Coaguila, M., Agredo-Delgado, V., Rondon, D., & Chicana-Huanca, S. (2023). Influencing factors in creativity in learning, mediated by self-confidence in university students in the post-pandemic. CEUR Workshop Proceedings,

Henseler, J., Ringle, C. M., & Sarstedt, M. (2015). A new criterion for assessing discriminant validity in variance-based structural equation modeling. Journal of the Academy of Marketing Science, 43(1), 115-135. https://doi.org/10.1007/s11747-014-0403-8

Sadallah, M., Bin-Nashwan, S. A., & Benlahcene, A. (2025). ChatGPT: a transformative role in academia–insights into academic staff performance since adoption. Journal of Information, Communication and Ethics in Society, 23(1), 32-53. https://doi.org/10.1108/JICES-07-2024-0097

Sandu, R., Gide, E., & Elkhodr, M. (2024). The role and impact of ChatGPT in educational practices: insights from an Australian higher education case study [Article]. Discover Education, 3(1), Article 71. https://doi.org/10.1007/s44217-024-00126-6

Tomaylla-Quispe, Y., Gutierrez-Aguilar, O., Chicana-Huanca, S., & Duche-Pérez, A. B. (2023). Exploring Visual Creativity and Its Dimensions in Students of Online Training Program. CEUR Workshop Proceedings,

Wang, L., & Ren, B. (2024). Enhancing Academic Writing in a Linguistics Course with Generative AI: An Empirical Study in a Higher Education Institution in Hong Kong [Article]. Education Sciences, 14(12), Article 1329. https://doi.org/10.3390/educsci14121329

Wang, X., Zhong, Y., Huang, C., & Huang, X. (2024). ChatPRCS: A Personalized Support System for English Reading Comprehension Based on ChatGPT [Article]. IEEE Transactions on Learning Technologies, 17, 1762-1776. https://doi.org/10.1109/TLT.2024.3405747

Watkins, M. W. (2018). Exploratory Factor Analysis: A Guide to Best Practice. Journal of Black Psychology, 44(3), 219-246. https://doi.org/10.1177/0095798418771807

Yan, D. (2025). Posthuman Creativity: Unveiling Cyborg Subjectivity Through ChatGPT [Article]. Qualitative Inquiry, 31(2), 253-264. https://doi.org/10.1177/10778004241231923

RETHINKING EDUCATION: DESIGNING MINDS FOR INDUSTRY 5.0

Suzana Sampaio[1], Kathryn Cormican[1], Pablo Alvarez[2] and Deirdre Green[2]
[1]Enterprise Research Centre and LERO—The Irish Software Research Centre,
School of Engineering, University of Galway, Ireland
[2]Learnovate, Trinity College Dublin, Ireland

ABSTRACT

The shift to Industry 5.0 means that education must embrace a more human-centred, interdisciplinary, and sustainability-oriented approach. Traditional models that prioritise rote learning and standardised assessments are no longer sufficient in an era where human creativity is increasingly important. Design thinking as a viable approach shifts learning from passive knowledge acquisition towards active experiential learning where students engage in real-world problems, capture user needs and collaborate to create innovative solutions. We argue that it is essential to capture the voice of the user to understand the challenges faced by educators and students and identify good practices. Context-specific data collection is important, as educational settings vary widely. A one-size-fits-all model ignores nuances, potentially leading to ineffective or even counterproductive outcomes. Therefore, the goal of this paper is to uncover key challenges for educators and students to better understand their needs and requirements as we transition to Industry 5.0. To do this, we conducted six focus groups involving 38 educators working in vocational education and training settings in Spain, Slovenia and Portugal. Our analysis uncovers many challenges for educators and students and identifies effective teaching approaches as well as some skills and competencies needed to integrate innovative approaches into education. The results of this analysis offer insight into further advancements needed to develop and integrate innovative user-centred learning methods to accommodate respondents' concerns. Our work also supports evidence-based reforms that can enhance learning outcomes and better align with the demands of Industry 5.0.

KEYWORDS

Design Thinking, Focus Group, Challenges, Skills, Competencies

1. INTRODUCTION

Education systems worldwide face unprecedented challenges in preparing students for an increasingly complex and rapidly evolving world. As we transition to Industry 5.0, it is evident that we must move away from a focus on knowledge acquisition towards equipping our students with relevant skills and competencies. Industry 5.0. emphasises a synergy between humans and technology. It is characterised by human-centric automation, artificial intelligence, and sustainable innovation. Unlike Industry 4.0, which focuses on digital transformation and automation, Industry 5.0 emphasises the role of the human and advocates for user-centric solutions, personalisation, ethics and environmental responsibility in innovation. The emergence of Industry 5.0 necessitates a shift from traditional knowledge-based learning to a skills-oriented educational approach that fosters problem-solving, critical thinking, agility and interdisciplinary collaboration (Bathla et al., 2025; Jones, 2024; Oberer and Erkollar, 2024).

Traditional lecture-based teaching methods, often referred to as "chalk-and-talk" methods, although still prominent in most classrooms, are no longer adequate (Faregh et al., 2023; Lin et al., 2020). This approach relies heavily on rote learning, passive absorption of information, and standardised assessments. While it may be effective for basic knowledge acquisition, it fails to develop higher-order cognitive skills, such as critical thinking, adaptive reasoning, and creativity, required for the demands of Industry 5.0. There is an urgent need to transition towards an agile design thinking approach that encourages user-centred innovation, experimentation, and real-world problem-solving (Bathla et al., 2025; Jones, 2024; Scheer et al., 2012). Educators must integrate experiential learning, creativity-driven curricula, and digital fluency to prepare students for an era where human ingenuity complements intelligent machines (Wolf et al., 2024; Faregh and Jourabchi Amirkhizi, 2023).

While design thinking approaches have gained traction in educational settings, particularly in universities, in recent years, little empirical work has been conducted in this domain (Bathla et al., 2025; Lattemann et al., 2020; Panke, 2019). There is a pressing need for scholarly research to understand the challenges in education, the most effective teaching methods to use, whether design thinking approaches can meet the needs of students and educators and if so, what skills and competencies are required to integrate innovative approaches into education. Furthermore, researchers are calling for studies to capture and analyse context-specific data to address the unique challenges and requirements in different educational environments (Wolf et al., 2023; Razali et al., 2022; Panke, 2019). Therefore, the goal of this paper is to answer this call and uncover insights relating to challenges for students and teachers, determine effective approaches for teaching and learning and identify skills and competencies required to integrate innovative approaches into existing systems. To do this, we conducted focus groups in six schools in Spain, Portugal and Slovenia. We captured qualitative data and elicited user stories from 38 vocational education and training (VET) teachers in six different schools. The findings of this paper shed light on an important but neglected area.

2. RESEARCH METHODOLOGY

The research process employed in this study comprised three stages as illustrated in Figure 1.

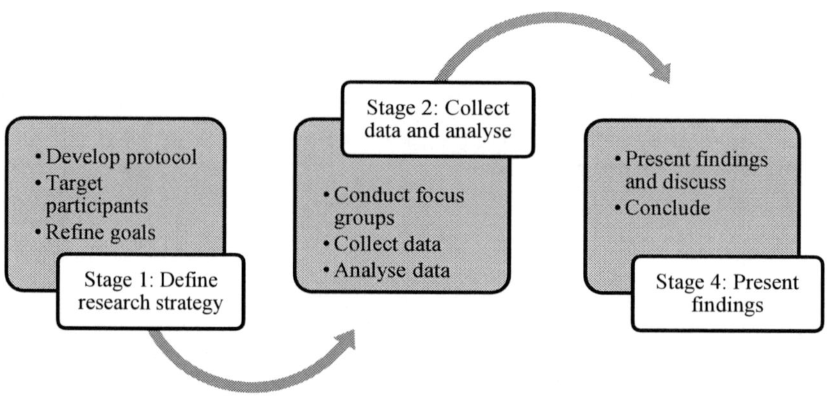

Figure 1. Research methodology

The first stage in the research process involved defining a research strategy. As this study aims to explore user stories, uncover deep insights and understand context-specific data, exploratory qualitative methods were deemed appropriate. The study employed a high-variation sampling strategy (Flick, 2014) comprising vocational education and training (VET) educators in Spain, Slovenia and Portugal. Table 1 presents the participants involved in the focus group, their area of teaching, years of experience and the country in which they are based. A total of 38 VET teachers participated in our study from six schools in Spain, Slovenia and Portugal. Focus groups were used to explore key issues in depth and helped to foster discussions among participants. This approach allowed for the emergence of diverse perspectives, as participants built on each other's insights, leading to a more comprehensive understanding of challenges and issues. An interview schedule was developed to guide the questions and to ensure consistency in the data collection process. The questions posed centred on the challenges for educators and students in VET education, the most effective teaching approaches and the skills and competencies needed for integrating innovative approaches into education. The focus group protocol and supporting script were pretested with academics (n=3) and pilot-tested with VET teachers (n=9), and amendments were made based on the feedback provided. A total of 38 VET teachers participated in our study from six schools in Spain, Slovenia and Portugal. Each focus group comprised between 5 and 9 teachers and lasted between 50 to 80 minutes, with an average time of 70.33 minutes.

Table 1. Profile of participants

Code	Country	Experience	Courses
T1	Spain	+20	Computing science
T2	Spain	15	Health care
T3	Spain	+20	Administration and economics
T4	Spain	3	Computing Science
T5	Spain	10	Computing Science
T6	Spain	11	English
T7	Spain	25	Information technology
T8	Spain	10	Business
T9	Spain	7	Business
T10	Spain	6	Business
T11	Spain	17	Business
T12	Spain	30	Business
T13	Spain	25	Electronics/information technology
T14	Spain	20	Information technology
T15	Spain	12	Business/electronics
T16	Slovenia	26	Economics
T17	Slovenia	32	Foreign languages (English, German)
T18	Slovenia	25	Mathematics
T19	Slovenia	5	Library and information education, Entrepreneurship club
T20	Slovenia	21	Counselling, Psychology
T21	Slovenia	9	Media production
T22	Slovenia	23	Sales and economics
T23	Slovenia	31	Practical education
T24	Slovenia	29	Business mathematics and statistics
T25	Slovenia	20	Business foreign language
T26	Slovenia	7	Business communication
T27	Slovenia	21	Business Finance
T28	Portugal	5	Physics and Chemistry
T29	Portugal	17	Philosophy
T30	Portugal	2	Electricity
T31	Portugal	6	Child Education
T32	Portugal	5	Portuguese
T33	Portugal	2	Mathematics
T34	Portugal	43	English
T35	Portugal	29	Portuguese
T36	Portugal	30	Information technology and Electrotechnics
T37	Portugal	26	Physics and Chemistry
T38	Portugal	20	History of culture and arts

The second stage of the research process involved collecting and analysing the data. Data were collected in the local language to ensure a deep understanding of the challenges and issues. The focus groups were recorded, and the data were transcribed and translated into English. Content analysis was used to generate themes and identify relationships for explaining the data. Miro boards were used to organise and validate the data. The data underwent an initial analysis at the national level, followed by a consolidation and further iteration of data from the three countries to identify shared and distinct challenges and insights.

3. FINDINGS FROM THE ANALYSIS

This section presents the findings from our analysis. An overview of the results and a sample of the Miro boards used are presented in Figure 2 (for illustrative purposes only). The challenges for educators and students are presented and discussed, followed by an overview of effective teaching approaches and skills and competencies required to implement innovative approaches in practice.

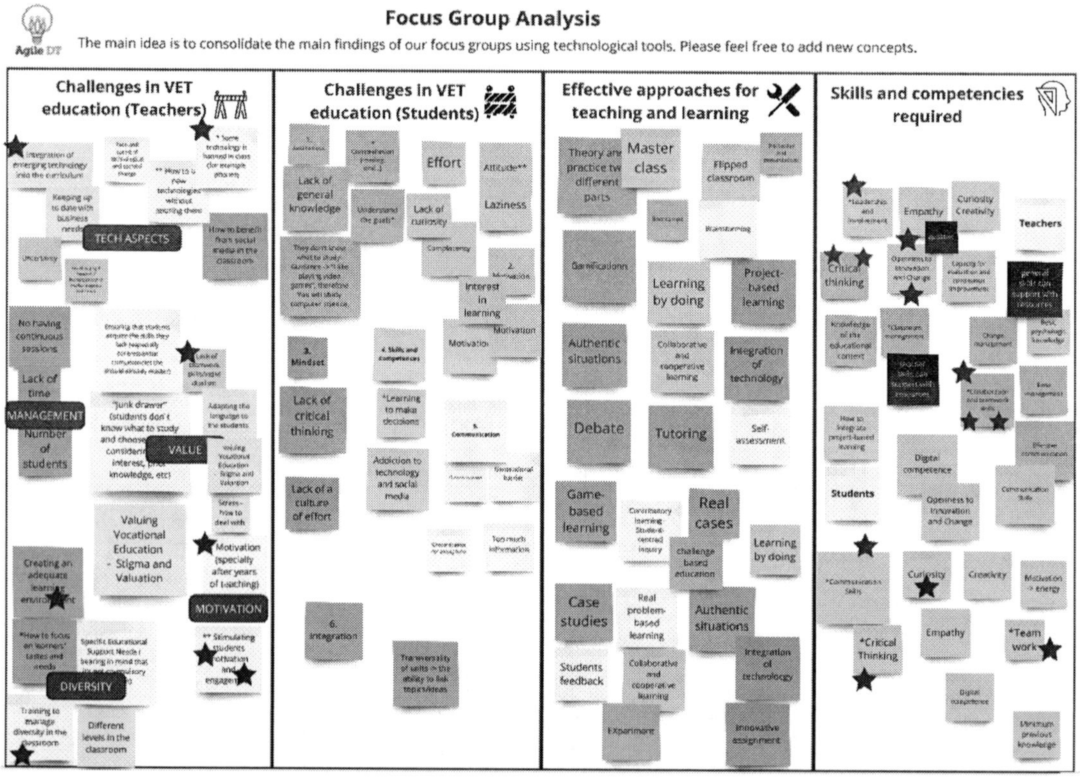

Figure 2. Overview of results

3.1 Challenges for Educators and Students

In this section, we present a range of challenges faced by both educators and students in the current educational landscape. By understanding these obstacles, we aim to shed light on the difficulties educators encounter in engaging and supporting their students, as well as the struggles students face in adapting to the demands of modern education. Table 2 presents the challenges for educators.

Table 2. Challenges for educators

Challenges	Description	Importance	Context
Inclusive education	Difficult to make learning inclusive Struggle to manage diversity Difficult to adapt the material for different students Difficult to deal with students with special needs Difficult to deal with different levels in the classroom Difficult to deal with neurodiversity Lack of a personal approach	High	Portugal Slovenia Spain
Motivation	Difficult to make the process interesting Unable to stimulate students Unable to motivate students Unable to engage students Unable to create an adequate learning environment	High	Portugal Slovenia Spain
Content	Outdated content Poor alignment with business needs Poor alignment with market needs	Medium	Portugal, Slovenia Spain

Technology	Poor alignment with future careers Difficult to update content quickly Difficult to integrate AI Difficult to integrate emerging technologies into the curriculum Difficult to use new technologies without abusing them Difficult to leverage social media in the classroom Difficult to deal with the speed of technologies	Medium	Portugal Slovenia Spain
Lack of time	Lack of time	Low	Spain
Number of students	Too many students in class	Low	Spain

The findings highlight several challenges for educators. Inclusivity remains a major concern, with difficulties in adapting materials for diverse student needs, managing varying levels of ability, and addressing neurodiversity. Motivation also poses a significant challenge, as educators struggle to engage students, create an interesting learning environment, and stimulate interest in the process. Content-related issues include outdated materials and a lack of alignment with both business and market needs, making it difficult to prepare students for future careers. Technological integration is another hurdle, with challenges in incorporating artificial intelligence and emerging technologies without overwhelming or misusing them. Furthermore, the fast-paced evolution of technology and a lack of time, along with large class sizes, further complicate the teaching process, requiring more efficient and adaptable solutions to meet the needs of today's students. Table 3 presents the challenges faced by students, according to the teacher's perception.

Table 3. Challenges for students

Challenges	Description	Priority	Country
Motivation	Lack of motivation Lack of interest in learning Lack of curiosity	High	Portugal Slovenia Spain
Learning capacity	Students do not know how to connect theory and practice Students don't know what to study Lack of ability to learn Lack of learning capacity Lack of ability to comprehend Gap in knowledge	High	Portugal Slovenia Spain
Organisation	Poor organisation Difficult to set goals Difficult to follow lessons and complete assignments in class Students do not complete assignments at home	High	Portugal Slovenia
Behaviour	Lack of a culture of effort Poor behaviour Laziness	Medium	Portugal Slovenia Spain
Critical thinking	Inability to concentrate Inability to think critically Do not know how to critically evaluate the information	Medium	Slovenia Spain
Teamwork	Difficult to engage in group activities Lack of motivation to work together	Medium	Slovenia
Stress	High level of stress Inability to cope with stress	Medium	Slovenia
Confidence	Lack of confidence	Low	Portugal

The findings reveal several challenges for students. Motivation is a major barrier, with students showing a lack of interest, curiosity, and overall drive to learn. Learning capacity is also a concern, as many students struggle to connect theory with practice, lack direction on what to study, and face difficulties in comprehending or retaining information. Organisational challenges are evident, with students struggling to set goals, stay on track with lessons, and complete assignments both in class and at home. Behavioural issues such as a lack of effort, poor behaviour, and laziness further hinder progress. Additionally, critical thinking skills are underdeveloped, with students unable to concentrate or critically evaluate information. Teamwork is another area of difficulty, with students lacking motivation to collaborate effectively. High levels of stress, inability to manage stress, and low confidence further exacerbate these challenges, creating barriers to academic success and personal growth.

3.2 Effective Approaches for Teaching and Learning

To identify the most effective teaching methodologies, educators participating in the focus groups shared insights reflecting the need for a student-centred, dynamic approach to learning. The findings reveal a strong consensus on the importance of active participation, collaboration, and engagement-enhancing strategies such as gamification and technology integration.

3.2.1 Challenge-Based Projects

Several participants emphasised the importance of challenge-based projects. Real-world projects serve to provide students with opportunities to explore authentic scenarios and develop practical solutions. By making learning more tangible and relevant, these methods foster greater student involvement and motivation. Educators emphasised the importance of simulating real-life challenges, with statements from Portuguese participants, "*Projects that simulate real-life challenges are crucial as they prepare students not just academically but practically* (T36)." Also, "*Approaching content by always linking it to real life gives them a more motivating vision, involvement in task preparation... that motivates them more to work*" (T37). The preference for practical learning is evident across multiple testimonies, as one Spanish educator explained, "*The teaching approach is 100% practical. The problem is that for practice, you have to know theory concepts. They are not interested in anything that is theoretical...*" (T05). Moreover, teachers from different schools in Slovenia pointed to student interest in concrete experiences offered by the challenge-based projects "*They want to have concrete experiences and are often very bored if they sit still all morning and copy from a whiteboard*" (T21), and *"Project work is the future, the basis for work" (T25).*

3.2.2 Collaboration

Collaborative project work was lauded for its ability to promote interaction, communication and teamwork. An educator highlighted this by noting, "*Group projects help foster teamwork and communication skills, which are essential in the modern workplace*" (T36). The role of collaborative learning, including cooperative group work and peer discussions, was emphasised in Portugal and Slovenia. Educators noted its effectiveness in developing essential problem-solving and communication skills. One participant shared, "*If we talk about projects in relation to their interests, they are able to identify with certain projects. And since there's a conversation, a collaboration, there's the need to solve problems, there's critical thinking*" (T31). However, it is important to note that collaboration is not easy and some educators acknowledged challenges in group work dynamics, particularly in assessment: "*There are two problems with group work: The first is that the more successful students want to claim all the credit and the good marks for themselves; the second is that such work is very difficult to mark and the criteria have to be very detailed*" (T16). Despite these challenges, the findings from our analysis reinforce the value of collaborative learning in fostering engagement, problem-solving, and teamwork.

3.2.3 Gamification

Gamification techniques, including game-based learning and competitions, were highlighted in Slovenia and Spain as effective tools for increasing student participation. Teachers shared their experiences with digital tools such as Kahoot and Socrative, e.g., "*I use a lot of game-based learning to review all the concepts we have learnt*" (T13). Educators also underscored the effectiveness of simple interactive games such as Scrabble for vocabulary enhancement, highlighting how gamified learning experiences make education more engaging and interactive. For example, one educator remarked, *"We use games like Scrabble to enhance vocabulary skills, which makes learning more engaging and interactive for students"* (T34).

3.2.4 Flipped Classroom

The flipped classroom model was endorsed in Portugal and Spain. One educator illustrated its benefits in subjects requiring higher-order thinking skills: "*In my subject, which requires a certain amount of thinking and which they don't like right from the start, I use a lot of simulations that exist online and I have to use mobile phones...*" (T28). This approach allows students to engage with theoretical concepts independently before applying them in the classroom, leading to deeper comprehension and more meaningful discussions. In addition, encouraging discussion and presentation of work was identified as a valuable strategy to cultivate critical thinking and public speaking skills among students. As one educator pointed out, "*Encouraging students to discuss current events and present their findings helps them develop critical thinking and public speaking skills*" (T38). It can also simulate real life "*The flipped-classroom format is very realistic with the reality they are going to have in the company, the boss is not going to come to you and explain in too much detail how they want things to be*" (T15).

3.2.5 Integration of Technology

The role of technology in modern education was a recurring theme in the focus group discussions, particularly in Portugal. The integration of technology, particularly through smartphone applications and simulations, was recognised for its practical applications in technical subjects, enhancing hands-on learning experiences. One educator emphasised this by stating, "*We incorporate apps in our lessons, like using smartphone applications that simulate traditional oscilloscopes*" (T37) Teachers recognised the importance of utilising mobile devices productively within the learning environment: "*Since we have mobile phones in the classroom, let's use them productively*" (T37). By integrating technology effectively, educators can create more interactive and personalised learning experiences.

3.3 Skills and Competencies Required

According to our analysis, the successful integration of innovative approaches into education requires several important skills and competencies. Educators should consider strategies to create engaging learning environments, foster student participation, and effectively implement innovative teaching strategies. For example, the results of our analysis reveal that effective leadership is essential to creating an engaging learning environment. A participant emphasised the importance of "*leadership, involvement, and empathy*" (T36), highlighting the crucial role of educators in guiding and supporting students. This sentiment was echoed by others who noted that fostering a "*sense of responsibility and the desire to do well*" (T34, T37) is vital for student engagement and the successful implementation of innovative approaches. Creativity also emerged as a critical component in innovative educational approaches. One participant stated that "*educators must think outside the box and develop innovative teaching strategies to captivate students' interest and enhance learning outcomes*" (T28). Teachers must be creative in designing learning experiences that are engaging and effective. Effective communication skills were deemed essential for both teaching and student involvement. This skill enables educators to convey ideas effectively and engage students in meaningful learning experiences. Moreover, communication skills facilitate collaboration and teamwork, essential elements highlighted by educators across the focus group discussions. Finally, the importance of being receptive to innovation and adaptable to change was also highlighted. This enables teachers to respond to the diverse needs and learning styles of students. Moreover, flexibility is crucial for educators to embrace new methods and educational approaches, ensuring that teaching practices remain relevant and effective in today's dynamic learning environment.

4. CONCLUSION

The evolution of industry has always shaped education, driving the need for new skills and competencies. As we transition into Industry 5.0, new approaches are required to prepare students for the challenges and opportunities of this new era. Design thinking is lauded to offer a potential solution for education in this new era. It is an experiential, student-centred approach that fosters active learning by encouraging learners to define problems, ideate solutions, prototype, test, and iterate. Empirical studies can provide valuable insights into how educators can integrate design thinking into curricula, ensuring that it fosters the creativity, adaptability, and problem-solving skills necessary for Industry 5.0. This study captures voices and insights from 38 vocational education and training teachers working in Spain, Slovenia and Portugal. The findings of our analysis highlight the pressing challenges faced by both educators and students while identifying key teaching approaches and competencies necessary for integrating innovative educational methods. By shedding light on these issues, we provide valuable insights into the advancements needed to develop user-centred learning strategies that address respondents' concerns. Ultimately, our findings support evidence-based insights that ensure education evolves in alignment with the demands of Industry 5.0.

ACKNOWLEDGEMENT

Our work is funded by the European Commission and Leargas by AgileDT: VET schools solve real-life challenges with innovation techniques. Grant number: 2023-1-IE01-KA220-VET-000157588. We would like to extend our gratitude to our school project partners, whose contributions enriched this research. Special thanks to Centro Privado Plurilingüe Liceo 'La Paz' and CEBEM Formación Profesional SL from Spain, AEVA - Associação para a Educação e Valorização da Região de Aveiro and Agrupamento de Escolas de Sertã from Portugal, as well as Ekonomska šola Novo Mesto and SIC Brežice from Slovenia

REFERENCES

Bathla, A., Chawla, G. & Gupta, A., (2025). Design Thinking in Education: Reviewing the Past for Setting Future Research. *Journal of the Knowledge Economy*, pp.1-39.

Faregh, S.A. & Jourabchi Amirkhizi, P., (2023). Design Thinking As An Effective Tool In Education. *Journal of Design Thinking*, 4(1), pp.69-86.

Flick, U., (2014). Mapping the field. In: *The SAGE Handbook of Qualitative Data Analysis*, 1, pp.3-18. Available at: [https://methods.sagepub.com/hnbk/edvol/the-sage-handbook-of-qualitative-data-analysis/toc].

Jones, P.A., (2024). Design Thinking in Action: Fostering 21st-Century Skills Alongside Subject Specific Knowledge at Key Stage 3 in D&T. *Design and Technology Education*, 29(2), pp.219-247.

Lattemann, C., Arntsen, E., Flaten, B.T., Fürst, N., Holen, J. & Cappelen, B., (2020). Is there a proper way to teach design thinking? Empirical evidence from design thinking in education. *Journal of Design Thinking*, 1(1), pp.35-48.

Lin, L., Shadiev, R., Hwang, W.Y. & Shen, S., (2020). From knowledge and skills to digital works: An application of design thinking in the information technology course. *Thinking Skills and Creativity*, 36, p.100646.

Melles, G., Anderson, N., Barrett, T. & Thompson-Whiteside, S., (2015). Problem finding through design thinking in education. In: Inquiry-based learning for multidisciplinary programs: A conceptual and practical resource for educators, pp.191-209. Emerald Group Publishing Limited.

Oberer, B. & Erkollar, A., (2024). Education 5.0: Using the Design Thinking Process-An. *Journal of Systemics, Cybernetics and Informatics*, 22(1), pp.1-17.

Panke, S., (2019). Design thinking in education: Perspectives, opportunities and challenges. *Open Education Studies*, 1(1), pp.281-306.

Razali, N.H., Ali, N.N.N., Safiyuddin, S.K. & Khalid, F., (2022). Design thinking approaches in education and their challenges: A systematic literature review. *Creative Education*, 13(7), pp.2289-2299.

Scheer, A., Noweski, C. & Meinel, C., (2012). Transforming constructivist learning into action: Design thinking in education. *Design and Technology Education: An International Journal*, 17(3), pp.8-19.

Wolf, P., Cormican, K., Frederiksen, M.H., Wilhøft, A., Ulus, H.E., Kunz, C. & van Leeuwen, M., (2024). I think they just logged on and fell asleep: Challenges to facilitating creativity online in higher engineering education. *Creativity and Innovation Management*, 33(3), pp.438-460.

GENERATIVE ARTIFICIAL INTELLIGENCE IN SCIENCE EDUCATION AND TEACHER TRAINING: DEVELOPMENT OF A DATA COLLECTION INSTRUMENT

Juliana C. C. Monteiro, Margarida M. Marques and Lúcia Pombo
Research Centre on Didactics and Technology in the Education of Trainers (CIDTFF),
Department of Education and Psychology (DEP), University of Aveiro (UA)
Campus Universitário de Santiago, 3810-193 Aveiro, Portugal

ABSTRACT

Generative Artificial Intelligence (GAI) is rapidly transforming various sectors, including science and education. In the educational sector, GAI offers opportunities, such as personalized learning and the reduction of teacher workload, alongside challenges, like balancing the advantages of GAI with students' independent problem-solving skills, but also the potential exacerbation of the digital divide. Consequently, educational stakeholders must critically assess GAI's capabilities while remaining cognizant of its potential risks. This article aims to detail the development of a data collection instrument specifically designed to qualitatively explore the perspectives of key stakeholders regarding the integration of GAI in Science Education. The developed instrument is grounded in established theoretical frameworks, notably the Intelligent Technology, Pedagogy, and Content Knowledge model (Intelligent TPACK) and the Technology Acceptance Model (TAM), which are instrumental in understanding the multifaceted dimensions of technology integration in educational settings. The instrument aims to capture: i) indicators of respondents' literacy in GAI in science education, ii) the GAI perceived ease of use in science education and GAI tools learning curve, iii) perceived usefulness of GAI tools in enhancing science education, iv) the perceived trust in GAI tools for science education content creation and factors influencing it, v) perceptions on knowledge needed for GAI integration in science education, vi) perceptions on the acceptance of GAI tools in science education, and vii) perceptions regarding GAI ethical issues in science education. In contrast to the prevailing emphasis on the technical deployment of AI in educational contexts, this study focuses on the nuanced perspectives of science education stakeholders, to inform the development of strategies for effective GAI integration in science education and science teacher education.

KEYWORDS

Artificial Intelligence in Education, Science Teacher Education, Pedagogical Practices Innovation, Qualitative Data Collection Instrument, Intelligent TPACK, TAM

1. INTRODUCTION

Artificial intelligence (AI) involves computer systems performing tasks typically requiring human intelligence, such as learning, problem-solving, decision-making, and perception. AI emulate human cognitive abilities, enabling them to execute repetitive, rule-based tasks with enhanced precision, speed, and cost-effectiveness (Dima et al. 2024). Generative Artificial Intelligence (GAI), a subset of AI, employs algorithms to generate new content, such as text, images, code generation, or music, based on training data, often performing complex tasks indistinguishably from humans (Banh and Strobel 2023). Its rapid evolution, with technologies, such as ChatGPT or DALL·E, has ushered in a transformative era in Education.

Within Education, GAI tools can potentially enhance, e.g., pedagogical design, scientific modeling, and content creation. Indeed, the integration of GAI into adaptive and personalized learning environments marks a significant shift, building upon conventional AI methods, potentially catalyzing equity and innovation within educational paradigms (Guettala et al. 2024). For example, GAI's capacity to tailor educational content to the specific requirements of individual learners has the potential to significantly diminish cognitive burden on educators, concurrently fostering enhanced learning outcomes (Bura and Myakala 2024).

Realizing GAI's transformative potential in Education demands careful attention to attitudes, readiness, and ethical issues of concern for those who are central to its implementation: teacher educators, in-service teachers, teacher trainees, and pre-service teachers. Furthermore, the limited availability of AI-learning opportunities in initial teacher education programs highlights the importance of focusing on this issue. This enables educators to make well-informed decisions about integrating GAI into their teaching practices, highlighting the need for further research to deepen understanding and best practices. These could potentially enhance the effective incorporation of AI instruction and educational outcomes (Marques 2024).

The effective integration of GAI in education demands robust conceptual models and validated instruments to gather empirical data regarding stakeholders' perspectives. One of the most recognized models is the *Technological Pedagogical Content Knowledge* (TPACK), by Mishra and Koehler (2006), that conceptualizes the need for the intersection of: a) Technological Knowledge (TK), or knowing how to use technology effectively, b) Pedagogical Knowledge (PK), or knowing teaching methods and learning theories, and c) Content Knowledge (CK), or knowing the subject matter being taught. Thus, teachers with strong TPACK can design educational experiences that explore technology to enhance student learning in specific content areas, addressing contextual nuances and diverse learner needs. Recently, Mishra, Warr, and Islam (2023) revisited the model to address the specific GAI challenge, considering these tools as active, adaptive participants rather than instruments. The *Intelligent TPACK* calls for critical teacher knowledge of AI's capabilities, limitations, and ethical implications, reframing the role of technology in education towards dynamic and responsible engagement. Another framework that is widely used in this context is the *Technology Acceptance Model* (TAM), which aims to predict how individuals accept and use new technology, focusing on the perceived usefulness and ease of use of the technology by the potential user.

Recent studies have explored these theoretical frameworks and produce data collection tools. For instance, Chiu et al. (2024) developed the *Teacher AI Competence Self-Efficacy* (TAICS) scale, grounded in Falloon's (2020) teacher digital competence framework and TPACK elements. The TAICS includes 24 items distributed in: a) AI Knowledge (AIK), b) AI Pedagogy (AIP), c) AI Assessment (AIA), d) AI Ethics (AIE), e) Human-Centered Education (HCE) and f) Professional Engagement (PEN), incorporating dimensions that reflect the ethical, pedagogical, and professional complexities introduced by AI tools. The authors reported a high reliability and strong model fit of the scale, with invariance across gender and teaching disciplines, and K–12 settings. Al-Abdullatif (2024) developed and validated a TAM-based questionnaire to examine factors influencing university teachers' adoption of GAI technologies. The tool demonstrated that AI literacy, perceived ease of use, intelligent TPACK, and perceived trust are key constructs in understanding and supporting teachers' integration of GAI in higher education. These two studies exemplify distinct but complementary approaches to understanding GAI in education. Nevertheless, both instruments can inform the design of professional development, curriculum innovations, and policy interventions aimed at fostering AI tools thoughtful analysis by educators. While valuable, these instruments primarily offer quantitative insights.

In this context, the project GAI-SciTeach aims to foster a GAI integration in teacher training that empowers and not overpowers educators and learners. The project follows a design-based research approach with iterative cycles of design, implementation, analysis, and redesign, involving active collaboration among researchers, teachers, and other stakeholders to develop a pedagogical framework tailored for the effective integration of GAI in science teacher initial education. The goal is to bridge theory and practice by generating both practical solutions and theoretical contributions. Recognizing the gap in the literature regarding tools designed for in-depth qualitative exploration of the topic, our project follows a stakeholder-driven approach to develop a qualitative data collection instrument specifically designed to explore the perspectives of researchers, teachers, future-teachers and other stakeholders regarding the integration of GAI in science education. Hence, this paper outlines the process of developing a focus group guide to collect qualitative data on stakeholders' views. The instrument addresses the research question: "What are science education stakeholders' perspectives about the integration of GAI in science education and in science teacher education?". It is structured using the PICo framework (Hosseini et al. 2024), referring to Population (stakeholders in science education, such as researchers and teacher educators, teachers, and future teachers), Intervention/phenomenon (educational integration of GAI), and Context (science teaching, science teacher training).

This article is structured to provide a comprehensive analysis of the development and design of the above proposed instrument. The next section, titled *Method for instrument development*, begins with a justification of the research question and overall approach to address it. Following, in the *Instrument description* section, each dimension of analysis is presented. The article concludes with *Final Considerations*, offering a reflection on the study, including its limitations and future work.

2. METHOD FOR INSTRUMENT DEVELOPMENT

The instrument was developed to gather qualitative data of perspectives on GAI integration via focus group (Williams and Katz 2001). The guide's development was grounded on existing quantitative instruments, the TAICS scale (Chiu et al. 2024) and an extended TAM questionnaire (Al-Abdullatif 2024), which are theoretically sound and empirically sustained tools. This strategy aims for qualitative insights, exploring the nuances and reasoning behind stakeholder perspectives on GAI integration more deeply than quantitative scales permit. The development process involved interpretation of the questionnaires' items, which were then systematically transformed into open-ended questions for the focus group. This adaptation sought to elicit authentic and non-induced viewpoints, ensure coherent yet concise discussions, and traceability to the original constructs.

To ensure the content validity of the semi-structured interview guide, expert validation was conducted involving a panel of three specialists with experience in science education and technology. These experts critically reviewed the instrument for clarity, relevance, and comprehensiveness of the questions in relation to the research objectives (Monteiro and Marques 2025). Their *feedback* informed iterative revisions, with particular attention to the alignment between the interview prompts and the conceptual framework underpinning the study. This process helped to enhance the guide's capacity to elicit rich, meaningful data, while minimizing ambiguity and potential bias.

3. INSTRUMENT DESCRIPTION

The semi-structured focus group guide is organized in three main sections, namely *Introduction*, *Development* (with the analysis dimensions), and *Closure*. according to Amado's (2014) orientations

The *Introduction* section introduces the GAI-SciTeach project and interviewer, outlines the focus group's objectives, acknowledges the value of participants' contributions, and addresses informed consent, data protection (in accordance with GDPR), and ethical procedures. Following, the core of the guide resides in the *Development* section, which explores seven key themes related to GAI integration in science education. This section utilizes a structured four-column format to ensure systematic information organization: a) Dimension/Objective, detailing the specific goal or competency area within the theme; b) Base Studies Questions, identifying the specific items from the quantitative instruments (Al-Abdullatif 2024; Chiu et al. 2024) that informed the corresponding developed qualitative questions; c) Questions, presenting the primary, open-ended questions adapted for the focus group; and d) Probing Questions, listing follow-up sub-questions designed to probe deeper into participants' initial responses, if needed. This structured approach within the *Development* section facilitates a comprehensive exploration of stakeholder perspectives while maintaining a clear link to the foundational research and ensuring depth through targeted probing. Following, each study dimension is described.

3.1 GAI Literacy

This dimension evaluates foundational competencies regarding GAI, focusing on conceptual understanding, tool recognition, productivity enhancement, and evaluative use, as depicted in Table 1. Questions probe how respondents define GAI, identify AI-powered tools, and apply GAI tools in science education settings. Items are derived from validated AI literacy constructs from both Chiu et al. (2024) and Al-Abdullatif (2024) and are supplemented with probing sub-questions that deepen inquiry into user knowledge and experience.

Table 1. Section of the instrument dedicated to the dimension of GAI Literacy

Dimension / Objective	Base Studies Questions	Questions	Probing Questions
GAI Literacy / To collect indicators of respondents' literacy in GAI in science education	I can explain what AI is [AIK3] (Chiu et al. 2024)	1. Can you explain what Generative Artificial Intelligence is?	1a. Please define the concept as you understand it. 1b. What examples would you use to illustrate what Generative Artificial Intelligence is?
	I can distinguish whether a tool is AI-based or not. [AIK1] (Chiu et al. 2024)	2. How can people determine if new tools are based, or not, in Generative Artificial Intelligence?	2a. Can you illustrate with an example? 2b. What characteristics indicate that a tool uses Generative Artificial Intelligence?
	I can use AI applications or products to improve my work efficiency (Al-Abdullatif 2024) I can create content with AI [AIK2] (Chiu et al. 2024)	3. How can people use GAI tools in science education for greater efficiency?	3a. Illustrate, by referring to concrete tasks and objectives (in science education) that can be accomplished/achieved more efficiently through the use of GAI tools.
	I know how to choose the right AI tools to effectively complete a task. [AIK4] (Chiu et al. 2024) I can evaluate the capabilities and limitations of an AI application or product after using it for a while (Al-Abdullatif 2024)	4. How can people choose GAI tools that are adequate for carrying out a given task in science education?	

3.2 GAI Perceived Ease of Use

This dimension is dedicated to discussing GAI perceived ease of use in science education and GAI tools learning curve, as prompted by the questions in Table 2. Adapting the Technology Acceptance Model (TAM), this section investigates users' subjective ease of engaging with GAI tools. It includes queries about learning curves, clarity of use, and usability in teaching. The questions emphasize personal and vicarious experiences with learning GAI tools, exploring facilitators and barriers to skill acquisition.

Table 2. Section of the instrument dedicated to the dimension of GAI Perceived Ease of Use

Dimension / Objective	Base Studies Questions	Questions	Probing Questions
Perceived Ease of Use of GAI tools / To assess the GAI perceived ease of use in science education and GAI tools learning curve	My interaction with GenAI tools is clear and simple (Al-Abdullatif 2024)	5. How easy or difficult is it to use IAG tools and what factors influence this?	5a. Can you illustrate with an example? 5b. What made it easy or difficult for you or for others you know?
	Learning how to use GenAI tools is easy for me (Al-Abdullatif 2024) It is easy for me to become skillful in using GenAI tools (Al-Abdullatif 2024)	6. How easy or difficult is it to learn to use GAI tools?	6a. Can you describe your experience of building skills with GAI? 6b. What kinds of support, experiences, or strategies have helped you feel more competent using them?
	I find GenAI tools easy to use for my teaching (Al-Abdullatif 2024)	7. How easy or how difficult is it to use GAI tools for science education?	7a. Can you illustrate with an example?

3.3 Perceived Usefulness

This dimension seeks to assess the functional value educators assign to GAI tools in achieving instructional goals, as depicted in Table 3. Original items explore GAI utility in enhancing teaching performance, increasing productivity, and reducing effort. Sub-questions guide respondents to articulate concrete examples.

Table 3. Section of the instrument dedicated to the dimension of GAI Perceived Usefulness

Dimension / Objective	Base Studies Questions	Questions	Probing Questions
Perceived Usefulness / To assess the perceived usefulness of GAI tools in enhancing science education	I find GenAI tools useful for performing my teaching duties (Al-Abdullatif 2024) Using GenAI tools increases my chances of achieving high job performance (Al-Abdullatif 2024) Using GenAI tools helps me accomplish my teaching tasks effortlessly (Al-Abdullatif 2024) Using GenAI tools increases my productivity (Al-Abdullatif 2024)	8. How useful can GAI tools be in science education?	8a. Can you illustrate with an example? 8b. Can you elaborate on its usefulness for achieving higher quality results/outputs or to reduce the time needed for a certain task? 8c. How does it affect the effort required to complete a task?

3.4 GAI Perceived Trust

Trust is a critical determinant of tool adoption. This dimension of the instrument aims to address the perceived trust in GAI tools for content creation in science education and factors that influence it, as shown in Table 4. Questions investigate whether users believe GAI tools provide reliable, secure, and expectation-aligned outputs. Additional probes explore concerns around data privacy, content veracity, and psychological comfort with AI-driven recommendations.

Table 4. Section of the instrument dedicated to the dimension of GAI Perceived Trust

Dimension / Objective	Base Studies Questions	Questions	Probing Questions
GAI Perceived Trust / To assess the perceived trust in GAI tools for content creation in science education and factors influencing it	I will use the GenAI tools if I feel that the content is trustworthy (Al-Abdullatif 2024) I will use the GenAI tools if I feel that these tools provide reliable information (Al-Abdullatif 2024) I will use the GenAI tools if I feel that these tools meet my expectations (Al-Abdullatif 2024)	9. What factors affect people's trust and will to use GAI tools for content creation?	9a. Do you trust it? In what aspects do you (not) trust it? Can you illustrate with examples? 9b. Does knowing whether something is created with GAI affect your opinion or the way you use it? Why?
	I will use the GenAI tools if I feel that these tools are secure (Al-Abdullatif 2024) I will use the GenAI tools if the data use poses no risks to teacher and learner privacy (Al-Abdullatif 2024)	10. How does users' sense of security and data privacy—or lack thereof—influence their willingness to use GAI tools?	10a. What kinds of security concerns come to mind about these tools? 10b. How do you weigh the potential benefits of GAI tools against the possible risks (e.g., privacy issues)?

3.5 Intelligent TPACK: Knowledge for GAI integration

This section assesses perceptions on knowledge that science teachers need to effectively integrate GAI tools in science teaching and learning, as shown in Table 5. This knowledge includes the ability to properly combine specific disciplinary content (Content Knowledge), pedagogical strategies (Pedagogical Knowledge) and GAI functioning (Technological Knowledge) tools, to pedagogically integrate GAI in lesson planning or adaptive instruction, for example. The questions in this section also probe the knowledge necessary for effective AI integration, such as understanding LLMs (large language models) for feedback or assessments.

Table 5. Section of the instrument dedicated to the dimension of Intelligent TPACK

Dimension / Objective	Base Studies Questions	Questions	Probing Questions
Intelligent TPACK: Knowledge for GAI integration / To assess perceptions on knowledge needed for GAI integration in science education	I can teach lessons that appropriately combine my teaching content, GenAI tools, and teaching strategies (Al-Abdullatif 2024) I can choose an AI tool to use in my classroom that enhances what I teach, how I teach, and what students learn. [AIP1] (Chiu et al. 2024) I can choose an AI tool that enhances my teaching subject content for a lesson. [AIP2] (Chiu et al. 2024) I can teach lessons that appropriately combine my teaching subject, AI tools, and Teaching approaches. [AIP3] (Chiu et al. 2024)	11. What type of knowledge science teachers need to appropriately combine their teaching content, GAI tools, and teaching strategies to promote learning?	11a. Can you illustrate with an example? 11b. What teaching content is suited to be addressed in teaching strategies that include GAI? 11c. What GAI tools can be successfully used in science education? 11d. What teaching strategies can be explored in this context?
	In teaching my field, I know how to use different GenAI tools for adaptive and real-time feedback (Al-Abdullatif 2024) In teaching my field, I know how to use different GenAI tools for personalized learning (Al-Abdullatif 2024) I can select various GenAI tools to monitor students' learning in my teaching process (Al-Abdullatif 2024) I can use AI tool to foster assessment for learning. [AIA1] (Chiu et al. 2024) I can design an assessment approach to improve student learning in an AI-based environment (e.g., learning with ChatGPT). [AIA2] (Chiu et al. 2024) I can choose an AI tool to foster student self-assessment. [AIA4] (Chiu et al. 2024)	12. What type of knowledge science teachers need to appropriately use GAI tools to foster assessment for learning and to assess processes?	12a. E.g., what GAI tools support monitoring students' learning, to provide real-time feedback and personalized learning to students, or to foster students' self-assessment? 12b. What challenges or limitations can happen when using GAI tools this way? E.g., could it be relevant to know how to use Large Language Models (LLM) to fill in a draft of students' assessment report?
	In teaching my field, I know how to use different GenAI tools for generate quiz and assessments (Al-Abdullatif 2024)	13. What type of knowledge science teachers need to use GAI tools to evaluate students' learning?	13a. What GAI tools can work well in this area (e.g., to generate quizzes)? 13b. What challenges or limitations can happen when using GAI tools this way?

3.6 GAI Acceptance

This dimension captures perceptions on the acceptance of GAI tools in science teaching and learning, namely affective and behavioral intentions toward GAI adoption, including enthusiasm, future use plans, and advocacy, as depicted in Table 6. It details how this dimension also directs to broaden the scope beyond educators to include other stakeholders (e.g., students, parents, administrators), and examines perceived value and barriers to institutional uptake.

Table 6. Section of the instrument dedicated to the dimension of GAI Acceptance

Dimension / Objective	Base Studies Questions	Questions	Probing Questions
GAI Acceptance / To assess perceptions on the acceptance of GAI tools in science education	I look forward to use GenAI tools in my teaching (Al-Abdullatif 2024) I intend to use GenAI tools in my future teaching (Al-Abdullatif 2024) I plan to use GenAI tools in my future teaching (Al-Abdullatif 2024) I support the adoption of GenAI tools in higher education (Al-Abdullatif 2024)	14. What are the perspectives among science teachers and students, parents and school heads about integrating GAI in teaching practice?	14a. E.g., how important is it to integrate GAI in science teaching and why? 14b. Can you illustrate with examples of situations when people revealed reservations or excitement about using these tools in science education? 14c. Should GAI adoption in science education be supported? How? Why?

3.7 GAI Ethics

This dimension focuses on perceptions regarding GAI ethical issues, as depicted in Table 7. Ethical considerations encompass responsible use, safety, well-being, and information security. Questions investigate the ethical literacy of educators and their capacity to teach these principles to students, emphasizing compliance, vigilance, and content protection.

Table 7. Section of the instrument dedicated to the dimension of GAI Ethics

Dimension / Objective	Base Studies Questions	Questions	Probing Questions
GAI Ethics / To assess perceptions regarding GAI ethical issues in science education	I can teach students ethics [AIE1] (Chiu et al. 2024) I teach students how to behave safely and responsibly when learning with AI tools. [AIE4] (Chiu et al. 2024) I can ensure my health and well-being while using AI tools. [AIE3] (Chiu et al. 2024)	15. What topics do you consider important to include when discussing GAI ethics with students and colleagues?	15a. What kind of guidelines or expectations should be set for responsible and safe GAI use, e.g., to ensure people's health and well-being while using it?
	I always comply with ethical principles when using AI applications or products. (Al-Abdullatif 2024) I am alert to privacy and information security issues when using AI applications or products. (Al-Abdullatif 2024) I can protect sensitive content from AI tools (e.g., exams, students' grades and personal data). [AIE2] (Chiu et al. 2024)	16. What kinds of privacy and information security issues are important to consider when using AI tools in science education?	16a. What steps should be taken to ensure that sensitive materials — like exams, grades, or student data — are protected when using GAI tools?

4. FINAL CONSIDERATIONS

The developed data collection instrument constitutes a rigorous, multidimensional tool for investigating GAI role in science education in qualitative studies. This theoretically anchored approach ensures construct validity, while the qualitative design, featuring open-ended and optional probing questions, supports rich, contextual data collection suitable for thematic analysis. Its comprehensive structure facilitates a nuanced understanding of how educators and other stakeholders perceive the integration of GAI in science education.

This tool aims to identify key barriers to GAI adoption, as well as explore factors that could facilitate its effective and thoughtful use. These insights could inform educational policy, professional development programs, and curriculum design, particularly in the context of GAI integration in an era of rapid technological change. Moreover, this work aims to provide a foundational tool for further qualitative research in educational technology adoption, enriching the current literature on GAI in education.

Despite the contribution this work is expected to bring to the field of teacher education and GAI integration, the instrument's limitations must be acknowledged. As a qualitative tool, it relies on participants' self-reported experiences, which may be influenced by subjective biases or social desirability. Additionally, while the instrument was developed with strong theoretical and empirical foundations, its application in different cultural or educational contexts may require further adaptation to ensure relevance and validity. Thus, although the interview guide was submitted to face validation by a panel of three experts, the authors expect to refine further the tool after a piloting study with educational stakeholders.

Overall, this research contributes to the growing body of literature on educational technology adoption, offering practical implications for science educators, policymakers, and stakeholders in the field of science teacher education. Future research will include the application of the instrument to further investigate GAI tools integration in science education, considering their relation with teaching practices and student outcomes, while exploring additional factors that may influence the widespread acceptance of GAI in education.

ACKNOWLEDGEMENT

The GAI-SciTeach project is funded by National Funds through the FCT - Fundação para a Ciência e a Tecnologia, I.P., under the project 2023.13203.PEX. The work of the first author is funded by FCT, under the research grant BIPD/UI57/12159/2025. The work of the second author is funded by FCT, under the Scientific Employment Stimulus - Individual Call [2022.02153.CEECIND].

REFERENCES

Al-Abdullatif, A.M., (2024). Modeling Teachers' Acceptance of Generative Artificial Intelligence Use in Higher Education: The Role of AI Literacy, Intelligent TPACK, and Perceived Trust [online]. *Education Sciences*, 14 (11), 1209. Available at: https://www.mdpi.com/2227-7102/14/11/1209.

Amado, J., (2014). *Manual de investigação qualitativa em educação* [eBook]. Imprensa da Universidade de Coimbra. Available at: https://ucdigitalis.uc.pt/pombalina/item/54493.

Banh, L. and Strobel, G., (2023). Generative artificial intelligence. *Electronic Markets*, 33 (1), 63. 10.1007/s12525-023-00680-1.

Bura, C. and Myakala, P.K., (2024). Advancing Transformative Education: Generative AI as a Catalyst for Equity and Innovation [online]. Available at: https://arxiv.org/abs/2411.15971v1 [Accessed 17 April 2025].

Chiu, T.K.F., Ahmad, Z. and Çoban, M., (2025). Development and validation of teacher artificial intelligence (AI) competence self-efficacy (TAICS) scale. *Education and Information Technologies*, 30 (5), 6667–6685. 10.1007/s10639-024-13094-z.

Dima, J. et al., (2024). The effects of artificial intelligence on human resource activities and the roles of the human resource triad: opportunities and challenges. *Frontiers in Psychology*, 15, 1360401. 10.3389/FPSYG.2024.1360401/XML/NLM.

Falloon, G., (2020). From digital literacy to digital competence: the teacher digital competency (TDC) framework [online]. *Educational Technology Research and Development*, 68 (5), 2449–2472. Available at: https://link.springer.com/article/10.1007/s11423-020-09767-4 [Accessed 28 April 2025].

Guettala, M. et al., (2024). Generative Artificial Intelligence in Education: Advancing Adaptive and Personalized Learning. *Acta Informatica Pragensia*, 13 (3), 460–489. 10.18267/j.aip.235.

Hosseini, M.-S. et al., (2024). Formulating research questions for evidence-based studies. *Journal of Medicine, Surgery, and Public Health*, 2, 100046. 10.1016/J.GLMEDI.2023.100046.

Marques, M. M., (2024). Generative artificial intelligence in teacher education: systematic literature review [online]. *EDULEARN24 Proceedings*, 1, 2503–2511. Available at: https://ria.ua.pt/handle/10773/42297 [Accessed 26 April 2025].

Mishra, P. and Koehler, M.J., (2006). Technological Pedagogical Content Knowledge: A Framework for Teacher Knowledge. *Teachers College Record: The Voice of Scholarship in Education*, 108 (6), 1017–1054. 10.1111/j.1467-9620.2006.00684.x.

Mishra, P., Warr, M. and Islam, R., (2023). TPACK in the age of ChatGPT and Generative AI. *Journal of Digital Learning in Teacher Education*, 39 (4), 235–251. 10.1080/21532974.2023.2247480.

Monteiro, J. and Marques, M. M., (2025). *Guide for Expert Review of the GAI-SciTeach Focus Group Semi-Structured Interview Script* [online]. Zenodo. Available at: https://doi.org/10.5281/zenodo.15691541.

Williams, A. and Katz, L., (2001). The use of focus group methodology in education: Some theoretical and practical considerations. *International Electronic Journal for Leadership in Learning*, 5.

FORESTED: INTERACTIVE TOOL FOR COMPETENCY CENTERED CURRICULUM VISUALIZATION

Sorana Cimpan[1], Romain Ferrer[1] and Hinata Yokoyama[2]
[1]LISTIC – University Savoie Mont-Blanc
5 Chemin de Bellevue, 74940 Annecy-Le-Vieux, France
[2]Kagawa University
2217-20 Hayashi-cho, Takamatsu, Kagawa 761-0396, Japan

ABSTRACT

This paper addresses the topic of curricula visualization in the context of competency-based approaches. We introduce ForestED, an interactive tool for competency centered curriculum visualization, highlighting how visualizing the links between competencies and the courses helps understand the interplay between the two. The visualizations concern one computing engineering curriculum. While the visualization reflects the static curriculum definition, it also allows inference of competency development over time through the course timeline.

KEYWORDS

Curriculum Visualization, Competency-Based Curricula

1. THE WHY AND HOW OF CURRICULUM VISUALIZATION

Curriculum design approaches vary considerably. The most common is the *course-centered curriculum* which focuses on individual courses and their content. Other approaches include to name but a few the *competency-based curriculum* (focusing on the development of specific competencies or learning outcomes - LO), *program-based curriculum* (ensures coherence and alignment between courses, LO and assessments), *learner-centered curriculum* (focuses on student needs, interests and learning styles). Regardless of the approach used in their design, curricula are carefully crafted complex systems (Aldrich, 2015), designed by highly qualified groups of experts in particular fields. A curriculum definition most often takes the form of a detailed document presenting the constituent elements and the interplay between them. As with any complex system represented textually, understanding the curriculum is challenging, even for those involved in its design. Visual representations of complex systems aid understanding by engaging the visual cortex in addition to the prefrontal cortex.

Educational data plays a key role in the field of Learning Analytics, which is defined as the measurement, collection, analysis, and reporting of data about learners and their contexts, with the goal of understanding and optimizing learning and the environments in which it occurs (Siemens, 2013). One of the most common applications in learning analytics is the design of student dashboards using visualizations created by using instructional data. These dashboards are intended to assist learners and enhance the learning experience (Susnjak et *al.*, 2022). The study of curricula with the aim of improving their quality, coherence, and design is the focus of the academic field of Curriculum Analytics (Barthakur et *al.*, 2024). One of its practical applications is curricula visualizations.

While some definitions of curricula include graphic representations, comprehensive curriculum visualizations remain relatively rare. The purpose of curriculum visualization is to enhance the understanding of educational offerings. Depending on the curriculum design approach, visualizations may focus on different aspects. For content-based curricula, visual representations often highlight relationships between curricular elements. One common approach involves using heat maps (Joyner, 2016; Plaza et *al.*, 2007; Siirtola et *al.*, 2013; Vuong et *al.*, 2011) which visually represent correlation matrices. In these matrices, both the rows and columns correspond to courses, and the cell at position *(i, j)* indicates the degree of relationship between *course i* and *course j*.

These relationships can take various forms. In some cases, prerequisites are not defined between modules but rather among LO (Auvinen, 2011; Auvinen et *al.,* 2014). The relationship values may be binary (i.e., in the set *{0,1}*), as is common in prerequisite structures, where a value of *1* indicates that *course i* is a prerequisite for *course j*. Alternatively, the values can range continuously between *0* and *1* to represent degrees of similarity, calculated using various methods—such as the proportion of shared topics between courses (Siirtola et *al.,* 2013).

The same information presented in correlation matrices is sometimes visualized as a graph, where nodes represent courses and edges (or arcs) denote the identified relationships between them (Akbaş et *al.*, 2015; Aldrich, 2014; Gestwicki, 2008; Kabicher & Motschnig-Pitrik, 2009; Kriglstein, 2008; Zucker, 2009). This graph-based representation not only offers a different visual perspective on course relationships but also enables further curriculum analysis through graph theory. For instance, courses (as graph nodes) can be characterized by their incoming and outgoing degrees, providing insights into their roles within the curriculum structure.

As with correlation matrices, the values used in these graphs are not limited to prerequisite relationships. They can also represent various forms of similarity between courses, calculated using different approaches. Such similarity may relate to shared topics—derived from course content—or from connections to the same competencies (Blasco Soplon et *al.*, 2014). When interpreted this way, the graph structure supports advanced analyses, such as cluster detection, to reveal groups of closely related courses.

In the case of competency-based curricula, the link between competencies and content can materialize in different forms. In the design of such curricula the targeted competencies are associated to each of the courses, in a declarative manner, like "*course A is related with competency X and Y*". The heat maps are the most common form of rendering visual relations. These matrices allow analysis like coverage of competencies in the curriculum, as you can easily spot the competency column that has no course associated to it. Some approaches provide visualizations for understanding the link between competencies and content by also constructing graphs where nodes are courses, an arc between two nodes indicates that the courses are related to the same competency.

We address in this paper the case of competency-based curricula and present ForestED – an interactive curriculum visualization tool. We explore how visualizing curricula as a set of trees can help to understand the interplay between the competencies targeted by the curriculum, and the content of curriculum (courses, projects, etc). As presented in the following section, we use trees to highlight the relations between courses and competencies but also exploit these relations to show how competencies (or rather resources related to them) develop over the curricula.

2. FORESTED – A VISUAL LINK BETWEEN CURRICULUM AND COMPETENCIES

We tested ForestED in the context of an engineering school in France, in which the curriculum design is competency-based. The definition of the competency referential follows the model proposed by Tardif (Tardif, 2003) and defines a competency as a *"complex ability to act within a family of situations relying on the effective mobilization and combination of a set of internal and external resources"*.

A competency referential has been defined for each engineering diploma, comprising 4 transversal and 4 specific competencies. As the goal of the paper is to present the visualization technique (and not the referential), we will refer to these competencies by their identifier only. Thus, the transversal competencies, shared by all diplomas in the engineering school are labeled as *TC-1* to *TC-4* in our visualizations, whilst the 4 specific competencies are labeled from *IDU-1* to *IDU-4*. Each competency has several core-components, or sub-competencies intuitively labeled *IDU-1.1, IDU-1.2* and so on. In the curriculum design, the link between modules and competencies is made by attaching each course learning outcome to one or several sub-competencies. A correlation matrix renders visual these links, an extract of which is presented in Figure 1 and contains one course with two LO. The first learning outcome is attached to 5 sub-competences, while the second one is attached to two sub-competences. Each learning outcome has a complexity level declared, the possible values being *Notion(N)*, *Advanced(A)*, *Mastered(M)* and *Expert(E)*. These levels are inspired by hierarchical model of development such as Bloom's Revised Taxonomy (Anderson et *al.*, 2001), which organizes learning objectives according to levels of complexity and mastery.

UE	EC	Acquis attendu de l'apprentissage Cette unité d'enseignement (UE) vise à rendre l'étudiant apte à :	Niveau	IDU-1			IDU-2			IDU-3		IDU-4		TC-1					TC-2			TC-3				TC-4			
				1.1	1.2	1.3	2.1	2.2	2.3	3.1	3.2	4.1	4.2	1.1	1.2	1.3	1.4	1.5	2.1	2.2	2.3	3.1	3.2	3.3	3.4	4.1	4.2	4.3	4.4
	ISOC731 - Économie et gouvernance de la donnée	Understand principles of data governance	M						✓		✓		✓			✓							✓						
		Understand principle of data economics	M												✓								✓						

Figure 1. The curriculum design makes the link between courses LO and sub-competencies

ForestED visualizes the correlation matrix as a set of trees (hence the name of the tool, ED standing for education). Each tree in the forest corresponds to a course, the branches are the LO, and the leaves are the competencies.

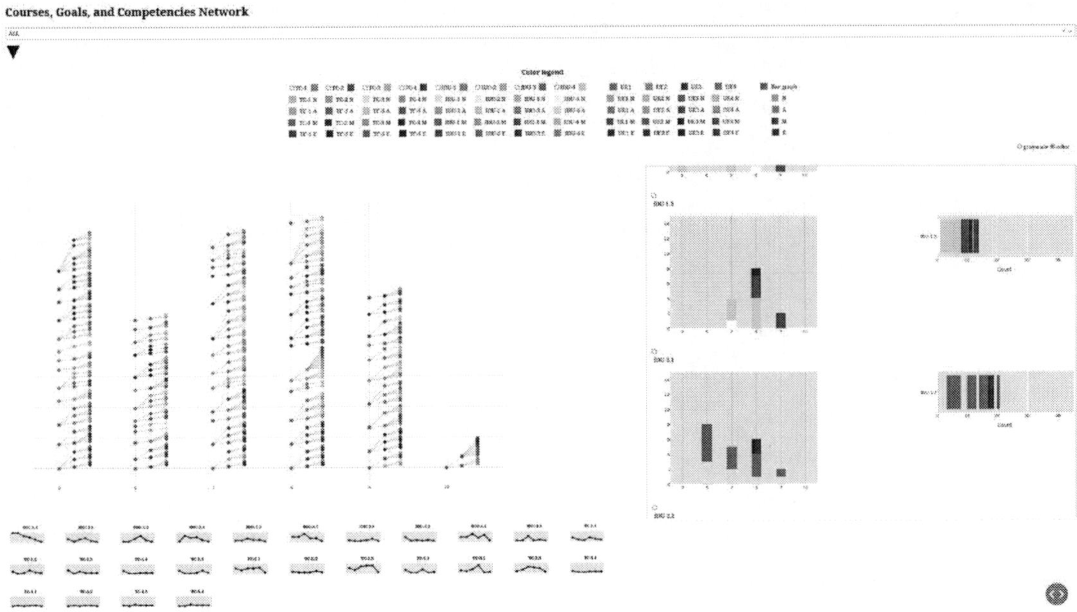

Figure 2. Overview of the interactive tool, when the IDU diploma has been selected

Once a diploma is selected, ForestED displays the curriculum trees, as presented in Figure 2, as well as some additional graphics: a set of spark-lines for competencies, and more-detailed visualizations for each competency, as well as the color scheme. The visualization is interactive, with the possibility of choosing which data is displayed – per competency group, competency of core-component of a competency.

The main component – the forest - provides an overall visualization of the curriculum, as a set of trees structures, presented in columns (see Figure 3). Each column from left to right represents a semester. In our case study the semesters go from 5 to 10. Thus, the number of trees in a column corresponds to the number of courses in that semester. Each tree structure represents a course and has three levels of depth. The roots represent the subjects, the intermediate nodes are the LO, and the leaves are the core components of the competencies. We deliberately made the choice not to display any analytic data on this visualization, keeping it purely structural. When the curriculum is displayed entirely, the main goal is to give a global view, and the colors employed were chosen to help this understanding.

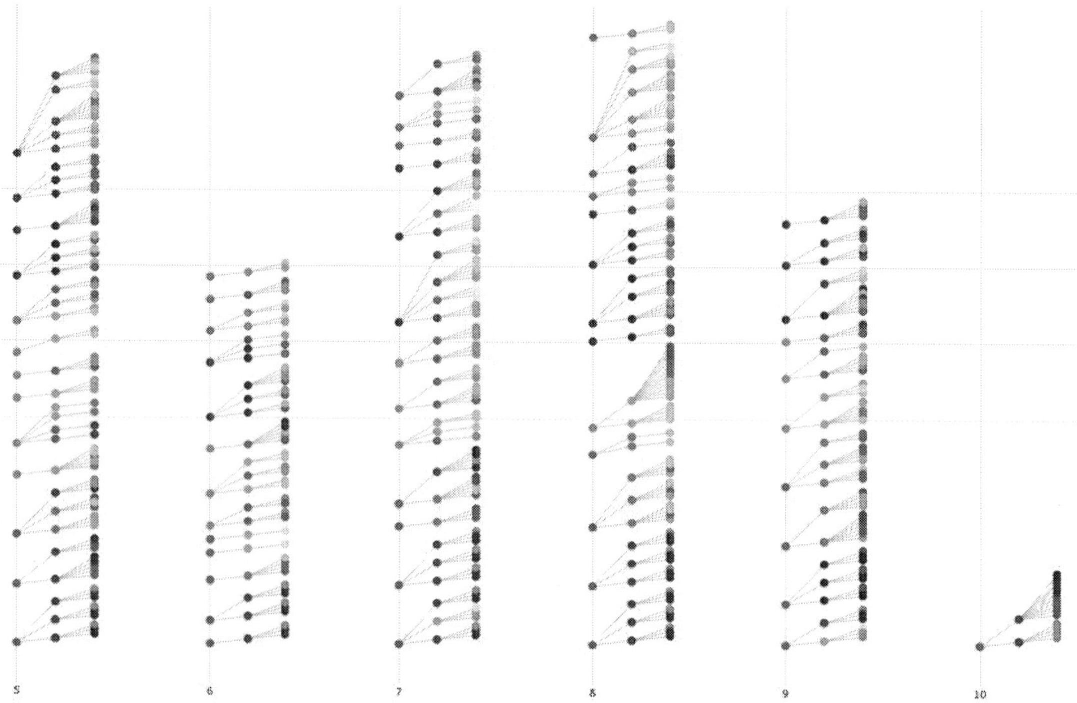

Figure 3. The curriculum viewed as forest

Each competency has a color attached to it, with warm colors for the specific competencies (IDU) and cold colors for the transversal competencies (TC). This global view of the forest allows us to make some analysis. In the example presented in Figure 3, the different columns, corresponding to the different semesters, vary in size. Nevertheless, semesters have the same length and weight in the diploma. The case of the last semester (10) is special, because there is just one module, corresponding to the internship. But the other columns still considerable vary in size. On closer inspection, we notice that the trees are of different shapes. In semester 6 (second column) the trees are smaller, which indicates that there were less learning objectives identified for the corresponding courses, with fewer competencies attached. This is normal as the designers (teachers) were not given specific directions as to the number of LO and competencies they should consider for the courses they oversaw, but it surprises, and it begs for further enquiry (undergoing).

As mentioned, the tool is interactive. When choosing one competency, the forest displays only the modules related to the select competency. Figure 4 presents the courses (trees) in the curriculum (forest) that are related to the core competence *IDU-1.1*. Nevertheless, the trees are entirely displayed, and we can spot which other competencies are addressed jointly with *IDU-1.1*. Hoovering over the nodes displays their label (core competency). Note that there is no course displayed in the last semester, as the only course in that semester did not address competency *IDU-1.1*.

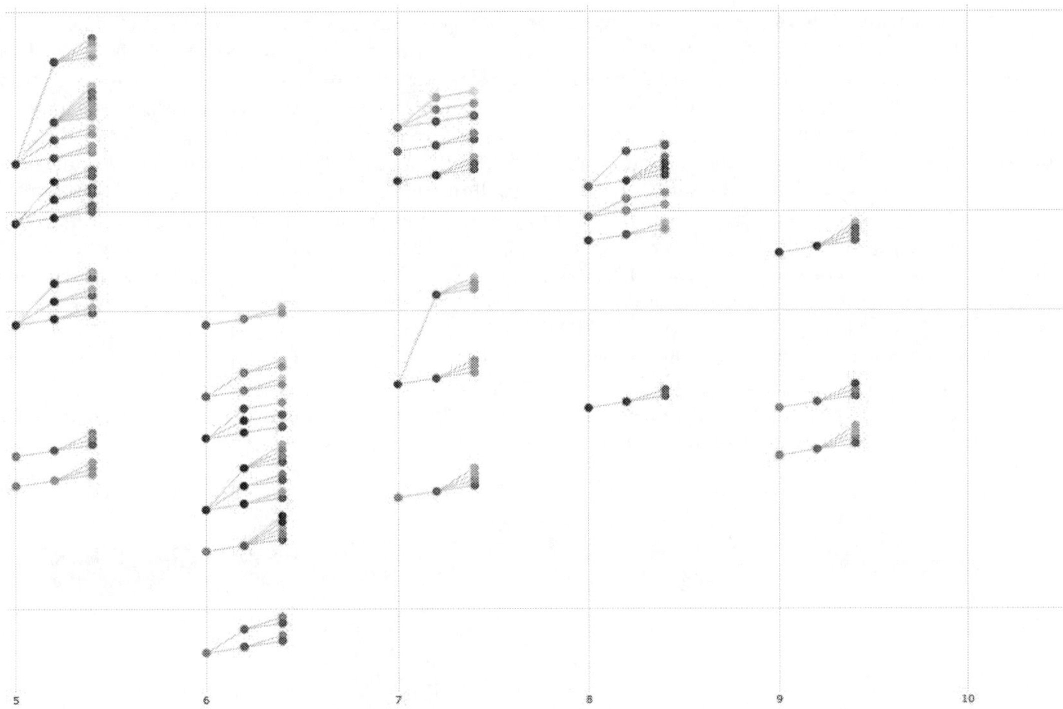

Figure 4. The trees (courses) that correspond to the IDU-1.1 competence

The color palette is chosen carefully as it supports better comprehension. A sufficient color distance between hues is ensured to avoid confusion due to similarity. The same color scheme is reused constantly across the different graphics. The different levels of acquisition are represented through shades of the same color, enhancing visual coherence. Recall that the root nodes of trees correspond to courses. The color of these nodes is given by the learning unit to which they belong, allowing to pre-attentively identify courses belonging to the same unit. For the nodes corresponding to the LO of a course, the hue is the one of the courses, but the shade corresponds to their maturity or difficulty level. As the visualization is interactive, a simple selection in the forest allows to zoom on specific modules, such as presented in Figure 5, where we show examples of two modules. The first tree has warm colorful leaves. This indicates it is a specialization course, some transversal competencies are also addressed, as there are green leaves. The second example tree has blue and green leaves, so it is related to transversal competencies. Also, as the root nodes for the two trees are of different colors, this indicates the fact that the courses belong to different learning units.

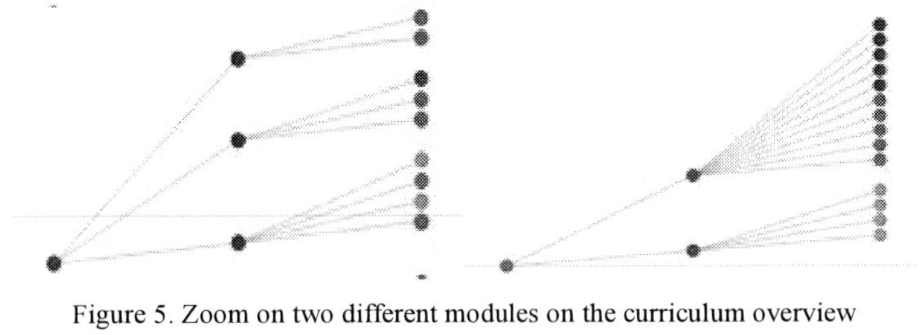

Figure 5. Zoom on two different modules on the curriculum overview

The two-remaining graphics of the interactive tool focus on competency core components. The graphics on the right side in Figure 2 present two displays for each core-component. Figure 6 presents these graphs for 4 different competencies. The stacked bar-charts on the left display how many times the core competency appears in the LO (LO) per semester by displaying also the complexity level of each learning outcome (the darker the shade, the higher the complexity). The horizontal bar-chart consists of counting the total number of LO helping to the acquisition of the core competency during the overall duration of the whole curriculum - this chart essentially sums up the values from the semester-based bar charts. Although the information may seem redundant, as our goal is to rely as much as possible on visual perception, the second graph (horizontally stacked chart) allows for quick comparison among competencies. It allows us to compare visually how many LO each competency has attached to it. Looking at the different competencies presented in Figure 4, we easily see which is specific (left hand side, warm colors), which is transversal (right hand side, cold colors). Also, we can easily see that IDU-1.2 has less coverage in the curriculum than the others, we can detect that pre-attentively – by comparing the size of the bars. We can also attentively see that IDU-1.2 has around 13 LO attached to it whereas TC-2.1 has around 35. Looking at the shades used, we can see that specific competencies (IDU) have fewer LO that address high complexity levels attached to them (the very dark shades of red and yellow) than the transversal competencies.

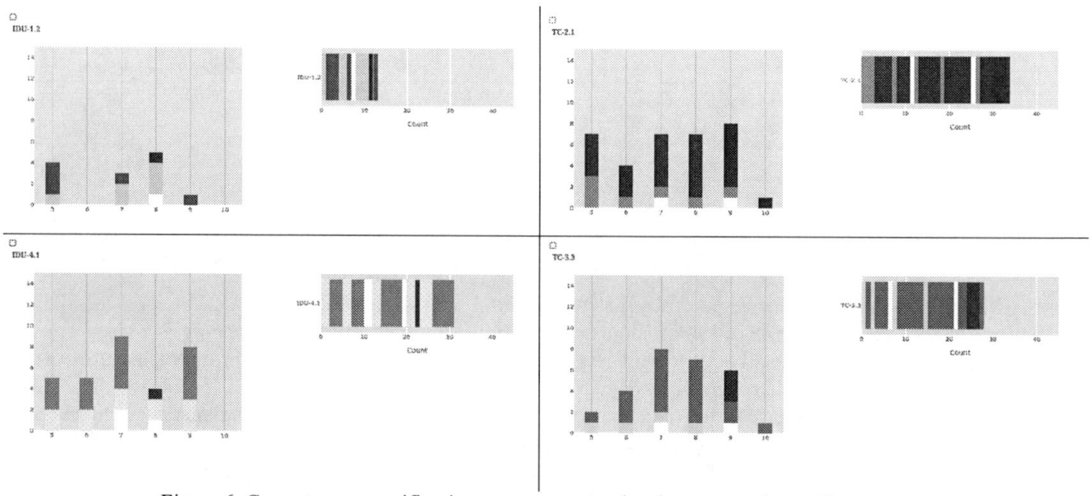

Figure 6. Competency-specific views: per semester development and overall coverage

The last graphic, consisting of spark-lines positioned side by side is intended to further ease the comparison between competencies (see Figure 7). The sparklines are minimalist versions of the stacked bar-charts from the previously presented competency graphs. They were thought to make a fast comparison of two core-components or more, as the bar-charts already give detailed information about every core-component. Using this graphic, it is easy to see that the way competencies develop over time is different from one competency to another. Note that each dot in the spark line corresponds to how many LO are attached to this competency in a particular semester. Each spark line provides thus a chronological view on how the curriculum addresses it. The shapes are different, indicating that the way competencies are covered in time varies from one competency to another.

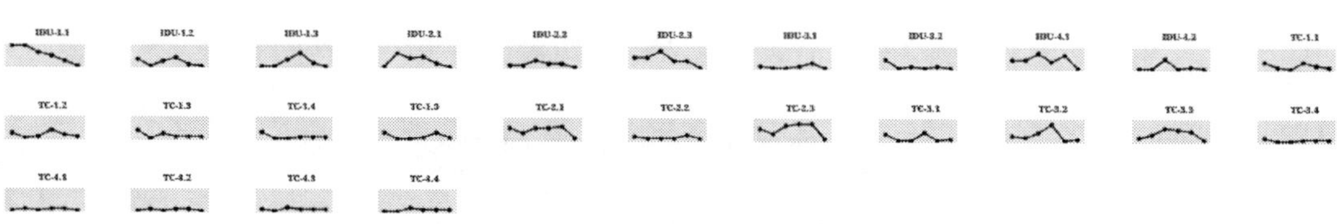

Figure 7. Spark-lines showing competencies temporal development

This kind of comparative analysis is easy to make using this visualization, but it will require immense effort without it. As mentioned, the ForestED is an interactive tool. If one competency is selected, the curriculum overview only displays the courses (trees) that are related to that competency.

3. CONCLUSION

The tool presented in this paper - ForestED - showcases the relevance of visualization as a means for understanding educational offers. Different diplomas in our engineering school have been visualized, as their curriculum design is based on the same model. In terms of architecture, ForestED relies on a CVS file where the curriculum and its relation to competencies have been stored. Further tests with diplomas outside our engineering school are envisaged and will allow us to test its genericity. ForestED was presented to the teachers involved in designing and delivering the curriculum and the feedback was very positive. The teacher panel included the person in charge of the curriculum updated – and the tool appears to be of help in this case.

ForestED focuses on the interplay between the aimed curriculum competencies and the curriculum content (such as defined in the curriculum design). At the current time, most of the evaluation of students following the curriculum is done traditionally, at the level of each course. The competencies are evaluated only in some specific modules - mostly transversal projects in each semester, but even these modules must deliver grades. So, the curriculum, although designed using this competency-based approach, delivers the diplomas on grades-based evaluation, related to each module. Thus, the two models' competency based (design) and content based (delivery and certification) co-exist. This is the case in most of the higher education institutions in France. The link made by the curriculum design between the content (course LO) and competencies corresponds to the development of resources (internal and external) that hopefully will be effectively mobilized by the learners when the corresponding competency is needed.

ForestED is part of a toolkit that aims supporting the competency-based approaches at different levels. Whilst we focus here on the curriculum visualization, other tools in the kit include the student's dashboard focused on the actual competency development.

ACKNOWLEDGEMENT

This work was supported by the @SPIRE (Accompagnement, Spécialisation Progressive et Individualisation pour la Réussite de tous les Étudiants) project, which we gratefully acknowledge. The @SPIRE project aims at transforming undergraduate education, particularly through the integration of the competency-based approach.

REFERENCES

Akbaş, M. I. et al. (2015). Curriculum GPS: An Adaptive Curriculum Generation and Planning System. *Interservice/Industry Training, Simulation, and Education Conference (I/ITSEC).* Orlando, USA. pp. 1-11.

Aldrich, P. R., (2014). *The curriculum prerequisite network: a tool for visualizing and analyzing academic curricula.* [online] arXiv.

Aldrich, P. R., (2015). The curriculum prerequisite network: Modeling the curriculum as a complex system. *Biochemistry and molecular biology education: a bimonthly publication of the International Union of Biochemistry and Molecular Biology*, pp. 168–180.

Anderson, LW. et al, (2001). *A Taxonomy for Learning, Teaching, and Assessing: A Revision of Bloom's Taxonomy of Educational Objectives.* Longman, New York, USA.

Auvinen, T. et al. (2014). STOPS: a graph-based study planning and curriculum development tool. *Proceedings of the 14th Koli Calling International Conference on Computing Education Research.* New York. USA. pp. 25-34.

Auvinen. T. (2011). Curriculum Development. *Full paper proceedings of the 1st EUCEET association.* Patras. Greece. pp. 27-36.

Barthakur, A. et al. (2024). The application of curriculum analytics for improving assessments and quality assurance in higher education. *Australasian Journal of Educational Technology.* Vol. 40. No. 4, pp. 20–37.

Blasco Soplon, L. et al, (2014). Curriculum visualization based on subject competences. *ICERI2014 Proceedings.* Seville, Spain, pp. 1439-1445.

Gestwicki, P. (2008). Work in progress – curriculum visualization. *2008 38th Annual Frontiers in Education Conference.* Saratoga Springs, USA. pp. T3E-13-T3E-14.

Joyner (Melito), H.S. (2016). Curriculum Mapping: A Method to Assess and Refine Undergraduate Degree Programs. *Journal of Food Science Education*, Vol. 15. No. 3. pp. 83-100.

Kabicher S. and Motschnig-Pitrik. R. (2009). "Coordinating Curriculum Implementation Using Wiki-supported Graph Visualization, *Ninth IEEE International Conference on Advanced Learning Technologies*, Riga, Latvia, pp. 742-743.

Kriglstein, S., (2008). Analysis of Ontology Visualization Techniques for Modular Curricula. *HCI and Usability for Education and Work.* Berlin, Heidelberg, pp. 299–312.

Plaza, C. M. et al. (2007). Curriculum mapping in program assessment and evaluation. *American journal of pharmaceutical education*, Vol. 71. No. 2, 20.

Siemens, G. (2013). Learning analytics: The emergence of a discipline. *American Behavioral Scientist.* Vol. 57, No. 10, pp. 1380–1400.

Siirtola, H. et al, (2013). Interactive Curriculum Visualization. *17Th International Conference on Information Visualisation.* London, United Kingdom. pp. 108-117.

Susnjak, T et al. (2022). Learning analytics dashboard: a tool for providing actionable insights to learners. *International Journal of Educational Technology in Higher Education.* Vol. 19. No. 1. pp. 12.

Tardif, J. (2003). Développer un programme par compétences : de l'intention à la mise en œuvre. *Pédagogie collégiale.* Canada. pp. 36-44.

Vuong, A et al. (2011). A Method for Finding Prerequisites Within a Curriculum. *Proceedings of the 4th International Conference on Educational Data Mining.* Eindhoven, The Netherlands, pp. 211-216.

Zucker, Ron. (2009). ViCurriAS: A curriculum visualization tool for faculty, advisors and students. *Journal of Computing Sciences in Colleges.* Vol. 25. No. 2. pp. 138-145.

INTEGRATING AI-BASED SEARCH FEATURES IN LMS: DESIGN AND IMPLEMENTATION NEEDS AND ISSUES

Andrea Molinari
Dept. of Industrial Engineering University of Trento, Trento, Italy
Lappeenranta University of Technology, Lappeenranta, Finland

ABSTRACT

The impact of artificial intelligence (AI) in the field of education is largely debated, but in most of the platforms that support educational processes (Learning Management Systems - LMSs), AI and large language models (LLMs) are mostly an external component. Present implementations primarily involve LMS platforms interfacing with AI engines to generate responses, suggestions, and content summaries, resulting in a limited and surface-level integration. This constraint is largely due to the design of widely used LMS platforms, which were conceptually and architecturally developed in the early 2000s. Indeed, enormous investments would be required to carry out a deep refactoring of the entire platform, and thus, the current choice to integrate the existing system with additional AI components is rather natural. This paper proposes a novel LMS architecture designed to natively integrate AI capabilities and take advantage of these technologies. Specifically, it presents the replacement of relational database models with graph-based structures, to facilitate the development of a native knowledge graph incorporating ontological frameworks for the description of the educational domain. Such a transformation would enable more sophisticated AI functionalities, including personalized learning pathways, semantic search capabilities, and AI-driven content recommendations. The article also addresses the obstacles to overcome and the open scientific challenges, attempting to outline a roadmap for evolution based on specific ongoing experiences.

KEYWORDS

Artificial Intelligence, Learning Management Systems, Knowledge GraphIntroduction

1. INTRODUCTION

Since their emergence in the late '90, Learning Management Systems (LMSs) have evolved into mature proprietary or open-source platforms, offering comprehensive solutions for various educational needs (Turnbull et al., 2020). However, the rise of cloud-based collaboration tools has overshadowed the advantages of LMS services and their pedagogical approaches. The COVID-19 pandemic underscored the critical role of technology-enhanced learning (TEL) within institutional information systems. More recently, the advent of generative AI and large language models (LLMs) (Wang et al., 2024) (Xu et al., 2024) (Hadi et al., 2024) has introduced significant opportunities for enhancing LMS functionalities. Despite this potential, AI integration within LMSs remains limited, often restricted to a basic interaction where the LMS requests AI-generated responses, summaries, or suggestions based on provided content.

This type of "external" integration stems from a fundamental software engineering challenge: major LMS platforms (Moodle, Sakai, Docebo, Blackboard Learn, ILIAS, Canvas, etc.) developed in the early 2000s, adhering to structural principles typical of that era and mainly based on the LAMP stack, with MySQL prevalently as the relational persistence layer of the application. While these platforms support extensive multimedia content for educational purposes, their content organization—primarily structured in folders corresponding to courses, modules, or lessons—limits their adaptability to AI-driven functionalities. Some advanced LMSs incorporate text-based interactions (e.g., forums, FAQs, blogs), but their overall design remains outdated in terms of AI native structure.

Based on our 25 years of knowledge in e-learning platform development, this paper examines the internal architecture of LMSs that proposes a novel software framework optimized for AI-driven education. The proposed approach entails a fundamental redesign of content creation and storage mechanisms, enabling the direct integration of AI functionalities. Specifically, transitioning from relational to graph-based data models would allow for the construction of a knowledge graph, which could serve as the foundation for advanced AI applications. By embedding ontological frameworks within LMS content (Villegas et al, 2020) (Molinari and Molinari, 2024), this paradigm shift could lead to a new generation of AI-enhanced LMSs, offering more sophisticated functionalities and services to users. However, transitioning to this model presents several challenges, including the need for a comprehensive redesign of platform persistence and business logic layers, the complexity of integrating domain-specific ontologies across educational institutions, and the technological demands of maintaining large language models capable of evolving in real-time, particularly during periods of intensive content generation (like lecturing periods).

The paper is organized as follows: chapter 2 explores the literature related to the integration of AI in LMS architecture, chapter 3 presents a review of searching functionalities in LMSs and Chapter 4 presents the new features guaranteed by a tighter integration of AI tools with LMS

2. INTEGRATION OF AI IN LMS ARCHITECTURE: BACKGROUND

The integration of artificial intelligence (AI) into LMSs has been extensively studied in recent years, particularly regarding its potential to enhance learning experiences through automation and personalization. Foundational research (Baker, 2016) highlights the role of AI in adaptive learning, demonstrating how data-driven approaches can tailor educational experiences. Similarly, Kose and Deperlioglu (2020) explores AI's capacity to create personalized learning paths, dynamically adjusting content based on student performance and engagement. By leveraging machine learning, LMS platforms can continuously assess learner needs, optimizing the delivery of educational material. The systematic review presented in Zawacki-Richter et al. (2019) provides an overview of AI applications in LMS, emphasizing the rise of language models like GPT for content generation and intelligent tutoring. This aligns with Holmes et.al (2019), which offers a critical analysis of AI's benefits and risks in education, highlighting both its potential to improve engagement and concerns about over-reliance on automation. The application of natural language processing (NLP) models, particularly for generating student feedback and instructional content, is detailed in Kumar and Vivek (2021), demonstrating how AI can enhance interactivity and support learners through automated assistance.

Looking ahead, Dillenbourg et al. (2022) discusses future AI-driven LMS developments, particularly in collaborative learning environments where AI fosters student interaction and real-time feedback. The transformative impact of adaptive AI systems is further explored in Alvarez and Conati (2020), which highlights their role in customizing learning based on individual progress and needs. Additionally, Popenici and Kerr (2017) examines AI's broader implications for higher education, suggesting that AI-enhanced teaching and assessment methods can improve efficiency and effectiveness. The potential of generative AI in content management is explored in Gonzales and Vicente (2021), emphasizing its ability to generate high-quality materials, freeing educators to focus on student-centered teaching strategies. Similarly, Roschelle and Teasley (2023) underscores AI's role in collaborative learning, enhancing communication and personalized feedback between students and instructors. As AI research intersects with educational technology across multiple disciplines, Turnbull et al. (2020) offers a comprehensive literature review on the topic. Meanwhile, even the new generation of LMS platforms remain rooted in architectures from the early 2000s, typically structured as relational database-driven web applications with multi-layered logic. While they provide extensive multimedia resources, their content is often organized in static, folder-based structures linked to courses or modules, limiting adaptability to AI-enhanced learning strategies. Historically, LMSs have included basic search and retrieval tools, but AI now offers advanced functionalities previously unavailable or experimental. Modern AI tools, such as GPT-based models, provide not only enhanced search and retrieval capabilities but also the ability to generate new, contextually relevant educational content. The gap between current LMS functionalities and AI's potential is particularly evident in multimodal generative AI applications.

To explore the potential of full, structural integration of AI tools into the LMS architecture, this study takes our in-house developed LMS platform called "Online Communities", developed in 1998 and later integrated into commercial systems. With a substantial knowledge base (2.2TB of textual data), it presents a manageable dataset for AI training compared to large-scale industry models. The research investigates the key points where AI integration could modify, if not revolutionize, an LMS structure, optimizing user experience and cost efficiency while transforming the LMS into an "intelligent" system tailored to educational needs (Dillenbourg et al., 2022). This transition requires significant architectural and technical changes, from the persistence modules to the user experience where conversational AI can modify significantly the user interaction. Rather than relying on external AI service APIs, our study proposes a profound restructuring of LMS knowledge bases. This involves, as the first pillar modification, a radical change in the storage mechanisms, shifting from relational databases to knowledge graphs, thus enabling more intelligent AI-driven functionalities.

As a broader goal of our study, we try to demonstrate that LMSs, thanks to this profound refactoring of the platform, can serve not only as educational tools but as fundamental components of the whole corporate information system, supporting collaboration in professional and vocational training contexts. Recent educational paradigms (Alvarez and Conati, 2020) emphasize peer-to-peer learning, flipped classrooms, and co-learning models, further reinforcing the importance of AI-driven adaptability in LMS. As part of this research, seven AI-driven pilot services have been developed as LMS components. Most of the work has been devoted to software architecture and design, particularly in the business and persistence layers. In this paper, we will focus specifically on the integration between LMS traditional functionalities and search capabilities provided by conversational, chat-based interfaces provided by LLM. Indeed, among the various services provided by LMSs, content search is the one most closely aligned with contemporary artificial intelligence capabilities. Increasingly, users are turning to AI systems rather than conventional search engines, with certain AI tools gaining substantial popularity as aggregators. These tools offer faster and more efficient alternatives to traditional search engines, enhancing the retrieval and organization of educational content. A recent interesting development of research in our direction is the LAMB framework (Bouquet and Molinari, 2016), which presents a similar vision to the one we propose. Here, a central core called LAMB Core acts as the orchestrator between the user interface of the LMS, the AI infrastructure, the knowledge base components (RAG Engine, Content Library, Augmentation functions, and Embeddings Database), and the LMS integration. It is more focused on creating learning assistants using AI tools and techniques, but the idea of radically approaching the LMS architecture differently is evident, and in line with our research hypothesis that a comprehensive, foundational redesign of LMSs is essential to fully harness AI capabilities, particularly those enabled by GPT and LLM technologies.

3. SEARCHING FUNCTIONALITIES IN LMS: AN OVERVIEW

To understand the structural changes required for an LMS to maximize interaction with artificial intelligence tools, we start with an analysis of the primary structure of LMS platforms, particularly focusing on the major open-source platforms where it is possible to review the software structure and the methods by which content is managed and presented to users. Learning Management Systems (LMS) offer search functionalities to help users navigate and quickly access relevant content. These features are crucial for improving efficiency in learning and information management. In this feature analysis, we have looked at the main search features provided by Moodle, Docebo, Sakai, Canvas, and TalentLMS, due to the availability of documentation and knowledge about the platforms.

Moodle is one of the most widely used open-source LMS platforms globally. Its search functionalities are quite powerful and flexible, with different implementations. We can access the "Global Search", where Moodle allows users to search across courses, documents, activities, and resources. Users can filter results by resource type, course, or user in the Search filter section, or they can use the full-text Search, which enables users to search for terms within course content, not just in titles. There are search functionalities in Forums and Quizzes, where teachers and students can search within discussion forums and quizzes. One of the most appreciated features (and also one of the most problematic) of Moodle is the possibility of taking advantage of the many plugins that the platform can integrate, among which the search Extension Plugins where Moodle supports plugins that extend search functionalities, making the platform more adaptable to specific needs.

Docebo is a cloud-based LMS platform, specifically aimed at companies for e-learning training. It provides various search features, like the AI-powered Search that utilizes an AI-based search engine to suggest relevant content based on the user's previous activities. Contextual Search is another functionality where users can search by categories, topics, or specific modules within a course. Docebo, in addition to search, suggests relevant content based on user behavior, and through the Advanced Filters feature, search results can be filtered by resource type, module, course, or creation date. There is also an Extended Search for Multimedia Content, where users can search for multimedia content, such as videos and audio files.

Sakai is an open-source LMS focused on the educational sector, primarily for universities and high schools. It presents classic features like full-text Search, which allows users to search for content within courses, including documents, wiki pages, shared resources, and more. A specific search feature, called Course Search Tool, allows users to search for instructional materials within individual courses or across multiple courses simultaneously. The search filters provide users with a filter feature where results are retrieved based on specific categories or materials. Finally, Sakai provides a search feature in Collaborative Tools, where it is possible to search for information within collaborative tools such as discussion forums, blogs, chat, and more.

Canvas is one of the most widely used and highly regarded LMS platforms due to its user-friendly interface and extensive course management features. Global Search is used to perform a global search across all courses, resources, and linked content. This includes files, discussions, quizzes, assignments, modules, and more. Canvas offers filters that allow users to narrow down search results to specific courses or types of resources (files, quizzes, discussions, etc.). Users can search for keywords not only in the titles of resources but also within the text of uploaded documents. Teachers and students can search within discussion threads to find specific comments or topics previously covered. Finally, search functionalities extend to evaluation rubrics and assignments, enabling quick retrieval of past activities and grading criteria. Recently, Canvas introduced its "Atomic Search", a smart, instructor-focused tool built to make finding and managing content in Canvas as easy as a Google search. Atomic Search provides cross-course access to everything in the LMS, using powerful filtering options, advanced analytics, find andreplace features to simplify content management.

TalentLMS is a cloud-based platform designed to offer a simple and fast learning experience, particularly suited for corporate training and small businesses. Users can search across the platform through courses, lessons, and training modules. The search is quick and allows for the rapid discovery of any educational content. Through advanced filters, TalentLMS allows users to filter search results by content type (courses, tests, documents), facilitating access to specific resources. Students can search for past quizzes or assessments, while administrators can search for created quizzes and tests to monitor the progress of training sessions. The Search in Reports feature enables searching within reports, such as student progress, course participation, and assessment results, simplifying performance monitoring, while with tagging and categorization, administrators can use tags and categories to make it easier to search for and organize courses and training modules.

As emerging from this brief analysis, except Docebo, which offers some AI functionalities in its main interface, the other platforms tend to standardize around key features, with some interesting additions such as tag-based searches or advanced filters. LMS platforms like Moodle, Docebo, and Sakai offer search functionalities that improve user experience by enhancing content accessibility and organization. These features vary in complexity, from Docebo's AI-powered intelligent suggestions to the full-text search capabilities of Moodle and Sakai, and Canvas contextual and global search. The search functionalities in Canvas and TalentLMS are designed to make content access simple and efficient. While Sakai focuses on social interaction and the search for shared resources within groups, Canvas provides a global and structured search across a wide range of content, including discussions and quizzes. TalentLMS, on the other hand, stands out for its speed and search features targeted at corporate training, such as searching within reports and managing assessments. These search capabilities are essential for optimizing users' time and enabling easy access to relevant information, making learning and educational project management more effective.

Artificial intelligence (AI) is reshaping Learning Management Systems (LMSs) by enhancing learning effectiveness, personalizing educational experiences, and streamlining the management of educational activities. AI integration within LMS platforms is transforming the learning process by improving accessibility, optimizing content delivery, and enabling adaptive learning pathways. From personalized instruction to automated assessment, AI fosters a more dynamic and targeted learning experience for both students and educators, ultimately enhancing the quality and effectiveness of education. However, when considering the integration of AI-driven, LLM-based solutions into an LMS, viable alternatives are limited. This integration typically consists of querying external engines via API calls to large language models (LLMs), which are then processed using the results presented to the end-user in a unified format. The confinement of structured and, in many cases, unstructured information within a relational database needs transferring this knowledge to an external AI engine for processing, after which the results must either be reintegrated into the persistence layer or directly presented to end users. Both approaches present significant drawbacks, as they do not constitute a native, structural integration. Specifically, they require pre-processing the entire LMS knowledge base (slides, forums, blogs, whiteboards, etc.) with the AI engine, a process that is largely infeasible due to performance constraints, effectively eliminating the possibility of real-time AI training on newly generated knowledge.

4. AI FEATURES AND LMS MODIFICATIONS

The integration of LLMs and generative AI presents both a compelling research and market opportunity, yet it also requires careful consideration due to its direct impact on individuals' personal and intellectual development. One of the critical risks is the phenomenon of 'hallucinations,' which is particularly concerning in educational contexts. Specifically, when searching for course materials associated with a particular topic, professor, or educational path, AI-generated responses may be overly generalized or derived from training data that is entirely different, potentially superior, or inferior, but ultimately not contextualized to the specific educational activity. We refer to this issue as 'local hallucinations,' distinguishing it from the well-documented problem of LLM-generated responses that are factually inaccurate or disconnected from reality. Instead, local hallucinations arise from the discrepancy between an institution's specific educational domain and the broader knowledge base on which the LLM has been trained. In the long term, this contributes to a well-recognized challenge in the widespread adoption of generative AI models: the homogenization of knowledge, a sort of knowledge "flattening" shaped by both the LLM's learned patterns and the conditioning imposed by its training data. This phenomenon can result in AI-generated responses that, while plausible, lack the necessary specificity and contextual accuracy for the particular educational materials or domain in question. Over time, the risk is that knowledge becomes overly standardized around the content that the LLM was trained on, potentially neglecting the richness and diversity of local educational contexts. This could dilute the uniqueness of course materials and reduce the specificity that teachers and students rely on for effective learning.

The structural changes proposed in this paper imply a complex but equally necessary step if we aim to fully exploit the capabilities of artificial intelligence, ensuring above all consistency in representation, service extensibility, and, in some ways, improved performance in the results. As said, at least in the first part of this long-term research, the persistence layer of the application is the one that will see the most important modifications, moving from the database-oriented structure to a more AI-oriented set of technologies for storage. The pipeline that will feed the storage of our LMS is presented in Figure 1. The need for this transformation is, in our opinion, mandatory if we want to use inference engines of any type, like LLMs. For instance, if we consider the native representation of a document in PDF format or its transformation via NLP systems into a graph that includes the mentioned entities connected and mapped to ontology concepts, we can grasp the radical difference between a database-oriented storage and an AI-oriented storage for the same knowledge. In fact, on one side, a set of relational tables, and on the other side an unconventional storage mechanism (like the one provided by NoSQL or MapReduce-based storage systems) where subject-predicate-object triples are stored to represent the content. Using the second example of representation, where knowledge is stored in the format of a knowledge graph, the system is inherently more suited to be processed by AI tools in a far more efficient and qualitative manner.

We expect that this new architecture not only will facilitate more coherent and sophisticated handling of data by AI but will also enhance the scalability of services and the potential for integrating advanced functionalities like entity recognition, relation mapping, and semantic reasoning. Such a shift would ultimately lead to a more profound and reliable AI-driven system, capable of delivering more relevant and contextually accurate results, particularly in domains like education, where precision and context matter greatly.

The areas for improvement and integration with AI services are numerous and highly intriguing, given the wide range of services that an LMS provides. These improvements target key functionalities such as personalized learning pathways, enhanced content search capabilities, automated assessments, and adaptive learning environments. Additionally, the integration of AI-driven tools for feedback, student performance tracking, and resource recommendation could significantly boost the overall user experience and efficiency of the platform, maximizing the potential of the LMS while ensuring scalability, coherence, and optimal performance in handling AI-driven tasks.

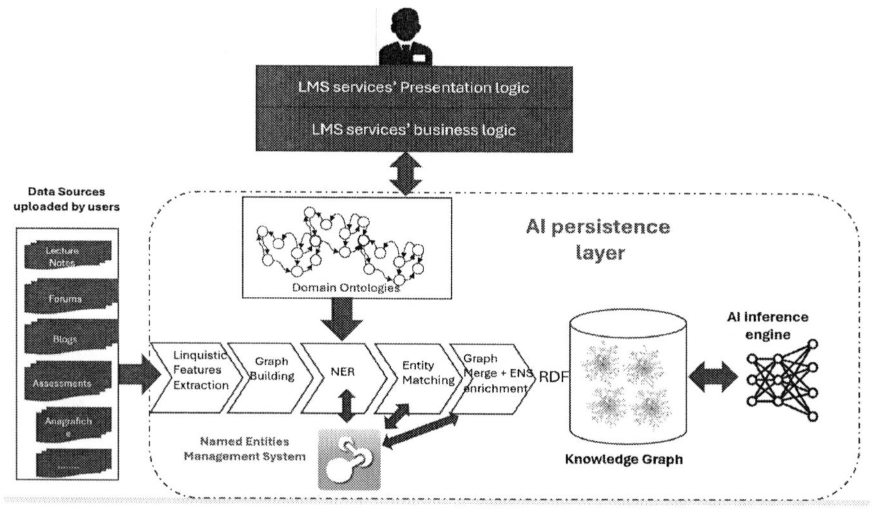

Figure 1. Conceptual view of the persistence layer of an AI-based LMS

The persistence layer consists of a pipeline that transforms the LMS's data sources (textual contents of various origins) into an entity-centric, ontology-driven knowledge graph. The role of ontology in software applications is traditionally to provide a structured, formal representation of knowledge within a specific domain. They define concepts, relationships, and constraints in a way that enables interoperability, semantic reasoning, and improved data integration across heterogeneous data sources and systems. By establishing a shared vocabulary and logical framework, ontologies facilitate enhanced machine understanding, enabling more effective data organization, retrieval, and analysis. They serve as a schema for structuring information in a way that reflects real-world entities and their interconnections, and this is particularly valuable in applications where artificial intelligence is needed to infer and increase knowledge: through the use of description logics and rule-based reasoning, systems can infer implicit relationships, detect inconsistencies, and enhance decision support mechanisms. This is a difficult but promising integration into an LMS, undoubtedly complex mainly due to the amount of different "contexts" to describe in educational contexts, where many different disciplines can be referred to inside the original data sources. For our ambition to provide machine learning and artificial intelligence services, ontologies provide explainability and interpretability by defining structured relationships between concepts. This enhances the trustworthiness of AI models, particularly in fields where transparency and accountability are critical, such as education.

The transformation pipeline presented in Figure 1 addresses a series of complex challenges from both a scientific and technical perspective, comprising the following key stages:

1. Linguistic Feature Extraction – This step involves identifying relevant textual elements and their corresponding relationships through a structured extraction process.
2. Graph Building – A specialized algorithm, LF2G (Linguistic Features to Graph), converts extracted tokens and lemmas into nodes within a directed graph.
3. Named Entity Recognition (NER) – Graph Neural Networks (GNNs) are employed to determine the semantics of each entity by analyzing adjacent nodes and their relational structures.
4. Entity Matching – The Entity Name System – ENS (Bouquet and Molinari, 2016)) is utilized to identify nodes that correspond to the same entity within the graph.
5. Graph Merging and ENS Enrichment – The ENS further refines entity identification, ensuring that corresponding nodes across different graphs are correctly linked.
6. Knowledge Graph Creation and Refinement – Finally, single document-based graphs are merged into a comprehensive knowledge graph, followed by an iterative refinement process to enhance accuracy and coherence.

At the end of this pipeline, the new content (for the moment, textual content like blog posts, slides, lectures' transcription, forum contribution, evaluations, students' tasks, etc.) will be stored in the proposed persistence layer: at the moment we are storing micro-graphs (single graph for single content) in a NOSQL repository, but for scaling and performance issues, we are evaluation to return to traditional, relational database using a single table, multi-column, fact-table like storage composed by the typical RDF triple (subject, predicate, object) with the addition of a graphID and a timestamp. From early experiments, the use of this storage structure seems to perform better than NOSQL solutions in the presence of ultra-large datasets.

5. CONCLUSION

The integration of AI into LMSs has the potential to revolutionize educational content management, delivery, and personalization. However, existing LMS platforms, primarily developed between the 1990s and early 2000s, are built on relational database structures that lack the flexibility needed to fully leverage modern AI technologies. As a result, AI-driven enhancements remain constrained, limiting their transformative impact on education. This paper presents an architectural shift in LMS design, particularly in the persistence layer, by transitioning from relational databases to graph-based systems. This transformation would facilitate the integration of knowledge graphs and ontological frameworks, enabling AI functionalities such as semantic search, intelligent content recommendations, and personalized learning pathways. Such advancements would enhance user experience and promote more adaptive, student-centered learning environments.

However, achieving this transformation presents several technical and research challenges. The proposed redesign necessitates a complete restructuring of both the persistence and business logic layers. Additionally, incorporating multi-domain ontologies to represent diverse and evolving educational content adds further complexity. A key obstacle is the development of adaptive AI models capable of real-time content generation, particularly during high-demand periods, which imposes significant computational and high-performance computing (HPC) requirements.

Currently, the project involves testing a linguistic recognition pipeline, graph construction, and named entity recognition (NER), applied to generic textual corpora not necessarily specific to the educational domain. In particular, the entity matching and graph construction components have been developed using an ontological domain derived from a parallel project (PAT-MAESTRO) focused on building a knowledge graph from pharmaceutical package leaflets. The ontologies employed—appropriately reduced and simplified—for the generation of a semantic, entity-centric knowledge graph include

- the Human Disease Ontology (DO), which provides standardized definitions of human diseases and disorders, available at the address: https://raw.githubusercontent.com/DiseaseOntology/HumanDiseaseOntology/main/src/ontology/doid.owl;
- the BRENDA Tissue Ontology (BTO) that provides descriptions for tissues, organs, and body parts, which are useful in annotating the anatomical sources of microbiome samples, available at the address https://www.brenda-enzymes.org/ontology.php?ontology_id=3

Using texts extracted from approx. 5,000 pharmaceutical package leaflets and leveraging the previously described pipeline integrated with the aforementioned ontologies, a navigable and searchable entity-centric graph has been constructed. The integration of this pipeline into a proprietary Learning Management System (LMS) is currently under experimentation. However, several challenges remain to be addressed, particularly in terms of performance and related issues, such as the indexing of newly ingested texts that require real-time search capabilities.

Despite these challenges, the potential benefits of an AI-native LMS architecture are substantial. A restructured system would empower educators and institutions to fully harness generative AI, creating more dynamic, personalized, and efficient learning experiences. To ensure sustainability, future LMS designs must not only integrate current AI technologies but also anticipate ongoing advancements, positioning educational systems at the forefront of innovation. While implementing AI in LMSs requires extensive structural modifications, the long-term advantages—enhanced learning adaptability, improved content management, and deeper engagement—justify in our opinion the effort.

ACKNOWLEDGMENT

The research presented in this paper is partially supported by the MAESTRO project, sponsored by the Autonomous Province of Trento – Italy, through L.P. n.6/99 and L.P. n.6/23

REFERENCES

Alier, M., Pereira, J., García-Peñalvo, F. J., Casañ, M. J., & Cabré, J. (2025). LAMB: An open-source software framework to create artificial intelligence assistants deployed and integrated into learning management systems. Computer Standards & Interfaces, 92, 103940.

Alvarez, A. I., & Conati, C. (2020). "Towards Using Artificial Intelligence for Adaptive Learning in Online Learning Environments." Educational Technology & Society*, 23(4), 33-45.

Baker, R. S. (2016). "Staying in Control: Automated Education with Learning Management Systems and Artificial Intelligence.", Journal of Educational Data Mining, 8(1), 12-25.

Bouquet, P., & Molinari, A. (2016). A new approach to the use of semantic technologies in e-Learning platforms, International journal of advanced corporate learning., 9(2), 5-12.

Dillenbourg, P., Schneider, D., & Synteta, P. (2022). "Virtual Learning Environments: Integrating AI Tools for Effective Collaboration and Learning.", AI and Education Journal, 39(3), 56-78.

Gonzalez, M. A., & de Vicente, M. J. (2021). "Generative AI in Learning Management Systems: Opportunities and Challenges.", AI and Learning Sciences Review*, 7(1), 41-58.

Hadi, M. U., Al Tashi, Q., Shah, A., Qureshi, R., Muneer, A., Irfan, M., ... & Shah, M. (2024). Large language models: a comprehensive survey of its applications, challenges, limitations, and future prospects. Authorea Preprints.

Holmes, W., Bialik, M., & Fadel, C. (2019). Artificial Intelligence in Education: Promises and Implications for Teaching and Learning, The Center for Curriculum Redesign.

Kose, U., & Deperlioglu, O. (2020). "Intelligent Learning Systems: Integration of AI into LMS for Personalized Education.", International Journal of Artificial Intelligence in Education*, 30(2), 183-203.

Kumar, V., & Vivek, S. (2021). "AI in Education: Application of Natural Language Processing (NLP) in Learning Management Systems.", International Journal of Learning Analytics and Artificial Intelligence for Education*, 4(1), 45-67.

Molinari, A., & Molinari, E. (2024). Designing a New Generation of Learning Management Systems, International Journal of Learning and Teaching, Vol. 10, No. 1, 2024

Popenici, S., & Kerr, S. (2017). "Exploring the Impact of Artificial Intelligence on Teaching and Learning in Higher Education.", Research and Practice in Technology Enhanced Learning*, 12(22), 1-14.

Roschelle, J., & Teasley, S. D. (2023). "Collaborative Learning with AI: Integrating GPT in Learning Management Systems for Enhanced Educational Outcomes.", Journal of Learning Analytics*, 10(2), 75-92.

Turnbull, D.; Chugh, R.; Luck, J. (2020) Learning management systems, an overview. In Encyclopedia of Education and Information Technologies; Tatnall, A., Ed.; Springer: Cham, Switzerland, 2020; pp. 1052–1058.

Vergara, D.; Lampropoulos, G.; Antón-Sancho, Á.; Fernández-Arias, P. (2024). Impact of Artificial Intelligence on Learning Management Systems: A Bibliometric Review. Multimodal Technol. Interact. 2024, 8, 75. https://doi.org/10.3390/mti8090075

Villegas-Ch, W., Román-Cañizares, M., & Palacios-Pacheco, X. (2020). Improvement of an online education model with the integration of machine learning and data analysis in an LMS. Applied Sciences, 10(15), 5371.

Wang, S., Xu, T., Li, H., Zhang, C., Liang, J., Tang, J., ... & Wen, Q. (2024). Large language models for education: A survey and outlook. arXiv preprint arXiv:2403.18105.

Xu, H., Gan, W., Qi, Z., Wu, J., & Yu, P. S. (2024). Large Language Models for Education: A Survey. arXiv preprint arXiv:2405.13001.

Zawacki-Richter, O., Marín, V. I., Bond, M., & Gouverneur, F. (2019). "Systematic Review of Research on Artificial Intelligence Applications in Higher Education – Where Are the Educators?", International Journal of Educational Technology in Higher Education*, 16(1), 1-27.

GEOMETRIC FIGURES IN STEREO BY CONICAL AXONOMETRIC PERSPECTIVE USING GEOGEBRA 2D SOFTWARE

Karla Bastos Guedes, Ronaldo da Silva Tavares, José Geraldo Franco Méxas
and Marcelle de Sá Guimarães
Fluminense Federal University
Brazil

ABSTRACT

The binocular or stereo perspective is important to improve our spatial reasoning. Studies point to the potential of digital systems for stereoscopic visualization, as a teaching tool, to improve the students' development of spatial reasoning. This paper addresses some of the main difficulties observed in traditional theoretical courses of Spatial Geometry, which refers to the abstract spatial visualization and the need to use technological resources to increase teaching methodologies. A new technique was proposed, to generate the stereo figures, from the orthogonal projections of the object, applying the axonometric conical perspective method and using GeoGebra software. The traditional method, used in digital computer graphics stereoscopic visualization software, generates figures very close to the screen, which is not appropriate in classroom's projections because it privileges an individual user. The proposed method can generate figures distant from the projection screen, 'floating' in the middle of the classroom, more resembling the cinema movies 3D method, where the images 'jump out' of the screen towards to different viewers.

KEYWORDS

Conical Axonometric Perspective, Stereoscopic Visualization, GeoGebra Software

1. INTRODUCTION

Spatial reasoning, including cognition and spatial intelligence is the ability to understand image variations and this concept was originally studied in the field of psychology. Thurstone (1938) defined spatial reasoning as the ability to memorize, twist, move and alter an image and compare it mentally with the original image. It is an extremely necessary skill for Spatial Geometry's students. The failure to develop spatial reasoning during school results in newcomers to undergraduate courses with recognized difficulties in understanding Spatial Geometry concepts. In addition, there is a lack of resources that could minimize these difficulties and support the abstraction process necessary for the development of visualization skills and spatial reasoning.

The great evolution of computing and electronics allowed the use of new technological resources in teaching and learning of graphic disciplines, such as Spatial Geometry, Technical Drawing, etc. Studies point to the potential of digital systems with stereoscopic visualization: Seabra et al (2005), Guedes et al (2012), Chu et al (2017) and augmented reality: Lima et al (2007), for the development of students' spatial reasoning. Both are virtual-reality techniques, used to create an immersive environment that aims to develop the visualization, understanding and solution of spatial geometry problems. Disciplines that need visualization and spatial reasoning, all over the world, from different areas of knowledge, have successfully used these digital teaching tools. Hoyek (2014) et al describe the effectiveness of a three-dimensional animation tool with stereoscopic vision, in teaching anatomy, which presents geometric shapes of human anatomy in a three-dimensional space. Garcia-Ruiz et al (2014) created a model for stereoscopic visualization by anaglyphs of spatial molecules, used in undergraduate chemistry courses. Joseph (2011) developed a stereoscopic visualization system to teach astronomy concepts, such as the behavior of the solar system, other galaxies and the interactions between them. Mexas et al (2015) developed a virtual system with stereoscopic figures by anaglyphs, using cylindrical-orthogonal axonometric perspective, applied to Descriptive Geometry teaching. In this work, a new technique was developed to create stereo figures, from the orthogonal projections of the object, applying the conical axonometric perspective method and using the free educational GeoGebra software. The conical axonometric perspective method was chosen because it is the drawing technique, within projective geometry, that generates more similar figures to the human eye vision. Section 2 presents the traditional method for generating stereo figures, used in computer graphics programs; Section 3 proposes a new methodology to generate stereo figures, from their orthogonal projections and using conical axonometric perspective in GeoGebra 2D; Section 4 shows a comparison between the methods and Section 5 the conclusion and future work's proposal.

2. STEREOSCOPIC VISUALIZATION IN TRADITIONAL METHOD

Stereoscopic vision consists of the perception of the depth of the scene by the human visual system Siscoutto et al. (2004). For this, it is necessary to capture the image of the object by both eyes, so that the brain can estimate the distance between them, that is, between the eyes and the object. To obtain a computerized image, in stereo, it is necessary to generate two computer images of the same object from slightly different view point and present them separately to each eye. In order to represent a point P in the projection plane, according to the observer's eyes, that is, in order to obtain a stereoscopic vision, we need to do the matrix calculations above twice, shifting the observer's eye slightly, in order to obtain P2, referring to the left eye (*le*) and the corresponding P_1, referring to the right eye (*re*), which must be on the same horizontal line. The distance between points P_1 and P_2 is called parallax and three situations can occur, depending on the position of point P in relation to the projection plane and the observer's eyes. At Figure 1 on the left, we have positive parallax because the image P is formed behind the Projection Plane (*PP*).

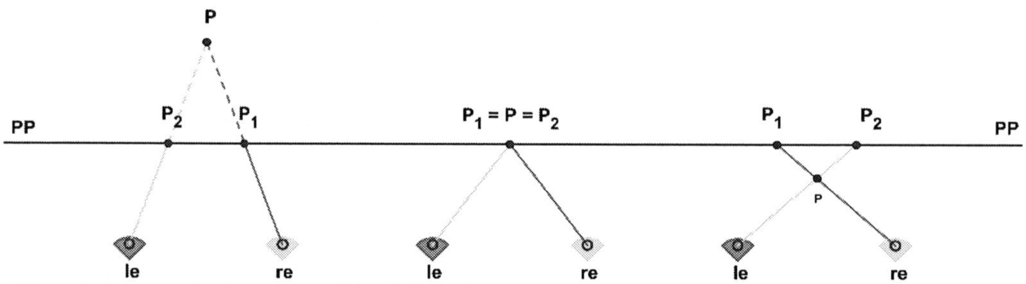

Figure 1. Stereo projections with positive (left), null (center) and negative (right) parallax for anaglyphic glasses

Figure 1 shows, on the center, zero parallax and on the right we have negative parallax when the image P is formed in front of the *PP*. In order to perceive the depth of the image, it is necessary that each eye sees only its respective image. This is done, in the anaglyphic method, using an image editor and coloring one of the images red and the other cyan, then, using anaglyphic glasses, with one lens red and the other cyan we can see in stereo.

The traditional matrix method, with homogeneous coordinates, used in computer graphics programs was explained at Méxas et al (2015). In this work, the GeoGebra 3D software was used to generate the figures by the traditional method.

3. THE PROPOSED METHOD

This paper proposes the use of a computational method, whose binocular conical axonometric perspective is obtained from the object's orthogonal projections. Its innovative nature comes from the implementation of stereoscopic vision from methods traditionally used in Projective Geometry. In addition to the fact that the new methodology allows controlling the magnitude of the figures' parallax, what means, allows controlling their distance to the projection screen. The proposed method can generate stereoscopic figures that are farther from the projection screen, therefore more suitable for projections in the classroom. The developed program uses the free educational software GeoGebra 2D. In an increasingly computerized world, it is necessary to innovate traditional methods of teaching spatial geometry which stimulate our students. The use of GeoGebra provides a lot of benefits to develop mathematic skills in all disciplines of STEM (Science - Technology - Engineering - Mathematics) education according to Tamam and Dasari (2020) and Kramarenko et al (2019).

3.1 Orthogonal Projections

The figures and objects in space are represented by their orthogonal cylindrical projections on two projection planes: a horizontal plane and a vertical plane, perpendicular to each other. One plane is rotated, around the intersection line between them, until both coincide, as shown in Figure 2 on the center. This process is called rotation of one plane over another.

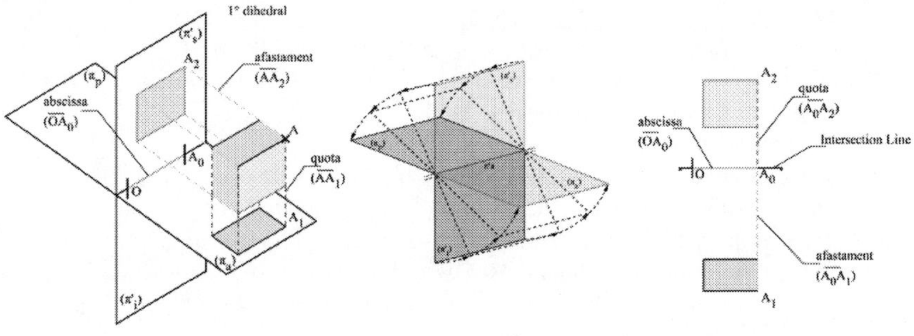

Figure 2. Rotation of the vertical plane over the horizontal plane Méxas et al (2015)

Figure 2 on the right shows the orthogonal projections of a parallelepiped, whose position in space is shown in Figure 2 on the left. In the proposed method, from the orthogonal projections of the object, two perspectives are drawn, under slightly different view point, in order to obtain the stereoscopy.

3.2 The Conical Axonometric Perspective Method

Perspective is a graphic representation that shows objects as they appear in our vision, with their three dimensions. The conical axonometric perspective, Cavallin (1976), was programmed in GeoGebra 2D and it is a conical projection on an oblique plane in relation to the three dimensions of the represented object. The word "axonometria" comes from Latin, where "axon" means "axes" and "metria" means "measure". According to the concepts of projective geometry, this perspective is essentially technical. It is a topic of particular interest to engineering and architecture students, as they need to develop the ability to perceive and create three-dimensional objects and be able to reproduce them in orthogonal projections.

The axonometric perspective, has its construction based on the inscription of the solid to be drawn, in a tri-rectangle trihedral (three right angles), where only the edges, called axes, are considered. The projection of this trihedral on the projection plane can generate the following types of axonometric perspective, shown in Figure 3.

- Conical: The projection lines depart from an objective point, called the center or projection vertex (*V*). Corresponding the letter "a" in Figure 3.
- Cylindrical: The projection lines start from a point at infinity, that is, the projection vertex is at infinity, which causes parallel projection lines (projectors), which can be oblique or orthogonal to the plane of projection. Letters "c" and "b" in Figure 3, respectively.
- Orthogonal: projection lines parallel and orthogonal to the projection plane corresponding to letter "b" in Figure 3.
- Oblique: projection lines parallel and oblique to the projection plane. Letter "c" in the Figure 3.

The conical axonometric perspective was chosen because it is the one that most closely resembles human vision.

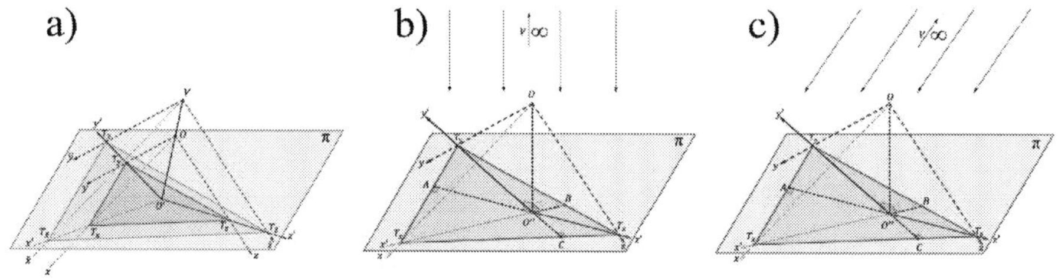

Figure 3. Axonometric Perspectives Conical (a), Orthogonal Cylindrical (b) and Oblique Cylindrical (c)

Figure 4 presents the conical axonometric perspective method, highlighting the elements used in its construction, where:

- *V*: Center or vertex of projections (observer's eye);
- *O*: Origin or vertex of the objective trihedral, as it is below the observer (between the observer and the projection plane), it is always visible;
- x, y and z: axonometric axes
- $\overline{x}$, $\overline{y}$ and $\overline{z}$: Vanishing point axes;
- x', y' and z': Axonometric axes projection;
- π: Projection plane, generally considered in the horizontal position. Keeping the trihedral fixed and moving the projection plane closer or further away, we obtain similar axonometric triangles, which result in the same perspective, since it depends only on the shape of the axonometric triangle and not on the size of its sides;
- Triangle T_x T_y T_z: Axonometric or fundamental triangle, whose vertices are the traces of the axonometric axes (*x*, *y* and *z*), in the projection plane (π);
- Triangle $T_{\overline{x}} T_{\overline{y}} T_{\overline{z}}$: Vanishing point triangle whose vertices are the traces of the vanishing point axes ($\overline{x}$, $\overline{y}$, and $\overline{z}$) in the projection plane (π); According to Basic Geometry, the two triangles are homothetic, that is, they have parallel and proportional sides.
- *O'*: Conical projection of vertex *O*;

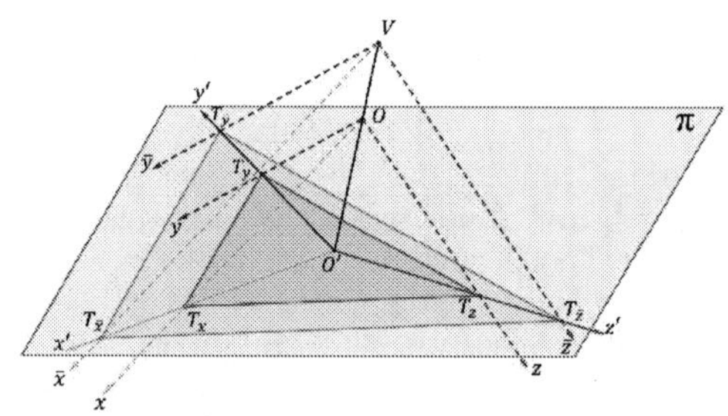

Figure 4. Elements for the conical axonometric perspective construction

The construction process of the conical axonometric perspective is similar to that of the orthogonal cylindrical axonometric perspective, described in Méxas et al (2015). The main difference resides in the fact that, in the conic, all lines from each of the object point must converge to the same point *V*, called 'point of view' or 'observer's eye'. In addition, two orthogonal reference systems are used: one called the objective or fundamental trihedral, with vertex *O* and another called the trihedral of the conical system, with vertex *V*, farther from the projection plane than the first. The two trihedral have parallel corresponding axes. The traces of the axes of the objective trihedral, on the projection plane, determine the vertices of the axonometric triangle (triangle T_x T_y T_z) and the traces of the axes of the conical system, on the projection plane, determine the vertices of the triangle of vanishing points (triangle $T_{\overline{x}} T_{\overline{y}} T_{\overline{z}}$). The two triangles are homothetic with respect to the point *O'*, which is the intersection of the projection plane with the projecting line passing through the points *V* and *O*, that is, the conical projection of the point *O*.

According to Figure 5, the grading process of the projected axes (x' y' and z') starts with the reflection of the axes of the conical system on the projection plane, obtaining the reflected point V. Then we reflect the axes of the objective trihedral on the projection plane, obtaining their true magnitudes, where the scales of units in true magnitude will be marked on the reflected objective axes. To grade the projected axes in conical perspective, just draw straight lines, all of them starting from the reflected V point, passing through the markings of the scales of the reflected objective axes. The intersections of these lines with the projected objective axes (x' y' and z'), determine the projected scales on these. In conical perspective, each graded unit will have a different size, according to Desargues' 'Fugitive Scales Theorem'.

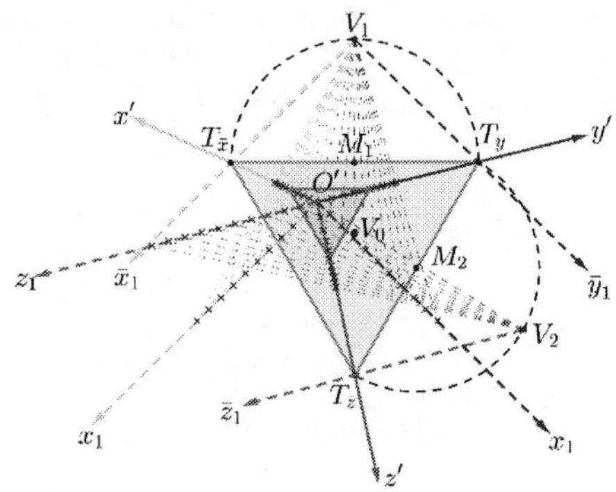

Figure 5. Axonometric axes graduation

Figure 6 shows the conical axonometric perspective of a cube, whose edges measure 10 units of measure. Its construction takes into account that all lines that would be parallel to the projected objective axes (x' y' and z') must converge to the vanishing points $(T_{\bar{x}} T_{\bar{y}} T_{\bar{z}})$.

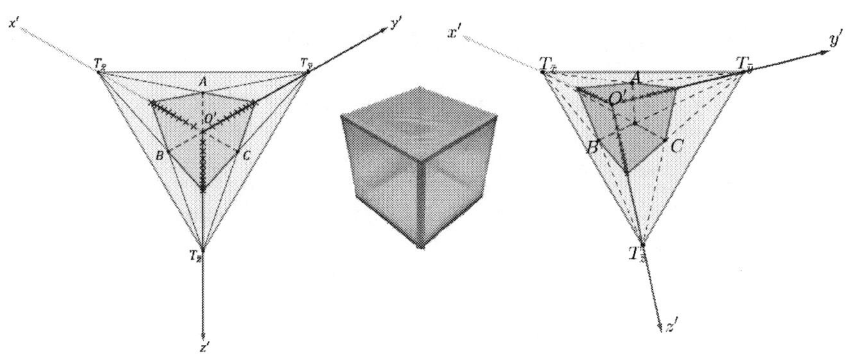

Figure 6. Cube's Conical Axonometric Perspective when *O'* is the triangle orthocenter (left) and when not (right)

Next it will be explained as two slightly different conical axonometric perspectives are constructed, one red and the other cyan blue, so that, when observed using anaglyphic glasses, the desired stereoscopic effect is obtained.

3.3 Stereoscopic Visualization by Anaglyphs Using the Proposed Method

Using the educational software GeoGebra 2D, two conical axonometric perspectives are constructed, on the same axonometric triangles and vanishing points, generated by the trihedral of the objective and conical systems. Each of the generated figures will be seen by one eye, in order to enable its vision in stereo. To obtain the second figure, slightly different from the first, we move the object in space and then construct its second conical axonometric perspective. In the proposed method, instead of displacing the axonometric system to obtain a second figure, as in the orthogonal cylindrical axonometric perspective method shown in Méxas et al (2015), only the object was rotated around the central axis that passes through its center and is perpendicular to the horizontal projection plane. Thus, we have the first and second perspectives with the same central rotation axis, but distinguished by a minimum rotation angle, which is chosen according to the best stereo perception of the object. The size of the rotation angle used and the distance between the two figures obtained determine how far or close to the projection screen the binocular or stereo image will be formed, that is, the magnitude of its parallax. One of the figures will be drawn in red and the other in cyan, to be viewed with anaglyphic glasses, thus enabling stereo vision. In this work we used anaglyph glasses because they are less expensive. Glasses with polarized lenses could be used, as in cinema stereo projections. For this, instead of generating the two figures in red and cyan colors, they would be generated in two different polarizations. Figure 7, on the left, shows the orthogonal projections of an F-117 plane and on the right, the stereoscopic visualization by anaglyphs, obtained using the proposed method.

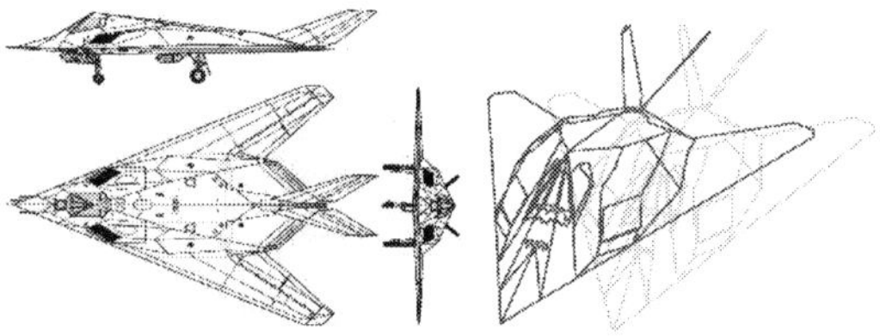

Figure 7. F-117 orthogonal projections (left) and conical axonometric stereo perspective, in GeoGebra 2D (right)

Figure 7, shows exactly how they appear on the GeoGebra screen, the orthogonal projections and the conical perspective projection, side by side, dynamic and interactive, that is, changing the orthogonal projections, the stereo perspective is automatically changed.

4. COMPARISON BETWEEN THREE METHODS

In this section we will make a graphic comparison between three methods to generate stereo figures, from the analytical view point, only to highlight the differences. The validity of the methods depends on the didactic objectives they propose. Figure 8 shows the F-117 plane and Figure 9 shows the five regular Plato solids, comparing the three methods.: the Orthogonal Cylindrical Axonometric Perspective method, shown in Méxas et al (2015), the Conical Axonometric Perspective method, proposed in this work and the traditional matrix method, used in computer graphics.

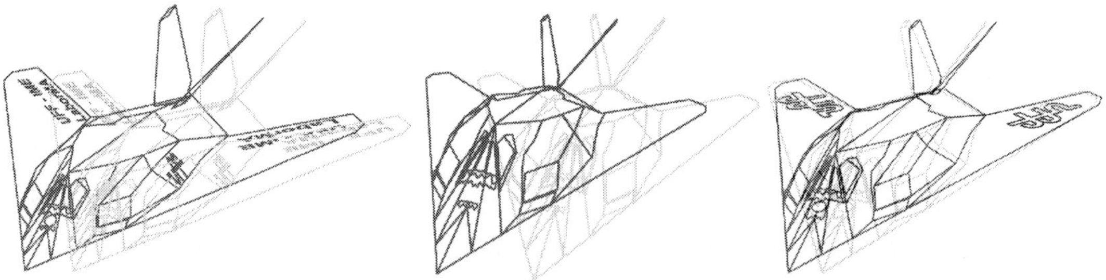

Figure 8. F-117: Méxas et al (2015) (left), Proposed Method (center) and Traditional Method (right)

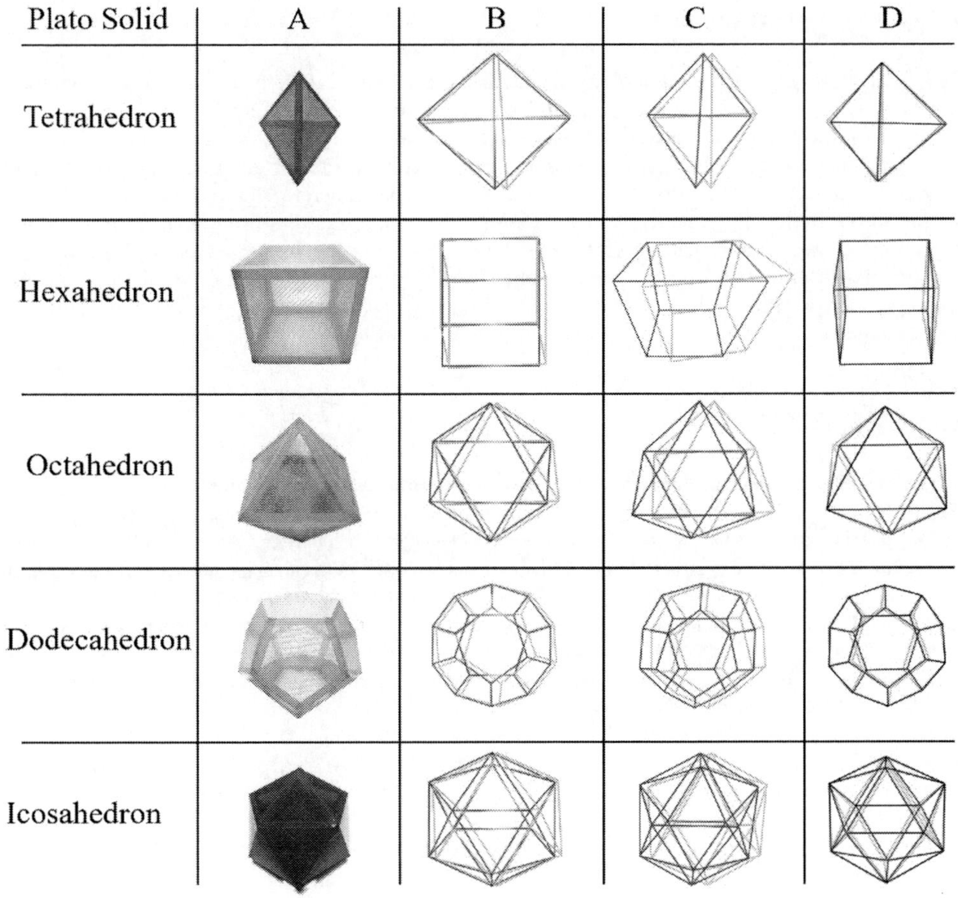

Figure 9. A) Photo of a real solid B) Méxas et al (2015) Method C) Proposed Method and D) Traditional Method

By the traditional method, column D, it was used the GeoGebra 3D. The figures formed have one half on the inside of the projection screen and the other on the outside, which is ideal for a viewer in front of the computer screen or smartphone device.

By Méxas et al (2015) method, on the column B, the figures formed can be away from the projection screen, as desire, but simulates a case with the observer's eye at infinity, therefore different from human vision.

By the proposed method, column C, our objective was achieved. The objective for which we propose our method is to promote stereoscopic visualization for a group of people gathered in a classroom, looking at a projection screen with figures similar to human vision. That way, students at the back of the room, away from the projection screen, only need to put on their anaglyph glasses and will also experience the stereoscopic vision.

5. CONCLUSION

This work presents a new methodology for stereo spatial visualization: the use of a computational method with anaglyph stereoscopy, whose binocular conical axonometric perspective is obtained from the object's orthogonal projections. Its innovative feature comes from the fact that stereoscopic vision is implemented from a technique used in Projective Geometry and Technical Drawing. In addition, the new methodology allows controlling the magnitude of the parallax and can generate stereoscopic figures farther from the projection screen than those generated by traditional computer graphics programs, therefore more suitable for projections in classrooms. The program was developed using the free GeoGebra 2D educational software. The figures generated by the proposed method are compared with the same figures obtained by other methods, such as the traditional method. For future work, we intend to investigate and implement the proposed method, using modeling from Bezier curves.

REFERENCES

Cavallin, J.,(1976). *Perspectiva Linear Cônica.* Curitiba, RS, Brazil: Ao Livro Técnico Press.

Chu, P. Y., Chien Y. H., (2017). The Effectiveness of Using Stereoscopic 3D for Proportion Estimation in Product Design Education. *Eurasia Journal of Mathematics Science and Technology Education*, 13(10), 6635-6647.

Garcia-Ruiz, M. A. et al, (2014). Using effective stereoscopic molecular model visualizations in undergraduate classrooms. *International Journal for Cross-Disciplinary Subjects in Education*, 5(1), 1593-1598.

Guedes, K. B. et al, (2012). Virtual Reality Using Stereoscopic Vision for Teaching/Learning of Descriptive Geometry. *In Proceedings of the Fourth International Conference on Mobile, Hybrid and On-line Learning*, 24-30.

Hoyek, N. et al, (2014). Effectiveness of three-dimensional digital animation in teaching human anatomy in an authentic classroom context. *Anatomical sciences education*, 7(6), 430-437.

Joseph, N; (2011). *Stereoscopic visualization as a tool for learning astronomy concepts* Master's thesis, Purdue University.

Kramarenko, T. H. et al, (2019). *Prospects of using the augmented reality application in STEM-based Mathematics teaching.*

Lima, A. J. R.; Haguenauer, C. J. L, (2007). Visualização das Seções Cônicas da Geometria Descritiva Através de Realidade Aumentada. *IX Symposium on Virtual and Augmented Reality,* Petrópolis, RJ, Brazil.

Méxas, J. G. et al, (2015). Stereo Orthogonal Axonometric Perspective for the Teaching of Descriptive Geometry. *Emerald, Interactive Technologies and Smart Education,* 12(3), 222-240.

Seabra, R. D., and Santos, E. T., (2005). Análise de Requisitos de uma ferramenta 3D para desenvolvimento da Cognição Espacial, *17° National Simposium of Geometry and Technique Drawing*, Recife, PE, Brazil.

Siscoutto, R.A. et al.; (2004). Estereoscopia - Realidade Virtual: Conceitos e Tendências. *Book of Pre-Simposium SVR 2004*, pp.179-201, São Paulo, SP, Brazil: Mania de Livro Press.

Tamam, B., and Dasari, D., (2021). The use of Geogebra software in teaching mathematics. In *Journal of Physics: Conference Series* (Vol. 1882, No. 1, p. 012042). IOP Publishing.

Thurstone, L. L., (1938). *Primary mental abilities.* Chicago, IL: University of Chicago Press.

EDUCATIONAL ASSESSMENT IN THE AGE OF AI: A NARRATIVE REVIEW ON DEFINITIONS AND ETHICAL-LEGAL PRINCIPLES FOR TRUSTWORTHY AUTOMATED SYSTEMS

Giannangelo Boccuzzi[1,2], Flavio Manganello[2] and Alberto Nico[3]
[1]*University of Bologna ALMA MATER STUDIORUM*
Via Zamboni, 33, 40126 Bologna BO, Italy
[2]*Italian National Research Council – Institute for Educational Technologies*
Torre di Francia, Via de Marini 6 - 16149 Genova, Italy
[3]*University of Bari "A. Moro"*
Piazza Umberto I, 1, 70121 Bari BA, Italy

ABSTRACT

This study presents a narrative review of scholarly literature at the intersection of education, explainable artificial intelligence (XAI), and automated assessment algorithms. Acknowledging the pressing need for a unified regulatory framework to enable the safe deployment of AI-based assessment technologies in educational environments, the paper investigates whether a coherent and consistent terminology is employed across academic research. It further compiles and systematizes the ethical and legal principles underpinning the use of such systems, from the Ethical "AI explainability" to the emergent "Right to Explanation" introduced by Article 22 of the General Data Protection Regulation (GDPR). The review finds that key concepts such as "assessment" and "explainability" are currently used with divergent meanings across disciplines. The analysis identifies transparency, accountability, fairness, privacy, and non-discrimination as the core principles most frequently emphasized in the literature, highlighting the need for interdisciplinary convergence to ensure trustworthy and equitable AI integration in Education.

KEYWORDS

Automated Assessment, Artificial Intelligence, Explainable AI, Ethics

1. INTRODUCTION

The integration of Artificial Intelligence (AI) into education is transforming how we understand and implement assessment processes. As automated systems increasingly take on evaluative function, from grading to predicting student outcomes, the very notion of "assessment" demands renewed scrutiny (Luckin et al., 2016; Holmes et al., 2019). In this evolving context, it is essential to establish univocal definitions of core constructs such as "assessment" and "explainability," which are often interpreted differently across technical, educational, legal, and ethical domains (Wang et al., 2023; Suresh & Guttag, 2021).

Assessment in education is not merely a tool for measurement; it is a fundamental mechanism through which learners receive feedback, understand their progress, and shape their learning pathways (Black & Wiliam, 1998; Boud & Falchikov, 2007). When such evaluations are performed by AI systems, they inevitably raise several ethical and legal concerns, potentially leading to biased or opaque decision-making, lack of recourse for contesting outcomes, and diminished agency on the part of the learner (Williamson & Eynon, 2020; Floridi et al., 2018; Selwyn, 2021). Here, transparency is not merely desirable but imperative, as students must clearly understand the methodological foundations and criteria upon which their academic outcomes are assessed (Wachter et al., 2017; Binns et al., 2018). Automated assessments must disclose the specific criteria and weighting schemes employed, as well as how these assessments align with learning objectives and competencies (UNESCO, 2021; Holmes et al., 2022). Such clarity ensures fairness, supports informed reflection by learners, and facilitates constructive interactions between students and educators (Hwang et al., 2020; Tsai et al., 2022).

A narrative review approach was taken here to add to the current understanding of transparency and the right to explanation in automated educational decision-making. The review critically examines existing scholarly literature on the implementation of algorithmic transparency and the right to explanation within educational contexts.

2. METHODOLOGY

This study formulates five research questions (RQs) situated within educational research, focusing on conceptual, methodological, ethical, and practical dimensions of AI-driven automated assessment. **RQ1** investigates whether a universally accepted definition of "automated assessment" exists in the literature, highlighting the need for terminological consistency. **RQ2** examines the conceptual coherence of "explainability" across disciplines, assessing its implications for transparency and interpretability in educational contexts. **RQ3** addresses how the ethical and legal principle of the "right to explanation" is represented in current research, reflecting scholarly engagement with regulatory frameworks. **RQ4** identifies the most frequently cited ethical and legal concerns in academic discourse on automated assessment. **RQ5** evaluates the presence and effectiveness of practical guidelines proposed in educational research to ensure transparency and explainability, considering their alignment with user needs and regulatory standards.

2.1 Search Strategy

A systematic search strategy was implemented using **Scopus** and **Web of Science** to identify peer-reviewed literature on *explainable AI*, the *right to explanation*, and their application in educational assessment. The **Scopus query** targeted documents mentioning "right to explanation" (field: ALL) or "explainable AI" (in TITLE-ABS-KEY), combined with "education" and further refined by the inclusion of "assessment." The search was limited to **Social Sciences (SOCI)**, included only **original research articles (ar)** and **literature reviews (re)**, and was restricted to **English-language publications** from **2023 to 2025** (PUBYEAR > 2022 AND PUBYEAR < 2026).

A **parallel search in Web of Science** mirrored these parameters, retrieving publications that referenced "right to explanation" or "explainable AI" as **topic keywords**, alongside "education" (in any field) and "assessment" (in all fields). Filters were applied for **publication years (2023–2025)**, **document type** (articles and reviews), and **language** (English), ensuring methodological alignment and high academic rigor across both databases.

2.2 Results

Applying the previously detailed search strategy yielded a targeted corpus of scholarly literature addressing the intersection of automated educational assessment, explainable artificial intelligence (XAI), and the right to explanation within educational research. Specifically, the Scopus database returned **90** publications that satisfied all predetermined inclusion criteria, reflecting a substantial engagement of the social science community with these emerging topics. The parallel query conducted on Web of Science resulted in **45** relevant publications. Upon merging the two datasets and systematically removing duplicates (**14** publications identified as common to both searches), a consolidated corpus of **121** unique scholarly contributions was established. This consolidated set provided a comprehensive overview of recent scholarly activity around transparency, explainability, and related ethical and regulatory issues in AI-driven assessment. Subsequently, each of these 121 publications underwent careful screening based on relevance and alignment with the research questions guiding this review. Through a rigorous assessment of their thematic and methodological pertinence, **25** publications were selected as particularly relevant for detailed analysis and synthesis. These studies demonstrated focused engagement with conceptual debates and empirical inquiries regarding the definition and operationalization of explainability, critical reflections on the right to explanation within educational contexts, and systematic investigations into ethical, legal, and technological dimensions of automated assessment systems. The resulting refined corpus thus offers a targeted collection of academic insights specifically oriented towards the critical examination of AI explainability, transparency, and associated rights and responsibilities within contemporary educational assessment practices.

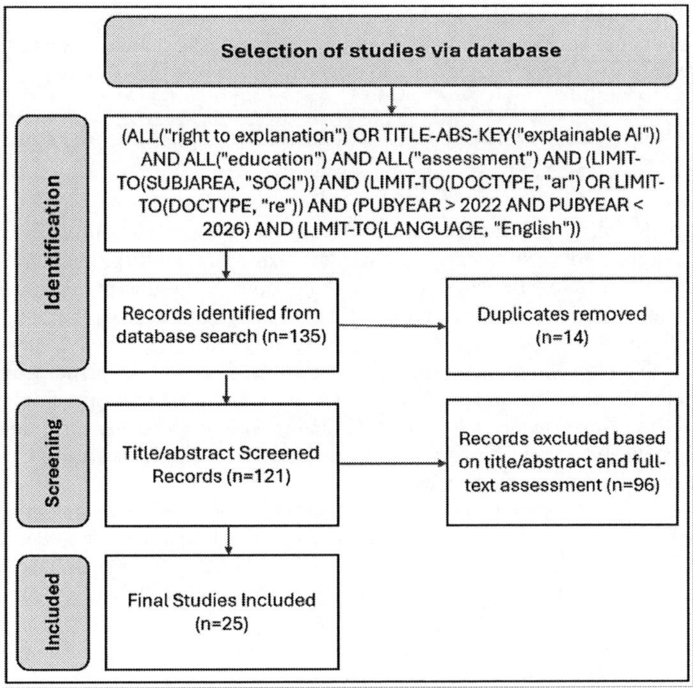

Figure 1. A chart depicting the search strategy implemented

3. FINDINGS

3.1 RQ1 – Is there a Universally Accepted Conceptualization of "Automated Assessment"?

The analysis of the selected literature indicates that, currently, there is no univocal or universally accepted conceptualization of "automated assessment" within educational research. Rather, the reviewed studies exhibit notable variation in terminology, definitions, and scope. While some researchers conceive automated assessment primarily as algorithmic grading or scoring of student performance, others adopt broader interpretations encompassing AI-driven analytics for predicting student outcomes, dropout risks, or automated decision-making in educational admissions and career counselling. Specifically, the review identified diverse interpretations, including **a)** Assessment as *information assessment* (Khan et al, 2024), emphasizing verification of information accuracy within educational contexts; **b)** Assessment as *Artificial Intelligence for Student Performance Prediction (AI4SPP)* (Fedele et al 2024) or similarly *student outcomes* (Gunasekara & Saarela, 2025) or *educational achievement* (Silva et al, 2023) or *learning achievement* (Duan et al, 2024) or *Academic Outcomes* (Albahli, 2025), focusing on predictive modelling of students' academic performance; **c)** Assessment as *guidance for career prediction* (Guleria & Sood, 2023; Boccuzzi & Manganello, 2024; Manganello et al, 2025), involving predictive analytics to support students' career decision-making processes; **d)** Assessment as *drop-out risk prediction* (Nagy & Molontay, 2024; Bulut et al, 2024), centred on identifying students at risk of leaving educational programs; **e)** Assessment as *Automated*

Decision-Making (ADM) (David & M.E.G., 2023) highlighting algorithmic systems making administrative or evaluative educational decisions without direct human intervention.

Conceptual variability reflects the diverse purposes and contexts in which automated assessment technologies are deployed. Consequently, the absence of a clearly established, shared definition complicates comparative analyses and potentially leads to ambiguities in regulatory discussions and ethical evaluations. Thus, future scholarship would benefit from explicit efforts to establish terminological clarity and consensus, facilitating more cohesive research agendas and effective policymaking.

3.2 RQ2 – Does "Explainability" Hold a Singular, Coherent Meaning Within Educational Research?

Based on the analysis of the selected literature, the concept of "explainability" in educational research does not exhibit a singular, coherent meaning; instead, it appears multifaceted and context-dependent. The reviewed studies demonstrate a range of conceptual and operational interpretations of explainability, varying significantly according to research objectives and methodological frameworks. Some studies focus explicitly on the technical dimension of explainability, evaluating how effectively various post-hoc techniques (e.g., SHAP, LIME) reveal the internal decision-making logic of machine learning models (Hooshyar & Yang, 2024; Gunasekara & Saarela, 2025). Such interpretations emphasize fidelity, reliability, and clarity in conveying feature importance, often distinguishing between intrinsically interpretable models (like decision trees or KBANNs) and more complex models (e.g., neural networks) that require supplemental methods to elucidate their reasoning processes.

Other studies approach explainability from the perspective of practical applicability in educational contexts, highlighting its role in supporting educators' understanding, trust, and decision-making capabilities. Here, explainability is linked directly to usability, as exemplified by research involving teacher co-design processes (Duan et al, 2024), personalized interventions (Nagy & Molontay, 2024), dropout risk predictions (Bulut et al, 2024; Mustofa et al, 2025), and career guidance systems (Guleria & Sood, 2023). This user-centric interpretation of explainability underscores the importance of explanations not only being accurate but also actionable, meaningful, and contextually relevant for educators and students alike.

Furthermore, certain authors critically analyze the ethical (Guizani et al, 2025; Fedele et al, 2024; Conjin et al, 2023) and socio-technical dimensions of explainability, asserting that superficial engagement with ethical theories has limited the depth and utility of XAI in educational policymaking and assessment. These studies advocate for a more nuanced, ethically grounded interpretation of explainability, calling for interdisciplinary collaboration and careful consideration of ethical complexity when designing and implementing XAI systems in educational settings.

The present literature highlights considerable heterogeneity in interpreting explainability, demonstrating that it cannot be treated as a uniform construct across educational research. Rather, explainability is variously conceptualized as a technical requirement, a pedagogical tool, and an ethical obligation, with its meaning shifting based on disciplinary perspectives, user needs, and intended applications within educational contexts. Multiplicity of interpretations underscores the need for researchers and practitioners to clearly articulate their definitions and operationalizations of explainability, fostering greater coherence and transparency in future educational research involving XAI.

3.3 RQ3 – How is the "Right to Explanation" Currently Articulated and Addressed as a Critical Ethical and Legal Principle within Existing Literature on Automated Educational Assessment?

The literature under analysis explicitly frames the "right to explanation" as an essential ethical and legal principle within automated educational assessment, although its operationalization remains inconsistent and context dependent. Central to this discussion is the European regulatory landscape, notably the General Data Protection Regulation (GDPR), the AI Act, and the guidelines such as the Assessment List for Trustworthy AI (ALTAI) (Fedele et al., 2024).

According to GDPR Article 22, individuals subject to automated decision-making processes that significantly affect them have the right to receive "meaningful information" regarding the logic, significance, and envisaged consequences of these automated processes. However, ambiguity persists concerning what constitutes an adequate explanation, especially in educational contexts involving vulnerable groups like students (Fedele et al., 2024). GDPR interpretative guidelines recommend caution in applying solely automated decisions to children unless strictly necessary, thus emphasizing the need for additional safeguards, including human oversight and clear consent procedures. Recent scholarly discourse further situates the "right to explanation" within broader ethical frameworks for trustworthy AI, particularly emphasizing the ALTAI checklist proposed by the High-Level Expert Group on AI (AI HLEG). ALTAI explicitly includes transparency and explainability among its foundational principles, mandating clear and accessible explanations of how AI-based educational assessment systems generate predictions or evaluations. These explanations must be understandable, actionable, and pedagogically meaningful to educators and students (Fedele et al., 2024). Empirical studies on Automated Essay Scoring (AES) suggest that providing detailed explanations does not necessarily enhance student trust or motivation; instead, discrepancies between student self-assessments and automated grades appear more influential (Conijn et al., 2023). This insight underscores a significant gap between regulatory intentions and practical educational outcomes, indicating that current explanatory mechanisms may not adequately address the psychological and educational needs of students (Conijn et al., 2023).

Technical solutions, notably Explainable AI (XAI) methods such as SHAP and LIME, are frequently suggested to fulfill transparency obligations. However, empirical evaluations highlight significant variability in the fidelity and reliability of these explanations in educational contexts (Albaladejo-González et al., 2024). Research indicates that explanations produced by XAI tools often misrepresent the decision-making logic of underlying models, inadequately capturing pedagogically relevant patterns or features, thus questioning their effectiveness in operationalizing the GDPR's "right to explanation" (Albaladejo-González et al., 2024; Nannini et al., 2024).

Further, ethical guidelines from international bodies such as UNESCO reinforce the critical importance of explainability, advocating for a comprehensive approach embedding explainability within broader ethical AI deployment strategies specifically tailored to education (Guizani et al., 2025). These guidelines recommend extensive capacity-building efforts to help educators fully understand AI operational logic and its ethical implications (Guizani et al., 2025). UNESCO's guidance emphasizes using explainability as a cornerstone for maintaining educational integrity, protecting student rights, and ensuring transparent governance of generative AI tools in higher education. It advocates for contextually meaningful explanations aligned with educational values and stakeholders' real needs rather than mere technical transparency (Guizani et al., 2025).

In conclusion, while the "right to explanation" is affirmed as a crucial ethical and legal principle within the literature on automated educational assessment, its practical realization remains fragmented across regulatory, ethical, and technical domains. Legal frameworks such as GDPR provide baseline obligations, ethical guidelines expand these into normative recommendations, and technical solutions partially operationalize explainability. However, considerable gaps persist in translating the conceptual "right to explanation" into effective educational practice, underscoring the need for ongoing interdisciplinary research and enhanced stakeholder collaboration (Fedele et al., 2024; Albaladejo-González et al., 2024; Conijn et al., 2023; Nannini et al., 2024).

3.4 RQ4 - Which Ethical and Legal Considerations are Most Frequently Emphasized Within Academic Studies of Automated Assessment?

Among the ethical and legal considerations most frequently emphasized within academic studies of automated assessment, issues related to **transparency**, **accountability**, **fairness**, **privacy**, and **non-discrimination** are prominent. Transparency is particularly critical, requiring that AI-driven assessment tools clearly disclose their decision-making processes to affected individuals, especially in contexts involving sensitive personal data or vulnerable populations (Fedele et al., 2024; Conijn et al., 2023). In educational settings, this implies the need to go beyond algorithmic traceability and provide meaningful, accessible, and pedagogically aligned explanations that students and educators can interpret and act upon. Transparency also intersects with the broader principle of educational fairness, as it enables scrutiny and contestability, which are central to protecting learners' rights.

Accountability frameworks emphasize the importance of clearly defined roles and responsibilities among educators, developers, and institutions using automated systems, ensuring clarity on liability, oversight, and mechanisms for addressing potential harms or inaccuracies (Fedele et al., 2024; Albaladejo-González et al., 2024). This includes the implementation of audit trails, external reviews, and the designation of ethics boards or data protection officers to supervise the use of AI in education. Accountability is increasingly seen not just as a legal necessity but as an ethical imperative to promote trust and ensure that harm can be adequately identified, prevented, and remediated.

Fairness and non-discrimination are highlighted due to the risks that AI systems may perpetuate existing biases or introduce new forms of inequality, necessitating rigorous ethical oversight, continuous monitoring, and iterative refinement of AI models and datasets (Nannini et al., 2024; Albaladejo-González et al., 2024). Several studies warn of algorithmic biases embedded in historical training data or produced by skewed data labeling practices, which may disproportionately affect marginalized student groups. Scholars increasingly advocate for participatory design practices and fairness audits that consider contextual vulnerabilities and intersectional risks.

Finally, safeguarding privacy and adherence to data protection standards, particularly GDPR, are consistently underscored as foundational for the ethical deployment of automated educational assessment technologies, ensuring the respect for students' personal data and autonomy (Fedele et al., 2024). This includes compliance with principles such as data minimization, purpose limitation, and data protection by design and default, as well as the mandatory execution of Data Protection Impact Assessments (DPIAs) when high-risk processing is involved. In educational contexts, where minors and sensitive educational records are frequently processed, the ethical responsibility to protect privacy is heightened.

3.5 RQ5 – Are Practical Guidelines Existing to Ensure Transparency and Explainability of AI-Based Assessment Systems in Education?

Although various practical guidelines exist to ensure transparency and explainability in AI-based educational assessment, their enforceability and implementation remain uneven. The most notable is the **Assessment List for Trustworthy AI (ALTAI)** by the European Commission's High-Level Expert Group, which operationalizes transparency, accountability, and explainability through a structured checklist (Fedele et al., 2024). ALTAI views transparency as a multi-dimensional duty involving documentation, stakeholder communication, and meaningful explanations—particularly critical in high-risk domains like education.

Legal frameworks such as the **GDPR** reinforce this obligation: Articles 13–15 and 22 establish the right to receive “meaningful information” about automated decisions. However, the ambiguity of what qualifies as “meaningful” and the lack of education-specific guidance remain problematic (Fedele et al., 2024; Conijn et al., 2023).

To bridge this gap, **UNESCO** has issued ethical instruments, including the *Recommendation on the Ethics of Artificial Intelligence* (2021) and *Guidance for Generative AI in Education and Research* (2023), emphasizing explainability as key to protecting learners' rights and pedagogical integrity (Guizani et al., 2025). These documents call for **contextualized and pedagogically relevant explanations** for all stakeholders.

Meanwhile, scholars advocate for using **Explainable AI (XAI)** methods like SHAP and LIME to enhance transparency. Yet, these techniques face limitations: they often produce explanations that are not educationally meaningful, can distort decision logic, and are hard for non-experts to interpret (Nannini et al., 2024; Albaladejo-González et al., 2024).

To move from principles to practice, proposals include forming **ethics committees**, promoting **teacher training**, and adopting **participatory design** to co-create AI tools with educators and learners. Complementary measures like **Data Protection Impact Assessments (DPIAs)** and **audit trails** are also recommended to ensure compliance and mitigate risks, especially for vulnerable populations like minors.

In sum, while regulatory (e.g., GDPR), soft-law (e.g., ALTAI, UNESCO), and technical (e.g., XAI) instruments offer a framework for transparent and explainable educational AI, actual implementation remains a challenge. Ensuring these principles are applied in context-sensitive, user-centered, and pedagogically appropriate ways is essential.

4. CONCLUSION

Taken together, the literature reviewed paints a picture of a field that is conceptually vibrant yet methodologically fragmented. The absence of shared definitions for both "automated assessment" and "explainability" unveils a semantic gap, hampers comparative scholarship, slows regulatory alignment, and leaves educators without stable reference points for practice. At the same time, the recurring focus on transparency, accountability, fairness, privacy, and non-discrimination signals an emerging normative core that can serve as a scaffold for future convergence. Translating that core into action, however, will require moving beyond high-level principles and checklist compliance toward context-sensitive, pedagogically meaningful explanations that genuinely empower learners and teachers. We therefore recommend that researchers prioritise *(i)* rigorous, interdisciplinary work to standardise key terminology; *(ii)* empirical studies that test how different explanatory formats affect student understanding, trust, and outcomes; and *(iii)* participatory design frameworks that embed legal and ethical obligations, such as the GDPR's right to explanation, directly into the lifecycle of educational AI tools.

REFERENCES

Al-Ansari, N., Al-Thani, D., & Al-Mansoori, R. S., (2024). User-centered evaluation of explainable artificial intelligence (XAI): A systematic literature review. *Journal of Healthcare Engineering.*

Albahli, S. (2025). Advancing sustainable educational practices through AI-driven prediction of academic outcomes. *Sustainability*, 17(3), 1087.

Albaladejo-González, M., Ruipérez-Valiente, J. A., & Gómez Mármol, F., (2024). Artificial intelligence to support the training and assessment of professionals: A systematic literature review. *ACM Computing Surveys*, 57(3), Article 56.

Amarasinghe, K., Rodolfa, K. T., Lamba, H., & Ghani, R., (2023). Explainable machine learning for public policy: Use cases, gaps, and research directions. *Data & Policy*, 5, e5.

Bai, J., Zawacki-Richter, O., & Müskens, W, (2024). Re-examining the future prospects of artificial intelligence in education (AIEd) in light of the GDPR and ChatGPT. *Turkish Online Journal of Distance Education*, 25, 20–32.

Binns, R., Veale, M., Van Kleek, M., & Shadbolt, N., (2018). 'It's Reducing a Human Being to a Percentage': Perceptions of Justice in Algorithmic Decisions. *CHI Conference on Human Factors in Computing Systems*, Montreal, Canada, 1-14.

Black, P., & Wiliam, D., (1998). Assessment and classroom learning. *Assessment in Education: Principles, Policy & Practice, 5(1), 7–74.*

Boccuzzi, G., Manganello, F., (2024). Integrating generative AI, synthetic data, and machine learning in a 5G-enabled metaverse for education orientation: the ITAVER5O project, *ICERI2024 Proceedings*, pp. 8613-8619.

Boud, D., & Falchikov, N., (2007). *Rethinking assessment in higher education: Learning for the longer term.* Routledge. New York, NY.

Bulut, O., Wongvorachan, T., He, S., et al., (2024). Enhancing high-school dropout identification: A collaborative approach integrating human and machine insights. *Discover Education*, 3, 109.

Chong, K. T., Ibrahim, N., Huspi, S. H., Wan Kadir, W. M. N., & Isa, M. A., (2025). A systematic review of machine learning techniques for predicting student engagement in higher education online learning. *Journal of Information Technology Education*: Research, 24(0).

Colonna, L., (2024). Teachers in the loop? An analysis of automatic assessment systems under Article 22 GDPR. *International Data Privacy Law*, 14(1), 3–18.

Conijn, R., Kahr, P., & Snijders, C., (2023). The effects of explanations in automated essay scoring systems on student trust and motivation. *Journal of Learning Analytics*, 10(1), 37–53.

Dara, T., Li, Y., Ukwuoma, C. C., & Dossa, J., (2024). Explainability of artificial neural network in predicting career fulfilment among medical doctors in developing nations: Applicability and implications. *Social Science & Medicine*, 360, 117329.

David, M. E. G., (2023). Government by algorithms at the light of freedom of information regimes: A case-by-case approach on ADM systems within public education sector. *Indiana Journal of Global Legal Studies*, 30(1), 105–171.

Duan, X., Pei, B., Ambrose, G. A., et al., (2024). Towards transparent and trustworthy prediction of student learning achievement by including instructors as co-designers: A case study. *Education and Information Technologies*, 29, 3075–3096.

Edwards, L., & Veale, M., (2017). Slave to the algorithm? Why a 'right to an explanation' is probably not the remedy you are looking for. *Duke Law & Technology Review*, 16, 18–84.

Fedele, A., Punzi, C., & Tramacere, S., (2024). The ALTAI checklist as a tool to assess ethical and legal implications for a trustworthy AI development in education. *Computer Law & Security Review*, 53, 105986.

Floridi, L., Cowls, J., Beltrametti, M., Chatila, R., Chazerand, P., Dignum, V., ... & Vayena, E., (2018). AI4People—An ethical framework for a good AI society: Opportunities, risks, principles, and recommendations. *Minds and Machines*, 28(4), 689–707.

Guizani, S., Mazhar, T., Shahzad, T., et al., (2025). A systematic literature review to implement large language model in higher education: Issues and solutions. *Discover Education*, 4, 35.

Guleria, P., & Sood, M., (2023). Explainable AI and machine learning: Performance evaluation and explainability of classifiers on educational data mining inspired career counseling. *Education and Information Technologies*, 28, 1081–1116.

Gunasekara, S., & Saarela, M., (2025). Explainable AI in education: Techniques and qualitative assessment. *Applied Sciences*, 15(3), 1239.

Holmes, W., Bialik, M., & Fadel, C., (2019). *Artificial Intelligence in Education: Promises and Implications for Teaching and Learning*. Center for Curriculum Redesign. Boston, MA.

Holmes, W., Porayska-Pomsta, K., Holstein, K., Sutherland, E., Baker, T., & Koedinger, K., (2022). Ethics of AI in education: Towards a community-wide framework. *British Journal of Educational Technology*, 53(2), 291–310.

Hooshyar, D., & Yang, Y., (2024). Problems with SHAP and LIME in interpretable AI for education: A comparative study of post-hoc explanations and neural-symbolic rule extraction. *IEEE Access*, 12, 137472–137490.

Hwang, G. J., Xie, H., Wah, B. W., & Gašević, D., (2020). Vision, challenges, roles and research issues of artificial intelligence in education. *Computers & Education: Artificial Intelligence*, 1, 100001.

Khan, U. A., et al., (2024). A system to ensure information trustworthiness in artificial intelligence enhanced higher education. *Journal of Information Technology Education: Research*, 23, 013.

Kieser, F., Tschisgale, P., Rauh, S., Bai, X., Maus, H., Petersen, S., Stede, M., Neumann, K., & Wulff, P., (2024). David vs. Goliath: Comparing conventional machine learning and a large language model for assessing students' concept use in a physics problem. *Frontiers in Artificial Intelligence*, 7.

Lu, Y., Wang, D., Chen, P., & Zhang, Z., (2024). Design and evaluation of trustworthy knowledge tracing model for intelligent tutoring system. *IEEE Transactions on Learning Technologies*, 17, 1661–1676.

Luckin, R., Holmes, W., Griffiths, M., & Forcier, L. B., (2016). Intelligence Unleashed: An Argument for AI in Education. Pearson, London, UK.

Manganello F, Rasca E, Villa A, Maddalena A and Boccuzzi G, (2025). Machine learning for post-diploma educational and career guidance: a scoping review in AI-driven decision support systems. *Front. Educ.* 10:1578979.

Mustofa, S., Emon, Y. R., Mamun, S. B., Akhy, S. A., & Ahad, M. T., (2025). A novel AI-driven model for student dropout risk analysis with explainable AI insights. *Computers and Education: Artificial Intelligence*, 8, 100352.

Nagy, M., & Molontay, R., (2024). Interpretable dropout prediction: Towards XAI-based personalized intervention. *International Journal of Artificial Intelligence in Education*, 34, 274–300.

Nannini, L., Marchiori Manerba, M., & Beretta, I., (2024). Mapping the landscape of ethical considerations in explainable AI research. *Ethics and Information Technology*, 26, 44.

Selwyn, N., (2021). AI and education: The importance of asking 'Why?'. *Learning, Media and Technology*, 46(2), 230–233.

Silva Filho, R. L. C., Brito, K. S., & Adeodato, P. J. L., (2022). A data mining framework for reporting trends in the predictive contribution of factors related to educational achievement. [*Journal name missing*].

Suresh, H., & Guttag, J. V., (2021). A framework for understanding unintended consequences of machine learning. *Communications of the ACM*, 64(1), 62–71.

Susnjak, T., (2024). Beyond predictive learning analytics modelling and onto explainable artificial intelligence with prescriptive analytics and ChatGPT. *International Journal of Artificial Intelligence in Education*, 34, 452–482.

Tsai, Y. S., Mello, G. D., & Williamson, B., (2022). Algorithmic accountability in education: Global policy and regulatory responses. *Learning, Media and Technology*, 47(2), 198–214.

UNESCO, (2021). Recommendation on the Ethics of Artificial Intelligence.

Wachter, S., Mittelstadt, B., & Floridi, L., (2017). Why a right to explanation of automated decision-making does not exist in the General Data Protection Regulation. *International Data Privacy Law*, 7(2), 76–99.

Wang, Y., Spector, J. M., & Yang, Y., (2023). Transparency and explainability in AI-supported learning environments: A review of the literature. *Educational Technology Research and Development*, 71(1), 1–24.

Williamson, B., & Eynon, R., (2020). Historical threads, missing links, and future directions in AI in education. *Learning, Media and Technology*, 45(3), 223–235.

UNDERSTANDING LEARNERS' PERCEPTIONS ON A GENERATIVE AI-ENABLE SYSTEM IN PERSONALIZED CONTEXTUAL VOCABULARY LEARNING

Mohammad Nehal Hasnine[1], Mirai Yamada[1], Hiroshi Ueda[1] and Kensuke Akao[2]
[1]Research Center for Computing and Multimedia Studies, Hosei University, Tokyo 184-8584, Japan
[2]Institute of Educational Science & Institute of Software Engineering, University of Stuttgart, 70174 Stuttgart, Germany
Institute of Informatics, Ludwigsburg University of Education, 71634 Ludwigsburg, Germany

ABSTRACT

Enhancing learners' vocabulary knowledge in foreign language learning is considered to have many positive effects. Artificial intelligence, particularly Large Language Models (LLMs), has opened up doors to generate content tuned to fulfill one's expectations. However, it is unclear in what ways and to what extent they could be used in personalized, contextual vocabulary learning. To explore this novel dimension, we developed a Generative AI (GenAI)-enabled system called Wordhyve-Glc for foreign language learners who are improving foreign vocabulary skills using learning contexts. Wordhyve-Glc provides personalized learning context support, including recommending contextual clues, correcting errors in learners-constructed sentences, and generating contextual images for the words to be learned. To understand foreign language learners' perceptions of how Wordhyve-Glc provides personalized learning context support, we conducted a study with native German speakers (N=9) learning Spanish, French, and Latin as a foreign language. They learned a total of 87 vocabularies using the Wordhyve-Glc system. A post-study survey was used to collect their perceptions of GenAI-enabled learning experiences. We found a positive outlook of learners towards GenAI-generated learning context support and favorable feedback on our app. We broadly discuss how these results support the significance of GenAI for personalized learning of foreign vocabulary.

KEYWORDS

Generative Artificial Intelligence (GenAI), Vocabulary, AI-Assisted Vocabulary Learning, Contextual Vocabulary Learning, Image Generation, Sentence Construction and Error Correction

1. INTRODUCTION

Understanding when, where, why, and how a foreign language learner learns a new vocabulary in context is essential. It can make learning more meaningful, engaging, and relevant. Also, this could lead to a deeper understanding of a new language and better retention. Contextual vocabulary learning or instruction is an effective strategy used in the classroom by the teacher and in informal learning by language learners. Using this strategy, a teacher can help a language learner acquire new words and infer meaning from both in-class learning materials and real-world encounters with unfamiliar words. For language learners, knowledge of foreign words will be of no use if they do not know how to use them in what context. Therefore, contextual vocabulary learning, where learners learn through contextual clues, establish personal connections to words and draw mental images, can facilitate and improve vocabulary acquisition. This strategy can effectively improve memory retention, better comprehension, and increased engagement. It has shown evidence of boosting learners' situational interest, value perception, and task effort when learning (Høgheim & Reber, 2015; Reber et al., 2018). Since each learner has different prior knowledge and interests, supporting personalized contextual vocabulary learning is exceedingly challenging. This is because they may quickly lose interest in learning and may have trouble establishing a connection with the learning contents (Krapp, 2002). Also, it is understandable that manually created learning materials could be immensely effortful and time-consuming.

The potential of artificial intelligence (AI) has opened the door to revolutionizing education and bringing new values to educational experiences. In particular, the introduction of large language models (LLMs) such as Llama, Gemini, Deepseek, and ChatGPT. LLMs use advanced and complex algorithms to generate human-like content, which opens up many possibilities. For example, well-known e-learning platforms such as Duolingo, Memrise, and Rosetta Stones are exploring various ways to adopt generative AI to tailor content to meet a learner's expectations. Generative AI has already been used for context personalization, where learning contents and tasks were customized to the individual learner.

Despite some efforts, research to understand learners' perceptions of AI-assisted systems that could personalize learning context is relatively few, particularly in the context of foreign vocabulary learning. Therefore, our research aimed to develop an AI-enabled learning app to support foreign vocabulary acquisition in context. Also, we aim to evaluate it with a user experience experiment to understand foreign language learners' perceptions of various AI-enabled app components. Moreover, we aimed to understand what the learners think of the opportunities and risks AI brings to the community. In this study, we address the following research questions:

- RQ1: Do AI-suggested contextual clues meet learners' expectations and help describe one's learning contexts?
- RQ2: Can AI-corrected errors be used to improve sentence construction?
- RQ3: Do AI-generated images meet learners' expectations and interests?
- RQ4: What do learners think of learning with AI interaction?

2. RELATED WORKS

Several chatbots and intelligent pedagogical agents have been developed in recent years to explore the concept of generative-AI-assisted personalized learning. @llegra (Bendel & Jabou, 2024) is a GPT-assisted learning app of Rhaeto-Romanic idiom, Vallader. This app can process the text and voice of the learners by which they can communicate and learn Rhaeto-Romanic idiom. Digital Learning Facilitator, Digital Learning Coach and Digital Learning Mentor are three GPT-enabled learning environments to enhance self-regulated learning (Hauske & Bendel, 2024). Digital Learning Facilitator is capable of replying to learners' questions and correcting misunderstandings. At the same time, the Digital Learning Coach, and Digital Learning Mentor can enhance learners' metacognition skills and acknowledge when a learner accomplishes some learning goals, respectively. To develop an AI-assistive tool for personalized English language instruction, (Lawrance et al., 2024) introduced an innovative tool where they primarily leveraged Natural Language Processing (NLP) techniques. The AI components of this tool can produce real-time feedback. It can guide learners to better use pronunciation, grammar, and vocabulary in context. AI has also been used to develop a tool for language learning assessment and feedback (Yesilyurt, 2023). This tool analyzes automatic scoring, speech detection, multidimensional analytics, and adaptive testing. This tool was able to enhance learners' learning experience and improve pedagogy.

In this line, we developed Wordhyve-Glc (Author et al., 2024)- a GPT-enabled language learning app that assists foreign language learners in learning foreign vocabulary using contexts. It embedded an innovative approach (hereafter, learning process) in it by which a learner can interact and learn with AI. In the learning process, the AI of the app recommends multiple contextual clues for a word to be memorized; it provides sentence construction support, including error detection and correction, and image generation for representing a learning context. By this, a learner can memorize new vocabulary with the help of multimedia annotations (text and image).

3. THEORETICAL FOUNDATIONS

Since the 1970s, Paivio's dual coding theory has been a prominent learning theory used in vocabulary learning (Clark & Paivio, 1991; Paivio, 2013). This theory assumes that our sensory system has two cognitive subsystems. One subsystem is specialized for the representation and processing of nonverbal objects or events (e.g., imagery), while the other subsystem is specialized for dealing with language. This theory sees learning occurring through verbal and non-verbal channels, and Paivio strongly asserts that two are better than one regarding memory aids (Yanasugondha, 2017). Paivio also introduced the phenomenon of the picture-superiority effect (Paivio & Csapo, 1973). He believed that the picture-superiority effect happens because any visual is stored in two ways in our memory: an image and a word or phrase that describes the image. On the contrary, words are processed in our memory in only one way- the word itself (Paul, 2024). When humans perceive an image, our mind generates a label (a word or phrase that describes it). Both the image and label are committed to memory.

Since the early 2000s, multimedia learning theory has been used to create vocabulary learning material. According to the multimedia learning theory by Mayer (Mayer, 2002; Mayer & Moreno, 1998), humans have two separate channels for processing information. One auditory and the other is visual (textual code or picture). This concept is well-known as the dual-channel assumption. The multimedia learning theory suggests that learners should actively engage in the learning process instead of being passive receivers of information. This theory has been practiced in foreign vocabulary learning research for many years, particularly in the era of computer-assisted language learning. In vocabulary learning, this theory suggests that learners learn the best from a combination of words and pictures. Therefore, developers and instructional designers should use words (text or narration) and visuals (images, animations, or videos) to create learning materials. Combining words and visuals helps language learners process and integrate information more efficiently.

Interactive learning theory focuses on active student participation and engagement in learning, moving beyond the passive reception of information to promote real-world application and a deeper understanding of the context. To elaborate this theory and to explain how we learn, Sims R associated interactivity with 1) interactive perceptions, 2) deriving interactive constructs, 3) interactivity, and learners (Sims, 1999). He also argued that learning should not just be linked with one single strategy or intervention; instead, it is viewed as a complex interaction between contexts, conditions, environment, circumstances, and culture. Therefore, emphasis is given to learner-to-learner, learning-to-teacher, and learner-to-technology in interactive learning theory. In this regard, we noted the Operant Conditioning theory of learning – a type of learning that focuses mainly on responses or actions that are controlled by consequences (Mulyana, 2017). This refers to actions or behaviors by learners that are rewarded and are likely to be repeated. In contrast, actions or behaviors that are punished or criticized are less likely to happen. As of now, few studies have used operant conditioning to increase learners' motivation while learning new vocabulary (Mulyana, 2017). One more learning strategy we can associate with interactive learning is adaptive learning. Many e-learning systems use adaptive learning techniques to adjust and tailor individual learning experiences. Adaptive learning provides real-time feedback and can assess learners' adaptive factors such as performance, behavior, and preferences (Polly et al., 2014).

The foundation of our study is to reflect the full range of three learning theories: dual coding theory, multimedia learning theory, and interactive learning theory. This work also coins components of operant conditioning and adaptive learning as the theoretical grounds. As a result, a learner can learn with the help of multiple multimedia annotations (i.e., a combination of text and image) by interacting with AI in our learning app. In the future, we aim to include audio (/pronunciation) as an annotation.

4. PERSONALIZED LEARNING WITH WORDHYVE-GLC

OpenAI's GPT Builder is used to develop Wordhyve-Glc. OpenAI was chosen for two reasons. First, previous studies (Bendel & Jabou, 2024; Hauske & Bendel, 2024) have used it for its flexibility and advanced features. Second, it offers- LLM models with API, chatbot, contextual clues recommendation, error detection and correction, and image generation functions. The GPT we used to build the app is configurable by prompt engineering. Prompt engineering is a process of crafting or structuring an instruction to produce the best result to fit one's expectations. Wordhyve-Glc supports learning foreign languages, including Japanese, Spanish, French, English, and Latin. Learners who wish to learn other languages can do so by prompt engineering. Prompt engineering is a method of guiding AI models to generate desired responses. In addition, we implemented one function in the app for the teachers by which a teacher can intervene in learners' learning styles. For example, if a teacher expects a learner to learn words related to the topic 'science' but the learner is learning words unrelated to 'science' (i.e., off-topic). A teacher can use this function to instruct the app to recommend more words related to the topic 'science.' We applied a Few-shot prompting approach to accomplish this task (*Few-Shot Prompting – Nextra*, 2025).

In designing the learning process, we assumed two common scenarios of informal learning: either the learner knows what the words are to be learned (e.g., taking the IELTS or TOEIC test) or the learner does not know the words to learn (e.g., traveling to a new place or learning a new culture). These scenarios are often observed in mobile or ubiquitous learning. To assist in both cases, we provide options to either input the word to be learned in Wordhyve-Glc or let the app decide the words to be learned based on a learner's profile. Hence, the learning process begins by inputting a word in the app, either by the learner or determined by the AI of the app. Once a word is entered into the app, the app recommends three contextual clues that could be used to describe a learning context. A learning context is typically one or two sentences written by the learner, leading to more meaningful learning and better retention. A learner is then expected to construct one or two sentences using those contextual clues suggested by the app. It is worth mentioning that sentence construction tasks as a technique to enhance vocabulary learning have been used by previous studies and found to be effective (Teng & Zhang, 2024). Then, the learner can use any other clues or words to write down a learning context without relying on the app's recommendations. Once the learner constructs a learning context in the form of one or two sentences, the app checks the mistakes made by the learner in writing it and eventually corrects them. The mistakes include typographical errors, grammatical errors, incorrect spelling and punctuation, and forgetting to use an apostrophe to form a contraction. Based on the corrections made by the app, the learner could reconstruct the learning context. A learner can skip doing it (i.e., rewriting the learning context) and go to the next phase, where the Wordhyve-Glc generates an image representing the word and its learning context. If the image generated by the app is unsatisfactory, the learner could instruct the app to generate another image. In getting the most appropriate image, the learner could have a natural conversation with the app. For example, 'My ethnicity is Asian (brown skin). So, can you generate an image closer to my race, or am I going to Spain in a week. So, can you generate images of taking Spanish trains'. Figure 1 shows a learning log of Wordhyve-Glc. Figure 1 shows the learning process of Wordhyve-Glc.

5. EVALUATION

To explore the idea of AI-enabled personalized contextual vocabulary learning, we conducted an onsite experiment with the participation of nine students. With this experiment, we aimed to understand foreign language learners' perceptions of the developed system, Wordhyve-Glc, including its contextual clue recommendation mechanism, error detection in learner-constructed sentences and correcting them, and image generation to represent the learning context. This research was approved by the ethical committee of the lead author's university. Moreover, the experiment was conducted by the EU's laws on personal data protection. Participation in this experiment was voluntary basis; hence, the participants received no gratuity. The experiment consisted of three steps: prescreening, study, and survey.

Nine participants took part in this experiment. The participants were native German speakers enrolled in the Teacher Training Courses at two German universities. They are expected to become teachers in the future. They are learning foreign languages for various purposes. Out of the 9 participants, 4 are learners of French as a Foreign Language (FFL), three are Latin as a Foreign Language (LFL), and two are Spanish as a Foreign Language (SFL). They are aged between 22 to 25 (M=23.11, SD=0.87). Of the 9 participants, seven were male, and two were female. They were asked to take part in a prescreening step to ensure their eligibility in the experiment.

Each participant was asked to complete a prescreening survey. This was to ensure that at least ten words in our word set were known to the participant and could, therefore, attempt to create a sentence (or describe a learning context) as part of our study. For prescreening, we prepared a set of 30 basic vocabulary words (with the combination of 10 nouns, 10 verbs, and 10 adverbs) recommended by Wordhyve-Glc. The prompts to Wordhyve-Glc for getting those words were:

- Can you recommend 10 nouns of Latin language to those foreign language learners who need to learn basic Latin?
- Can you recommend 10 verbs of Latin language to those foreign language learners who need to learn basic Latin?
- Can you recommend 10 adverbs of Latin language to those foreign language learners who need to learn basic Latin?

We used similar prompting for Spanish and French languages. We asked the participants to assign a ranking to each word on a 5-point adapted version of the Vocabulary Knowledge Scale (VKS) (1 = I have never seen this word. 2 = I have seen this word before but I don't know what it means. 3= I have seen this word before and I think it means ___. 4 = I know this word and it means _____ . 5 = I can use this word in a sentence.) (Brown, 2008). If a participant did not mark 10 or more words, the participant was not eligible for this experiment. The prescreening results revealed that all nine participants were familiar with at least 10 words out of the 30 words and could try to create a sentence (or describe a learning context).

Eligible participants from the prescreening step were invited to this step, where they learned vocabulary with the help of Wordhyve-Glc. After explaining the study and receiving their informed consent, participants were asked to study vocabulary using the app. Approximately 30 minutes were given to each participant to learn vocabulary with the contextual clues, context correction, and images generated by the Wordhyve-Glc. The instructions were:

i) study around 10 words (allowed to learn more or less; it's up to them)
ii) can use the words listed in the prescreening word set to learn from, and
iii) are allowed to learn any words they wish

During the study, the app provided three contextual clues for each word a participant wished to learn, pointed out and corrected the errors in participant-described sentences, and generated one (or more, upon the participant's request) image to represent the learning context. The nine participants created 87 vocabulary learning logs. Table 1 lists the words the participants learned during the study session. Following the study session, participants were asked to answer a set of survey questions related to the app's mechanism of personalized learning context support and their perception of the overall experience. The Wordhyve-Glc app logged participants' typed inputs during the study session. We used ChatGPT's shared links feature to record the conversation. However, we did not use any timestamps when a participant learned a word within 30 minutes of study time. We used the feedback function of Moodle to collect the data.

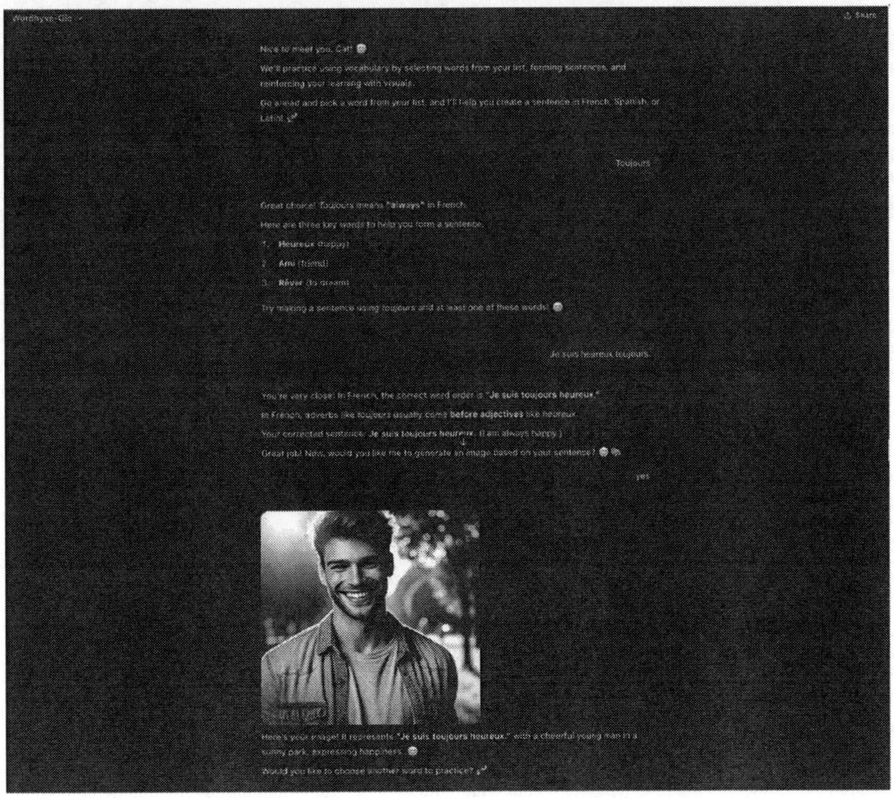

Figure 1. Learning with Wordhyve-Glc

Table 1. Vocabulary list

P. Id	Language	Words learned	No. of words
1	Spanish	Coche, Querer, Flor, Aquí, libro, escuela, perro, A veces, mal	9
2	Latin	Aqua, Fortuna, Stella, Mare, Luna, Flumen, Arbor, Pes, Urbs, Puer	10
3	Latin	Domus, Liber, Schola, Videō, Cibus, Hīc, Aqua, servus, anus, Sōl, Dō, Quārē, Tum, Legō, Legō	15
4	French	Là, parfois, prendre, retourner, temps, Voiture, attendre, apprendre, gateau, chateau	10
5	French	Avoir, maison, ici, aller, Toujours, Fleur, Aimer, ville, voiture, Être	10
6	French	Maison, chien, toujours, être, avoir	5
7	Spanish	Amor, Escuela, Siempre, Gato,Muy, Coche, Hacer, Vivir	8
8	French	Toujours, voiture, Livre, venir, pouvoir, ville, parfois, amour, mal, fleur	10
9	Latin	Videō, Aqua, Amō, Legō, Familia, Cibus, Male, Tum, Quārē, Bene	10

The post-study survey questionaries consisted of 10 questions. Those questions were about overall user experience (i.e., feedback on their experience of learning using our app), contextual clue recommendation (i.e., feedback on the three contextual clues recommended by the app to help them describe their personalized learning context), error detection and correction mechanism we implemented in the app, contextual image generation function of the app, and the overall app rating.

We used a 7-point Likert scale to understand their perceptions. We set two questions to know their overall user experience (i.e., their intrinsic motivation inventory towards learning apps). The 7-point Likert scale was: 1 = Very Playfully, 2 = Playfully, 3 = Somewhat Playfully, 4 = Normally, 5 = Somewhat Seriously, 6 = Seriously, and 7 = Very Seriously. The remaining questions (8 questions in total) were evaluated on 1 = Strongly Disagree, 2 = Disagree, 3 = Somewhat Disagree, 4 = Disagree, 5 = Somewhat Agree, 6 = Agree, and 7 = Strongly Agree. The survey also consisted of three open-ended questions to understand the opportunities and risks of AI-generated learning contexts. Those open-ended questions were: i) what do you think about this learning experience with generative-AI supported learning system, ii) is there anything you would like to change or improve about the vocabulary learning app, and iii) any final remarks or comments?

Participants rated two questions on a 7-point Likert scale (1= Very Playfully, 7 = Very Seriously), their overall user experience and attitude they adopted while using the app. The two questions were: 1) Generally I take any learning applications ___, and 2) What attitude did you adopt when using our app? With these two questions, we aimed to understand their intrinsic motivation towards any learning app with an emphasis on what they think about our app, Wordhyve-Glc. Participants' feedback on the first question suggested that their attitude towards any language learning app is normal (Median = 4, Normally). However, the answers to the second question suggested their attitude towards our app was Playfully (Median = 2). Participants were asked to rate 8 questions on a 7-points Likert scale (1 = Strongly Disagree, 7 = Strongly Agree) to understand their perceptions on our app. Table 2 shows the results.

Table 2. Results

Features	No	Question	Avg.	Mdn.	SD
Contextual clue recommendation	Q1	The words are relevant and aligned to the word that I wanted to learn	5.5	6.0	1.25
	Q2	The words were helpful to construct a sentence/describe my learning context	5.4	6.0	1.42
Error detection and correction	Q3	I think most of the learners would learn about the common errors (such as grammatical errors or spelling errors) they are making ir constructing a sentence using this app	4.6	5.0	1.56
	Q4	I would imagine that most of the learners would learn where they are making mistakes in constructing a sentence and how to corre them using this app very quickly	4.8	5.0	1.36
Contextual image generation	Q5	The image generation function was useful to remember the word and the learning context. This image will help me in rememberin the word after some time	4.2	4.0	1.54
	Q6	The AI of Wordhyve-Glc did a fantastic job in generating the image considering the complexity of the sentences	5.3	6.0	1.33
Overall app rating	Q7	I would like to use the app again to learn new vocabulary words	4.5	5.0	1.25
	Q8	I would recommend this app to a friend to learn new vocabulary words	3.8	5.0	1.36

The result indicates that they found the contextual clue recommendation feature the most important (M=5.45, SD=0.05) compared to the error detection and correction (M=4.7, SD=0.1) and the contextual image generation (M=4.75, SD=0.55) functions. Regarding the overall app rating, their perception of our app is favorable (M=4.15, SD=0.35). The lowest evaluation was noticed on Q8. This indicates that the participants disagreed to recommend this app to their friends. Participants were asked to write freely about their experiences with the Wordhyve-Glc app in the open-ended questions. We received a total of 12 comments. Mixed opinions were noticed. Positive comments for the question 'What do you think about this learning experience with Generative AI-supported learning system' include: 1) "I think that the app could be more helpful if it gave more grammatical context, like asking whether one could use extra help or practice when conjugating verbs or whether one might need help with other grammatical phenomena (for example articles and prepositions). The image creation did not particularly help me, but it might be a nice feature if someone does not know the words so far", 2) "It was fun and worked well", and 3) "It is a fun tool to train forming sentences because of the instant feedback". In contrary, negative comments include: "Not convinced this is helpful in its current state". We received several suggestions in the open-ended question 'Is there anything you would want to change or improve about the vocabulary learning app?' including: 1) "I would recommend to build an actual app that uses the API and not rely on custom GPTs", 2) "I was struggling with the grammar when building the sentences because it didn't show to correct form of the verb, or it would be useful to show more words so you have more choices to build a sentence. Because I could think of many sentences, but I could not write the sentences because i didn't knew all word", and 3) "The AI should explain why certain grammatical conjugation is needed in this context, otherwise there is no learn effect in the sentence arranging".

6. CONCLUSION

Contextual vocabulary learning is an effective strategy for learning new vocabulary quickly and utilizing it in various real-world contexts. However, contextualizing vocabulary, particularly context personalization, is a complex task. Because it involves adapting learning materials to be grounded in users' interests, preferences, and existing body of knowledge (Leong et al., 2024). To address this, this paper introduced Wordhyve-Glc, a language-learning app for enhancing foreign vocabulary in context. We used GPT-4o to build the app. The app facilitates learners with contextual clues, sentence construction, and image generation. We also conducted a user experience experiment in which nine native German students participated to understand their perceptions of the app. A recent study (Leong et al., 2024) by MIT Media Lab also introduced three variants of an app where they used gpt-3.5-turbo to develop it. They also conducted an extensive user study (N=272) to understand learners' motivational factors on generative AI-supported context personalization, concluding that it positively affected learning motivation. The findings of our study firmly suggest that interaction with AI can help construct a sentence/describing one's learning context; AI can also provide contextual clues that are relevant and aligned to a word that a learner wishes to learn. Our findings also suggest that AI did a fantastic job generating images, considering the complexity of the sentences. However, the participants disagreed with introducing the app in its current state to their friends and acquaintances. This result suggests some drawbacks and limitations the app has, which we aim to overcome in the future.

On the one hand, leveraging generative AI for personalizing learning contexts opens doors to exciting possibilities but raises many concerns. For instance, it has already raised concerns about cheating and an increasing amount of unemployment in educational institutions (Qadir, 2023). In vocabulary learning with AI, a dynamic partnership between a learner and the system is important. Therefore, user-centered and ethically aligned AI tools that can fulfill the real needs of learners and educators may lead to significant vocabulary learning over the long term.

ACKNOWLEDGEMENT

This work was supported by JSPS Grant-in-Aid for Early-Career Scientists Grant Number 24K16766.

REFERENCES

Bendel, O., & Jabou, D. (2024). @llegra: A chatbot for Vallader. *International Journal of Information Technology, 16*(4), 2035–2045. https://doi.org/10.1007/s41870-024-01779-0

Brown, D. (2008). Using a modified version of the Vocabulary Knowledge Scale to aid vocabulary development. *The Language Teacher, 32*(12), 15–16.

Clark, J. M., & Paivio, A. (1991). Dual coding theory and education. *Educational Psychology Review, 3*(3), 149–210. https://doi.org/10.1007/BF01320076

Few-Shot Prompting – Nextra. (2025, January 7). https://www.promptingguide.ai/techniques/fewshot

Hasnine, M. N., Yamada, M., & Akçapınar, G. (2024). How Can Generative AI be Adapted for Learning Context Support in Contextual Vocabulary Acquisition. *Proceedings of the 21st International Conference on Cognition and Exploratory Learning in the Digital Age (CELDA)*, 371–374.

Hauske, S., & Bendel, O. (2024). How Can GenAI Foster Well-being in Self-regulated Learning? *Proceedings of the AAAI Symposium Series, 3*(1), Article 1. https://doi.org/10.1609/aaaiss.v3i1.31234

Høgheim, S., & Reber, R. (2015). Supporting interest of middle school students in mathematics through context personalization and example choice. *Contemporary Educational Psychology, 42*, 17–25. https://doi.org/10.1016/j.cedpsych.2015.03.006

Krapp, A. (2002). Structural and dynamic aspects of interest development: Theoretical considerations from an ontogenetic perspective. *Learning and Instruction, 12*(4), 383–409. https://doi.org/10.1016/S0959-4752(01)00011-1

Lawrance, J. C., Sambath, P., Shiny, C., Vazhangal, M., S, P., & Bala, B. K. (2024). Developing an AI-Assisted Multilingual Adaptive Learning System for Personalized English Language Teaching. *2024 10th International Conference on Advanced Computing and Communication Systems (ICACCS), 1*, 428–434. https://doi.org/10.1109/ICACCS60874.2024.10716887

Leong, J., Pataranutaporn, P., Danry, V., Perteneder, F., Mao, Y., & Maes, P. (2024). Putting Things into Context: Generative AI-Enabled Context Personalization for Vocabulary Learning Improves Learning Motivation. *Proceedings of the CHI Conference on Human Factors in Computing Systems*, 1–15. https://doi.org/10.1145/3613904.3642393

Mayer, R. E. (2002). Multimedia learning. In *Psychology of Learning and Motivation* (Vol. 41, pp. 85–139). Academic Press. https://doi.org/10.1016/S0079-7421(02)80005-6

Mayer, R. E., & Moreno, R. (1998). A cognitive theory of multimedia learning: Implications for design principles. *Journal of Educational Psychology, 91*(2), 358–368.

Mulyana, S. N. (2017). *Applying Classical and Operant Conditioning Theory to Improve the Vocabulary of the First Year Students at MTS Madani Alauddiin Paopao* [PhD Thesis, State Islamic University]. https://core.ac.uk/download/pdf/198223578.pdf

Paivio, A. (2013). *Imagery and Verbal Processes*. Psychology Press. https://doi.org/10.4324/9781315798868

Paivio, A., & Csapo, K. (1973). Picture superiority in free recall: Imagery or dual coding? *Cognitive Psychology, 5*(2), 176–206. https://doi.org/10.1016/0010-0285(73)90032-7

Paul, S. (2024, April 26). *The Picture-Superiority Effect: Harness the Power of Visuals*. The Picture-Superiority Effect: Harness the Power of Visuals. https://www.nngroup.com/articles/picture-superiority-effect/#

Polly, P., Marcus, N., Maguire, D., Belinson, Z., & Velan, G. M. (2014). Evaluation of an adaptive virtual laboratory environment using Western Blotting for diagnosis of disease. *BMC Medical Education, 14*(1), 222. https://doi.org/10.1186/1472-6920-14-222

Qadir, J. (2023). Engineering education in the era of ChatGPT: Promise and pitfalls of generative AI for education. *2023 IEEE Global Engineering Education Conference (EDUCON)*, 1–9.

Reber, R., Canning, E. A., & Harackiewicz, J. M. (2018). Personalized Education to Increase Interest. *Current Directions in Psychological Science, 27*(6), 449–454. https://doi.org/10.1177/0963721418793140

Sims, R. (1999). The interactive conundrum I: Interactive constructs and learning theory. *Responding to Diversity. Proceedings of 16th Annual Conference of ASCILITE. Brisbane: QUT Teaching and Learning Support Services. [Verified 6 Apr 2002] Http://Www. Ascilite. Org. Au/Conferences/Brisbane99/Papers/Sims. Pdf.*

Teng, M. F., & Zhang, D. (2024). Vocabulary learning in a foreign language: Multimedia input, sentence-writing task, and their combination. *Applied Linguistics Review, 15*(5), 2123–2148. https://doi.org/10.1515/applirev-2022-0160

Yanasugondha, V. (2017). A Study of English Vocabulary Learning Using the Dual Coding Theory. *LEARN Journal: Language Education and Acquisition Research Network, 10*(1), Article 1.

Yesilyurt, Y. E. (2023). AI-enabled assessment and feedback mechanisms for language learning: Transforming pedagogy and learner experience. In *Transforming the language teaching experience in the age of AI*, 25–43. IGI Global.

EXPLORING THE PERCEIVED EFFECTS AND PERCEPTIONS OF USING NO-CODE AI EDUCATIONAL TOOLS IN AI LITERACY COURSES FOR NON-IT MAJOR UNDERGRADUATE STUDENTS

Hyunjin Cha
Soonchunhyang University
Chungcheongnam-do, 31538, Republic of Korea

ABSTRACT

As a small-scale empirical study, various No-code AI education tools, such as Machine Learning for Kids, Teachable Machine, Orange3, and AI ethic-related platforms, were applied to enhance AI literacy in the AI literacy courses during 15 weeks for non-IT major undergraduate students. To examine students' perceived effects and perceptions of the No-code AI tools, a survey was conducted in week 15, and 50 students participated. The questionnaire consisted of interest in the subject, perceived effectiveness, perceived usefulness, satisfaction, continued use intention, and perceived usability, including the 5-likert AI literacy scale. In terms of AI literacy scales, students scored highest in AI ethics (M=4.55, SD=0.45), followed by AI knowledge (M=3.84, SD=0.63), AI utilization (M=3.74, SD=0.62), and AI evaluation (M=3.57, SD=0.72). In terms of perceived effects and perceptions on the No-code AI education tools, students showed high scores in the following order: interest in subject (M=4.48, SD=0.69), intention to continue using (M=4.48, SD=0.72), perceived usefulness (M=4.42, SD=0.74), and perceived usability (M=4.16, SD=0.83). The results of this study revealed that No-code AI education tools can have various effects on non-IT major undergraduate students. In particular, in terms of AI ethics, the use of Moral Machine, Detectfake, and Video Generator showed that it was effective for students in terms of AI ethics. In addition, it showed that it has advantages such as interest, usefulness, and usability in AI literacy education. This study proposes future research such as comparing the use of Python coding, or expanding the sample size

KEYWORDS

No-Code AI Education Tool, AI Literacy, Effects, Perceptions, Non-IT Major Undergraduate Students

1. INTRODUCTION

Various studies in all sectors including education are being conducted on utilizing generative AI (Artificial intelligence) (Polyportis & Pahos, 2024), and AI is being discussed as an essential technology to solve social problems in all industry sectors (Chee, Ahn, & Lee, 2024). In this respect, AI literacy has emerged as one of the most crucial competencies that has had a major on future job markets (Ng et al., 2021). Therefore, many countries and non-profit organizations are developing and performing educational initiatives and making various efforts to help K-12 students enhance AI literacy (Chee, Ahn, & Lee, 2024)(UNESCO, 2024).

However, it is true that there are few discussions on systematic AI literacy education curriculum or diverse educational programs to develop competencies for undergraduate students who will enter society within the next five years. They are the generations who did not have opportunities with AI learning into official school curricula (UNESCO, 2024). The AI literacy is an essential basic skill for non-IT major undergraduate students as they need to create added value and solve global problems through creative problem solving using AI in each field engaged (Southworth et al., 2023)(Hilao-Valencia & Ruth, 2023). Therefore, this study explores how to make Non-IT major undergraduate students engaged into AI learning to build basic AI skills.

As no-code AI education tools have recently become more diverse, opportunities for non-IT major undergraduate students to easily and efficiently understand and apply AI and perform problem-solving projects using their career-related data are expanding (Chee, Ahn, & Lee, 2024). Even if the underlying concepts and AI machine learning processes are not easy (Chee, Ahn, & Lee, 2024), the no-code tools help to make students with difficulties and fears in programming engaged into the AI education more. However, there is few research on the effectiveness of these tools, or studies of the students' experiences has primarily focused on K-12 or kids (Pope et al., 2024)(Kim, Jeon, & Kang, 2024). In particular, it is difficult to find students' experiences of applying no-code AI educational tools such as Orange3 into data science-based problem-solving projects for non-IT major students in higher education contexts.

From this perspective, this study aims to examine the perceived effects and perceptions of non-IT major students based on their experiences of conducting data analysis and problem-solving projects to foster AI literacy using various tools, especially Orange 3, including other AI tools for one semester. This study especially tries to analyze the experiences and perceived effects based on AI competency frameworks suggested by Ng et al.(Ng et al., 2021) and Chee et al.(Chee, Ahn, & Lee, 2024).

2. RESEARCH METHODS

2.1 AI Literacy Course Design with No-Code AI Education Tools

In this study, an AI literacy course for the Non-IT major undergraduate students was designed to utilize no-code AI educational tools with various activities corresponding to AI literacy framework as shown in the Table 1.

Table 1. An AI literacy course with no-code AI education tools

Week	AI literacy framework	Contents	AI Education Tools
1	AI Knowledge & utilization [2][3]	- Creating their own pledge poster for the 2024-2 term using tools that incorporate AI technology	Autodraw
2		- Activities with objects recognition & trends comparison with keywords from their major	Google lens & trends
3		- Activities with the use of Chatgpt for their self-study by translating & summarizing foreign articles from their major	ChatGPT
4	Machine Learning (AI utilization & evaluation) [2][3]	- Distinguishing between good and bad posture - Understanding Machine Learning Using Titanic	ML4Kids
5		- Data Classifying recycles: plastic vs. bottle	Teachable Machine
6-12	AI evaluation & Problem-solving Projects with data science [3]	- Solving problems using AI in **various career-related fields (education, business, health, economy, environment, cultures, etc.)**	Orange3, Kaggle, Open data
13	AI ethics[2][3]	- Discussing AI ethics issues in autonomous Driving	MoralMachine: moralmachine.net/
14		- Discussing social issues through image creation and video creation using deepfakes	detectfakes.media.mit.edu, Stable Video
15	AI literacy	- Projects and reflection	- Questionnaire

The framework was based on Ng et al. (2021), and Chee et al.(2024). Ng et al. (2021) suggested the competency framework based on Bloom's taxonomy in cognitive domains and AI ethics in ethical domains. However, it is not focused on higher education contexts. Therefore, the framework analyzed by Chee et al. (2024) was combined in terms of data literacy, problem-solving, career-related competencies suggested for higher education.

For 1-3 weeks, students learned conceptual knowledge about AI, how AI works, and how AI tools are used through activities with various no-code AI educational tools that incorporated AI technology such as Autodraw, Google lens, Google trends, Chatgpt. For 4-5 weeks, students learned about machine learning (ML), how supervised learning works, the importance of data, AI model creation, and model evaluation with ML4kids and teachable machine. For 6-12 weeks, students conducted problem-solving projects using Orange (No-code Machine learning tools) to create and evaluate AI models with supervised and unsupervised ML by using data from 6 different sectors such as education, business, health, economy, environments, cultures. In the final two weeks, AI ethics were discussed by experiencing how well AI can create images and videos and by distinguishing between real and fake images. Students also experienced and discussed various issues related to self-driving cars through the Moral Machine.

2.2 Participants

A small-scaled empirical study was conducted in an AI literacy course for non-IT major students at a university located in a medium-sized city in South Korea. Fifty students from two classes with a total of 73 taught by the same professor based on the same curriculum voluntarily agreed to participate in the survey. Two classes were offered as a liberal arts course in the 2024-2 term, and Korean students from various non-IT majors participated. 37 (74%) female and 13(26%) male students participated and they are sophomore (56%), junior (38%), and senior (6%).

2.3 Research Instruments

The questionnaire mainly consisted of two parts: The AI literacy scale and perceptions of AI educational tools. The AI literacy scale (Kim, 2023) developed for high school students in Korea based on Ng et al. (2021) was utilized, since most of students do not have an experience to learn AI, the scale developed in Korean is easy to read for students, and the framework suggested by Ng et al. (2021) was referred on the design of the curriculum in the AI literacy course. Regarding the items for the perceived effects and perceptions of the no-code AI educational tools, questions from studies of technology acceptance were revised according to this study (Venkatesh & Davis, 2000; Venkatesh et al., 2003). The questions consist of learning interest, engaged in learning, help to understands, satisfaction, perceived effects, intention to continue using, perceived usability, etc.

3. RESULTS

3.1 Analysis of Students' Products Using No-code AI Tools

Depending on the AI literacy framework, undergraduate students might need different no-code tools. In the understanding AI phase, they tried to understand how AI works with the use of tools applied various AI technique.

The no-code AI educational tools helped non-IT major undergraduate students to focus on the authentic learning activities and problems without spending time learning and practicing programming. It helped students repeatedly think about real-world problems related to their careers, and provide them with experience integrating diverse thinking to solve problems with AI.

Table 2. Results and implications of the AI literacy courses

AI literacy framework	Tools	Students' product samples	Implications
AI knowledge & AI utilization	Autodraw, Google lens, Google trends ChatGPT +Discussion(Padelet)	 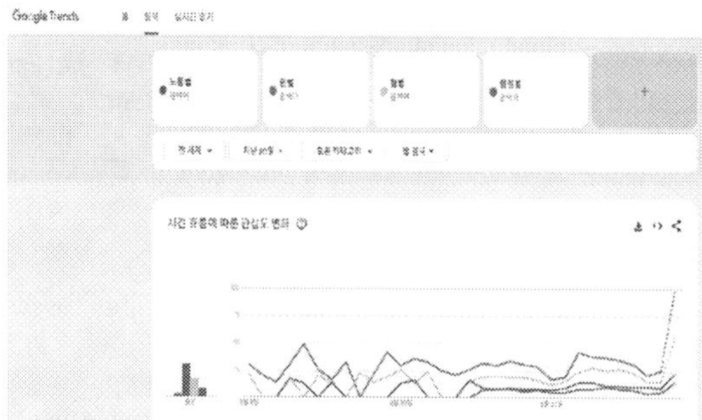	Students can understand how AI works, experiencing and applying their knowledge into authentic learning activities
AI utilization & AI evaluation	ML4Kids Teachable Machine + Quizzes (Kahoot)	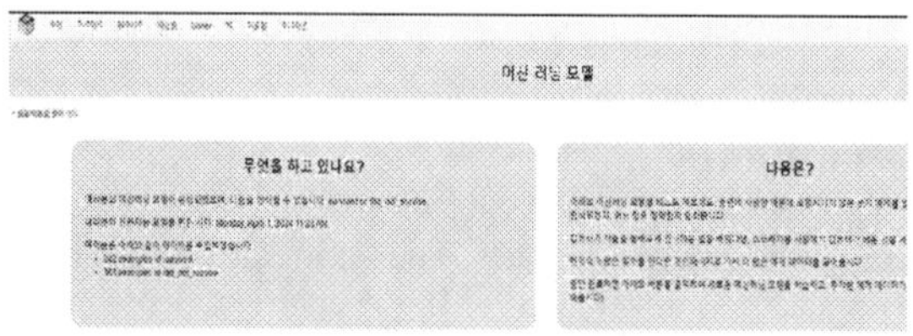  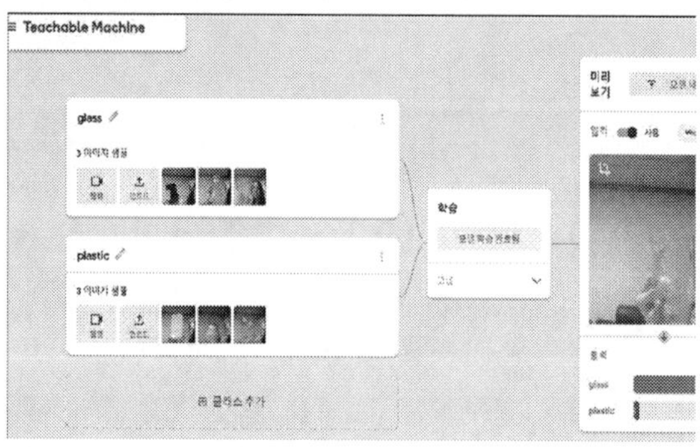	Students can utilize and evaluate AI models by incorporating AI knowledge and skills into learning contexts with authentic problems

AI literacy framework	Tools	Students' product samples	Implications
AI evaluation & problem-solving with AI	Kaggle, Orange3 +Report +Reflection	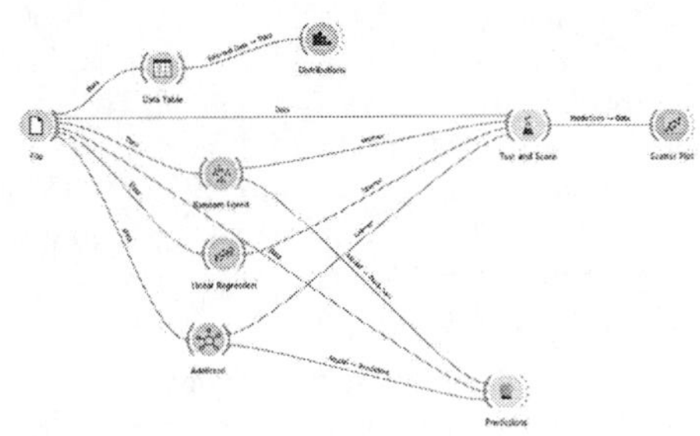 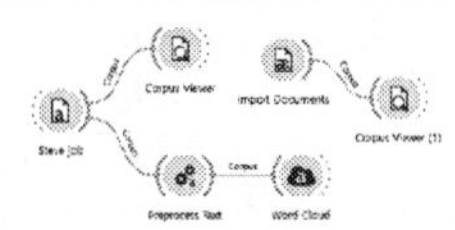 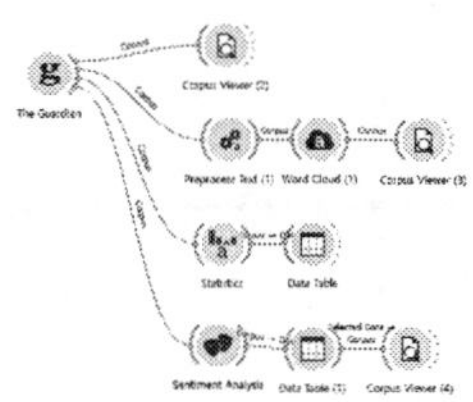	Students can evaluate various AI models by conducting problem-solving projects related to their career contexts in a repetitive way and can reflect how AI might influence their jobs.
AI ethics	MoralMachine Detectfake StableVideo +Discussions(Padlet)	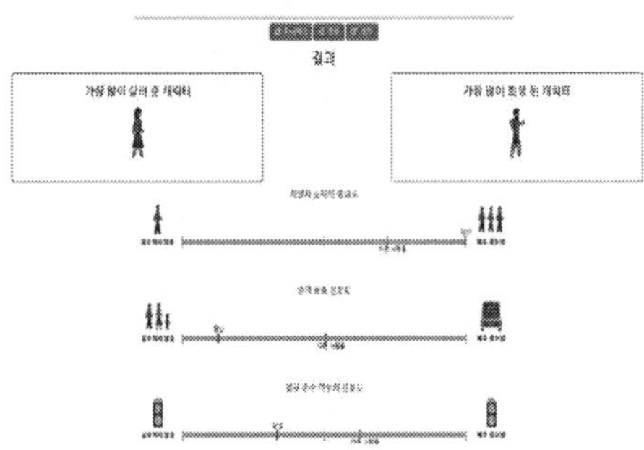 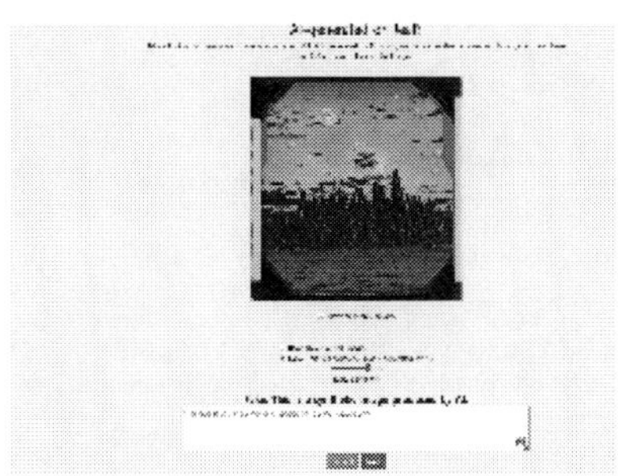	Students can feel and reflect how AI might have negative impacts on society and how complicated AI might be, experiencing and debating AI issues.

3.2 Effects of No-code AI Educational Tools

Regarding effect of AI literacy according to the tool use, although the no-code AI educational tools may not be a main direct factor to enhance student's AI literacy, the implications of AI tool use can be discussed through the results of AI literacy scale in that they were major learning methods to experience AI with various activities in each week. In this small-scale empirical study, in order to examine the effects of tools used in each learning objective and competency, we compared students' final mean scores in each component of AI literacy (Table 3). It found out that AI ethics had much higher mean score than others. On the other hand, AI evaluation where Orange 3 was mainly utilized had the lowest score. One of the reasons assumes that learning activities with Orange3 is to perform a project in each industry sector with critical thinking and a high-level of problem that require creating and evaluating models through machine learning techniques. As shown in Table 3, 'AI ethics (M=4.55, SD=0.45)' is the highest, followed by 'AI knowledge (M=3.84, SD=0.63)', 'AI utilization (M=3.74, SD=0.62)', and 'AI evaluation (M=3.57, SD=0.72)'.

Table 3. Results of AI literacy

AI literacy framework	Min	Max	M	SD
AI knowledge	2.07	5	3.84	0.63
AI utilization	2.33	5	3.74	0.62
AI evaluation	1.85	5	3.56	0.72
AI ethics	3	5	4.55	0.45

Regarding perceived effects and perceptions (Table 4.), all the mean scores are above 4.0, which means very high. The most positive effect students reported from the experience of the no-code AI educational tool use was 'interest (M=4.48, SD=0.69)', followed by 'perceived effectiveness (M=4.43, SD=0.75)', 'engagement (M=4.37, SD=0.75)', and 'help to understand (M=4.32, SD=0.72)'. These quantitative results are consistent with the qualitative comments. Analysis of open-ended questions about what was good about using tools in learning discusses that tools play a very positive role in stimulating students' interest, concentration, engagement, and participation. Some students also emphasized that it helped to increase their learning usefulness and understanding. In particular, it was discussed that the visualization of the results enhanced the sense of accomplishment and allowed for practical and authentic learning experiences.

Table 4. Results of Perceived Effects and Perceptions

Section	Perceived effects and Perceptions	Min	Max	M	SD
Perceived effects	Interest(4 items)	1.75	5.00	4.48	0.69
	Engaged(4 items)	1.50	5.00	4.37	0.79
	Help to understand(1 items)	3.00	5.00	4.32	0.71
	Perceived effectiveness(4 items)	2.00	5.00	4.43	0.75
Perceived perceptions	Satisfaction(1 items)	3.00	5.00	4.24	0.75
	Intention to continue using(4 items)	2.00	5.00	4.48	0.72
	Perceived usability	1.75	5.00	4.17	0.83

In terms of perceived perception, students expressed very highly the intention to continue using the tools (M=4.48, SD=0.72) and were satisfied with the experience of the no-code AI educational tool use (M=4.24, SD=0.75)'. In addition, they thought that the perceived usability of the tool was quite high (M=4.17, SD=0.83)'. In the open-ended questions, some students mentioned that they liked it because it was easy to use. However, in the open-ended questions about the difficulties or negative aspects of using the tool, some students expressed difficulties in using it. Interestingly, a difficulty expressed is their frustration at having to make decisions in such situations through Moral Machine. The comment can be translated that the students thought deeply and pondered about the moral machine. Other comments related to the challenges of the tool use were related to the Internet problems which means lag when connected at the same time.

4. CONCLUSION

This study is a small-scale empirical study to explore the effectiveness of AI no-code educational tools in AI literacy education for non-IT major undergraduate students. This study represents how an AI literacy course can be designed for the non-IT major undergraduate students with authentic learning contexts with AI no-code educational tools. In addition, it shows the perceived effects and perceptions of the tools with learning experiences by using the tools. It also reveals that the tools help students to focus on solving real-world problems by engaging into authentic learning activities as well as improving learning interest and helping to understanding AI. In the future, it suggests that further study should be conducted with a large number of students by comparing differences between experimental and control groups with use of Python coding.

ACKNOWLEDGEMENT

This work was supported by the Ministry of Education of the Republic of Korea and the National Research Foundation of Korea.

REFERENCES

Chee, H., Ahn, S., & Lee, J., (2024). A Competency Framework for AI Literacy: Variations by Different Learner Groups and an Implied Learning Pathway. *British Journal of Educational Technology*, Vol. 00, pp 1–37. https://doi.org/10.1111/bjet.13556.

Hilao-Valencia, P.M., et al., (2023). Appreciative Inquiry Approach and its Effects on English Oral Communication Skills of Teacher Education Students. *Mextesol Journal*.

Kim, H.G., (2023). *Development of AI Literacy Scale for High School Students*, Master thesis, Gacheon University.

Kim, S.Y., et al., (2024). Integrating Data Science and Machine Learning to Chemistry Education: Predicting Classification and Boiling Point of Compounds, *Journal of Chemical Education*, Vol. 101, No. 4, pp 1771-1776. doi:10.1021/acs.jchemed.3c01040

Ng, D.T.K., et al., (2021). AI Literacy: Definition, Teaching, Evaluation and Ethical Issues. *Proceedings of the Association for Information Science and Technology*, Vol. 58. pp 504-509. https://doi.org/10.1002/pra2.487.

Polyportis, A., & Pahos, N., (2024). Understanding students' adoption of the ChatGPT chatbot in higher education: the role of anthropomorphism, trust, design novelty and institutional policy. *Behaviour & Information Technology*, Vol. 44, No. 2, pp 315–336. https://doi.org/10.1080/0144929X.2024.2317364.

Pope, N., et al., (2024). A No-Code AI Education Tool for Learning AI in K-12 by Making Machine Learning-Driven Apps. *2024 IEEE International Conference on Advanced Learning Technologies (ICALT), Nicosia, North Cyprus, Cyprus*, pp. 105-109. doi: 10.1109/ICALT61570.2024.00037.

Southworth, J., et al., (2023). Developing a model for AI Across the curriculum: Transforming the higher education landscape via innovation in AI literacy. *Computers and Education: Artificial Intelligence*, Vol 4, https://doi.org/10.1016/j.caeai.2023.100127.

UNESCO, (2024). *AI competency framework for teachers*. UNESCO, Paris. https://doi.org/10.54675/ZJTE2084.

Venkatesh, V., & Davis, F. D., (2000). A Theoretical Extension of the Technology Acceptance Model: Four Longitudinal Field Studies. *Management Science*, Vol. 46, pp 186-204.

Venkatesh, V., et al., (2003). User acceptance of information technology: Toward a unified view. *MIS Quarterly*, Vol. 27, No. 3, pp 425–478.

DIGITAL TECHNOLOGIES AND HEALTHY LIFESTYLES: MOBILE APPLICATIONS FOR PHYSICAL ACTIVITY

Lilla Korenova[1], Marian Merica[1], Tomas Barot[2], Lucia Hlatka[1] and Katarina Zilkova[1]
[1]*Comenius University Bratislava, Faculty of Education*
Račianska 59, Bratislava, Slovakia
[2]*University of Ostrava, Faculty of Education*
Frani Sramka 3, Ostrava, Czech Republic

ABSTRACT

This study explores the relationship between a healthy lifestyle and the use of digital technologies, specifically mobile applications related to physical activity, among female university students. The research was conducted at the Faculty of Education, Comenius University in Bratislava, focusing on students enrolled in the Preschool and Elementary Education program. A questionnaire consisting of 15 questions was used to assess the attitudes of 71 female students towards physical activity, health, and digital tools that support an active lifestyle. The collected data were analysed using a quantitative approach, applying multidimensional regression analysis to evaluate interactions between selected variables. The findings indicate that while students recognize the benefits of physical exercise, regular participation in sports activities remains limited. However, a significant proportion of students are familiar with and actively use digital applications to track and improve their physical activity. The study highlights the potential of mobile applications and social media in influencing students' attitudes toward health and fitness. These insights can be instrumental in designing educational interventions and digital solutions to promote healthier lifestyles in academic environments.

KEYWORDS

Healthy Lifestyle, Digital Technologies, University Students, Physical Activity, Mobile Applications, Social Media Influence

1. INTRODUCTION

Adequate physical activity is an essential component of a healthy lifestyle for individuals, including university students. Both practical and theoretical findings from various authors emphasize the importance of maintaining regular physical activity in combination with a balanced diet as part of a healthy lifestyle. Increasingly, medical professionals recommend relaxation techniques, sports and recreational activities, lifestyle changes, and improvements in physical and mental fitness to mitigate stress. While not everyone has the potential to become an elite athlete, developing a lifelong relationship with sports and regular recreational physical activity—from university years to professional life and beyond—is far more valuable.

Several authors, including Horváth (2001, 2010), Belej et al. (2006), Šimonek (2012, 2018), Mačura et al. (2022), Petrovič (2022), and Gregor et al. (2023), have highlighted research findings from sports medicine and education experts, which confirm that physically active individuals exhibit higher vitality and overall well-being compared to those with insufficient fitness levels. Physically fit individuals are more capable of managing fatigue and sustaining higher energy levels, which positively impacts their professional performance and daily activities.

The benefits of sports and recreational activities extend beyond reducing fatigue or enhancing physical and mental fitness. The release of endorphins—often referred to as "happiness hormones"—positively influences mood, mental well-being, and overall sense of satisfaction. Additionally, increased blood circulation during exercise enhances oxygen supply to the brain, which is particularly beneficial for students and individuals engaged in intellectual work.

Physical activity is also closely linked to proper nutrition, hydration, and sufficient sleep. Clark (2000) highlights the significance of natural nutrition, emphasizing that fruits and vegetables, which contain 70–85% water, should be central to a healthy diet. Additionally, carbohydrate intake plays a crucial role in the rapid recovery of recreational athletes after physical exertion. On the other hand, Dahlke (2008) argues that nutrition occurs involuntarily, encompassing all physiological processes that food undergoes in the body until its complete transformation. In the absence of pathological conditions, proper nutrition manifests in overall well-being and health.

In the context of a healthy lifestyle and the use of digital applications for physical activity, it is essential to emphasize the role of educators, as their influence extends across all aspects of society, including students' leisure activities and engagement in sports. Various authors, such as Antala et al. (2018), Severini & Kostrub (2018), Duda et al. (2020), Antala (2021), Belešová (2021, 2022, 2023), Merica & Barnáková (2021), Rodin (2021), Severini, Kožuchová & Brezovská (2021), Benčuriková & Labudová (2022, 2023), Koreňová, Severini & Čavojský (2023), emphasize that teachers play a crucial role in shaping education, instruction, and student development. According to Gáborová & Porubčanová (2016), Bujdoš et al. (2022), Kostrub (2022), and Gunčaga & Belešová (2023), teachers influence classroom dynamics, stimulate student engagement—including attitudes toward sports and physical activity—and impact students' overall school experience, knowledge acquisition, and personal development. Other researchers, including Kampmiller & Vanderka et al. (2012), Gregor (2013, 2017), Sun (2013), Kaplánová (2018), Görner et al. (2021), Petrikán (2021), Rodin (2021), Skrypko et al. (2021), Duda, Stuła & Trzeciak (2023), Gregor et al. (2023), Harsa et al. (2023), and Ondrejková & Gregor (2023), advocate for the inclusion of physical activities as an integral part of a healthy lifestyle and stress the necessity of fostering physical activity from an early school age. Merica & Belešová (2022) particularly highlight the importance of recreational yet regular physical activity as an essential aspect of a healthy life for children, adolescents, and adults.

2. METHODOLOGY

2.1 Objective

The primary aim of this study is to assess and analyze responses from female university students at the Faculty of Education, Comenius University in Bratislava, regarding their attitudes towards physical activity, health, and the implementation of a healthy lifestyle. Additionally, the study seeks to determine how these students utilize mobile applications related to physical activity.

2.2 Research Tasks

To achieve this objective, the following research tasks were established: selecting a university in Bratislava as the study site, designing a 15-question survey, and subsequently identifying 8 key questions (Table 1) that best capture the knowledge and habits of the surveyed students concerning a healthy lifestyle and the use of mobile applications for physical activity. The collected data were subjected to statistical analysis for evaluation.

2.3 Hypotheses

Based on the research objective and tasks, 8 hypotheses were formulated, which are examined in the results section.

2.4 Study Population

The study sample consisted of 71 female university students enrolled in the Preschool and Elementary Education program at the Faculty of Education, Comenius University in Bratislava. Participants were from both the bachelor's and master's degree levels.

2.5 Data Collection Methods

A questionnaire comprising 15 questions was administered in person to the participants. The responses were analyzed using percentage-frequency analysis. For more detailed insights, 8 key questions were selected to assess the students' knowledge and practices related to a healthy lifestyle and the use of mobile applications for physical activity.

2.6 Data Processing and Analysis

The collected survey responses were analyzed using quantitative research methods, specifically statistical techniques based on mathematical induction—percentage-frequency analysis. To further explore the relationships among variables, multidimensional regression analysis was applied (Stockemer, 2019). Dependent and independent variables were determined to investigate potential statistically significant interactions. The results are presented in tables, where each of the 8 hypotheses is systematically evaluated. The testing of hypotheses using multidimensional regression analysis confirmed or refuted the presence of statistically significant dependencies at significance levels of 0.05 and 0.01.

3. RESULTS AND DISCUSSION

Clearly The study analyzed statistical hypotheses H1–H8 using multidimensional regression analysis. The primary objective was to determine interactions between health status (Q3) and various independent variables (Q1, Q5, Q11, Q12, Q13, Q14, Q15).

3.1 Hypotheses Proposal

The following items were considered in the frame of the quantitative research (Table 1):

Table 1. Description of Questions' Variables

Question Code	Description
Q3	health condition
Q1	age
Q5	smoking
Q11	approach to physical activities
Q12	assumption of physical activities after the university
Q13	knowledge of applications related to health style
Q14	utilization of applications related to health style
Q15	influence by social networks

The following hypotheses were proposed and tested in the software PAST Statistics (Hammer, 2001) as the multi-regression approach.

H_0: There are not statistically significant interaction between item Q3 and {Q1, Q5, Q11, Q12, Q13, Q14, Q15} on the significance level.

H_A: There are statistically significant interaction between item Q3 and {Q1, Q5, Q11, Q12, Q13, Q14, Q15} on the significance level.

H_0-Q1: There are no statistically significant interactions between items Q3 and Q1 on the significance level.

H_A-Q1: There are statistically significant interactions between items Q3 and Q1 on the significance level.

H_0-Q5: There are no statistically significant interactions between items Q3 and Q5 on the significance level.

H_A-Q5: There are statistically significant interactions between items Q3 and Q5 on the significance level.

H_0-Q11: There are no statistically significant interactions between items Q3 and Q11 on the significance level.

H_A-Q11: There are statistically significant interactions between items Q3 and Q11 on the significance level.

H_0-Q12: There are no statistically significant interactions between items Q3 and Q12 on the significance level.

H_A-Q12: There are statistically significant interactions between items Q3 and Q12 on the significance level.

H_0-Q13: There are no statistically significant interactions between items Q3 and Q13 on the significance level.

H_A-Q13: There are statistically significant interactions between items Q3 and Q13 on the significance level.

H_0-Q14: There are no statistically significant interactions between items Q3 and Q14 on the significance level.

H_A-Q14: There are statistically significant interactions between items Q3 and Q14 on the significance level.

H_0-Q15: There are no statistically significant interactions between items Q3 and Q15 on the significance level.

H_A-Q15: There are statistically significant interactions between items Q3 and Q15 on the significance level.

3.2 Results of Testing Hypotheses

In Table 3, the descriptive analysis can be seen in detail. Sample size N is equal to 71 with non-parametrical behaviour tested using the Shapiro-Wilk method (Table 2). Therefore, the descriptives are complemented by median M (explained by a confidence interval) and *IQR* (Inter Quartile Range).

The multidimensional regression analysis (Table 4) is bounded on the principle of p values with complementation of effect sizes supporting a case of rejecting a zero hypothesis in favor of an alternative hypothesis. Results of occurred p values are being corresponded as: $^{***}p < .001$, $^{**}p < .01$, $^{*}p < .05$. Across the independent variables, the nominal and cardinal types of variables have appeared. For each independent variable, the particular influence of each independent variable is denoted also using by p-value. The following effect sizes are considered as the Cohens' f: small > .02, moderate > .15 and large > .35; as the r: small < .3, medium from .3 to .5, large > .5 An adjusted multiple R^2 has been strengthened by consideration of the size of sample size inside the computations. Using regression coefficients β^+ or β^-, positive β^+ or negative β^- is being demonstrated the trend of the parameter influence in the frame of the regression analysis.

Table 2. Testing Normality of Considered Variables Using Shapiro-Wilk Test

	Q3	Q1	Q5	Q11	Q12	Q13	Q14	Q15
Shapiro-Wilk W	0.7541	0.891	0.5403	0.8361	0.8018	0.5817	0.6306	0.3668
p(normal)	1.44×10^{-9}	1.54×10^{-5}	1.44×10^{-13}	2.15×10^{-7}	2.26×10^{-8}	6.48×10^{-13}	4.43×10^{-12}	6.38×10^{-16}

Table 3. Descriptive Characteristics of Proposed Analyzed Variables

	Min	Max	Median	IQR	95 % CI	
Q3	2	4	3	1	3	3
Q1	21	25	23	2	23	23
Q5	2	5	5	1	5	5
Q11	1	4	3	1	3	4
Q12	0	4	3	1	2	3
Q13	0	1	1	1	1	1
Q14	0	1	0	1	-1	0
Q15	0	2	2	0	2	2

95 % CI: Lower and Upper Confidence Intervals for Obtained Medians

Table 4. Obtained Results of Multi-Regression Dependences Analyses

Related Hypothesis	
y Dependent Variable of Model	Q3
H: *p*-value for Multiregression	**2.45×10^{-5} *****
Effect Size Cohen's f^2	**.495 Large**
Adjusted Multiple R^2	.331
Independent Variable	
H-Q1:	.935
H-Q5:	.065
H-Q11:	**.001** (.420 Large) β^+**
H-Q12:	**.037* (.224 Moderate) β^+**
H-Q13:	.709
H-Q14:	.475
H-Q15:	.569

*** $p < .001$, ** $p < .01$, * $p < .05$; β^+ Positive Regression Coeffitient, β^- Negative Regression Coeffitient

3.3 Conclusion of Testing Hypotheses

According to Table 4, the statistically significant influence on the variable Q3(health condition) has variables Q11 (approach to physical activities) and Q12 (assumption of physical activities after the university).

The following hypotheses were accepted:

H_A: There are statistically significant interaction between item Q3 and {Q1, Q5, Q11, Q12, Q13, Q14, Q15} on the significance level.

H_0-Q1: There are no statistically significant interactions between items Q3 and Q1 on the significance level.

H_0-Q5: There are no statistically significant interactions between items Q3 and Q5 on the significance level.

H_A-Q11: There are statistically significant interactions between items Q3 and Q11 on the significance level.

H_A-Q12: There are statistically significant interactions between items Q3 and Q12 on the significance level.

H_0-Q13: There are no statistically significant interactions between items Q3 and Q13 on the significance level.

H_0-Q14: There are no statistically significant interactions between items Q3 and Q14 on the significance level.

H_0-Q15: There are no statistically significant interactions between items Q3 and Q15 on the significance level.

4. CONCLUSION

This study aimed to explore the health-related behaviors and implementation of a healthy lifestyle among female university students at the Faculty of Education, Comenius University in Bratislava, specifically those enrolled in the Preschool and Elementary Education program. A sample of 71 students was surveyed using a 15-question questionnaire to assess their attitudes toward physical activity and their use of mobile applications related to movement and fitness.

The study employed multidimensional regression analysis to test eight hypotheses, evaluating statistical significance at the 0.05 and 0.01 levels. The quantitative research methodology was chosen based on the study's analytical focus, with statistical techniques adopted from Stockemer (2019).

The findings indicate that while most students perceive their current health status positively, they report lower levels of physical fitness. The majority acknowledge the benefits of physical exercise for their health, yet they do not engage in regular sports activities. A noteworthy finding is that many students are familiar with and use mobile applications related to movement in their leisure time. Additionally, social media plays a positive role in influencing students' awareness of health and a healthy lifestyle.

An important observation from our study is that more than 24% of surveyed students do not engage in any physical activity beyond compulsory physical education classes at the university. However, a significant portion of students participate irregularly in skiing and hiking during the winter months. These findings highlight the need for targeted interventions to encourage consistent physical activity and increased utilization of digital health tools among university students.

ACKNOWLEDGEMENT

The research has been supported by projects:
- Erasmus+ 2023-1-CZ01-KA220-HED-000160664: "Accelerating STEAM-related Knowledge and Skills via 3D Modelling and 3D Printing"
- SGS06/PDF/2025 " Integration of e-learning, e-testing and GeoGebra software in mathematics education"
- KEGA 077UK-4/2025 Edu-Steam laboratory in primary education teaching

REFERENCES

Antala, B. et al., (2018). Pohybová aktivita žiaka v škole a jej ovplyvňovanie prostredníctvom nových technológií.Bratislava: SVS pre telesnú výchovu a šport.

Antala, B., (2021). Hodnotiaci proces v telesnej a športovej výchove, vzdelávacia oblasť Zdravie a pohyb. In: *Didaktika telesnej a športovej výchovy pre vzdelávaciu oblasť Zdravie a pohyb, vybrané kapitoly.* Bratislava: SVS pre TV a Š., pp. 310.

Belej, J. & Junger, J. et al., (2006). Motorické testy koordinačných schopností. 1. vyd. Prešov: Prešovská univerzita v Prešove, Fakulta športu. ISBN 80-8068-500-2.

Belešová, M., (2021). *Detské naivné teórie v kontexte edukácie.* Bratislava: UK.

Belešová, M., (2022). *Primárne vzdelávanie v teórii a v praxi.* Bratislava: Vydavateľstvo Univerzity Komenského. ISBN 978-80-223-5209-3.

Belešová, M., (2023). *Škola nie je výchovný tábor.* Bratislava: Grada.

Benčuriková, Ľ. & Labudová, J., (2022). Výučba plávania v primárnom vzdelávaní. In: *Telesná a športová výchova.*Bratislava: Nakladateľstvo Dr. Josef Raabe Slovensko, s.r.o. ISBN 978-80-8140-359-0.

Benčuriková, Ľ. & Labudová, J., (2023). Analýza úrovne plaveckých kompetencií žiakov mladšieho školského veku. In: *Žiak, pohyb, edukácia: vedecký zborník 2023.* Bratislava: Univerzita Komenského, pp. 29-36. ISBN 978-80-223-5702-9.

Bujdoš, G., Adamčák, Š. & Görner, K., (2022). Physical activities of grammar and vocational school students after returning to school following the fourth wave of Covid-19 in Eastern Slovakia. *Journal of Physical Education and Health,* vol. 11, no. 19, pp. 5-12.

Clarková, N., (2000). *Sportovní výživa.* Praha: Grada. ISBN 80-247-9047-5.

Dahlke, R., (2008). *Najlepšie tipy pre zdravie.* Bratislava: Eugenika. ISBN 978-80-8100-110-9.

Duda, H. et al., (2020). Reaction of the organisms of young football players to city smog in the sports training. *International Journal of Environmental Research and Public Health,* vol. 17, 5510.

Duda, H., Stuła, A. & Trzeciak, J., (2023). Kontrola dyspozycji wysiłkowych młodych piłkarzy nożnych z uwzględnieniem mechanizmów fizjologicznych – wymóg racjonalnego treningu. In: *Žiak, pohyb, edukácia: vedecký zborník 2023.*Bratislava: Univerzita Komenského, pp. 47-54. ISBN 978-80-223-5702-9.

Gregor, T., (2013). *Psychológia športu.* Bratislava: Mauro Slovakia. ISBN 978-80-968092-7-9.

Gregor, T., (2017). Osobnostné typy športovcov. *Telesná výchova & šport,* vol. 27, no. 4, pp. 39-42.

Gregor, T., Ondrejková, F. O., Zbončák, Ľ. & Harsa, P., (2023). Ideové východiská západných pohybových cvičení a športu. In: *Žiak, pohyb, edukácia: vedecký zborník 2023.* Bratislava: Univerzita Komenského, pp. 55-74. ISBN 978-80-223-5702-9.

Harsa, P. et al., (2023). Nové psychoaktívne látky v prostredí športujúcej mládeže. In: *Žiak, pohyb, edukácia: vedecký zborník 2023.* Bratislava: Univerzita Komenského, pp. 89-94. ISBN 978-80-223-5702-9.

Horváth, R. et al., (2010). *Diagnostika motorickej výkonnosti a genetických predpokladov pre šport.* Prešov: Vydavateľstvo Prešovskej univerzity. ISBN 978-80-555-0270-0.

Horváth, R., (2001). *Telesný vývin a pohybová výkonnosť rómskych detí mladšieho školského veku.* Prešov: Ústav národnostných štúdií a cudzích jazykov Prešovskej univerzity v Prešove. ISBN 80-89040-08-X.

Kaplánová, A., (2018). Individual differences of sensitivity of tennis players to injustice situations from the perspective of the five-factor model of personality big five theory. *Acta Gymnica,* vol. 48, no. 1, pp. 21-26. ISSN 2336-4920.

Kostrub, D., (2022). *Učiteľ – výskumník profesia založená na výskume: dizajny výskumu a premeny výučby.* Bratislava: UK v Bratislave. ISBN 978-80-223-5390-8.

Mačura, P., Habinák, A. & Krška, P., (2022). *Základné pohybové kompetencie detí mladšieho školského veku.* Ružomberok: Katolícka univerzita v Ružomberku, Pedagogická fakulta. ISBN 978-80-561-0959-5.

Šimonek, J., (2012). *Testy pohybových schopností.* Nitra: UKF. ISBN 978-80-970857-6-6.

Šimonek, J., (2018). Povinné testovanie pohybovej výkonnosti žiakov 1. ročníka ZŠ podľa nového zákona o športe. *Športový edukátor,* vol. XI, no. 2. ISSN 1337-7809.

Sun, H., (2013). Impact of exergames on physical activity and motivation in elementary school students: A follow-up study. *Journal of Sport and Health Science,* vol. 2, no. 3, pp. 138-145. https://doi.org/10.1016/j.jshs.2013.02.003

USE OF GENERATIVE AI BY PHD STUDENTS: IMPACT ON PRODUCTIVITY AND SKILLS DEVELOPMENT

Bruno F. Gonçalves
Transdisciplinary Research Center in Education and Development (CITeD), Polytechnic Institute of Bragança, Portugal

ABSTRACT

Higher education is constantly changing, especially with the growing use of generative artificial intelligence by doctoral students. This reality has led to significant changes in the way research activities are planned, organized, and executed, directly influencing productivity and skills development. This article aims to analyze the impact of the use of generative AI on productivity and the acquisition of skills by doctoral students at a university in Mozambique. The research, exploratory and descriptive, was based on the application of a questionnaire to all students on two postgraduate courses at a higher education institution. Sixty-two responses were validated, allowing for a robust analysis of the data. The results indicate that the use of AI has significantly changed the way students plan and carry out their research tasks, facilitating productivity. However, concerns were also identified about a possible over-reliance on these tools. In addition, participants recognized that certain skills were particularly impacted by the use of AI. Despite the risks, students reported a balance between productivity gains and skills development, highlighting the importance of conscious and strategic use of AI in the academic context.

KEYWORDS

Academic Productivity, Doctoral Students, Generative Artificial Intelligence, Higher Education, Skills Development

1. INTRODUCTION

Higher education is currently undergoing a period of intense transformation, driven by the rapid advance of digital technologies. Among these innovations, the growing adoption of AI by doctoral students stands out, which has led to profound changes in the way research activities are designed, organized, and executed. This new technological reality introduces a dynamic that, while challenging consolidated academic practices, also opens space for the reconfiguration of pedagogical and scientific processes in universities.

The use of AI, especially in advanced training contexts, has been significantly changing how knowledge is produced and disseminated. AI tools are now widely used in scientific writing, data analysis, bibliographic review, and time management, among other tasks relevant to the academic path. However, this widespread adoption raises crucial questions: to what extent are these technologies contributing to student productivity? Are they promoting the development of skills essential to scientific research, or, on the contrary, are they generating forms of dependence and cognitive superficiality?

In this scenario, it is essential to understand the real impact of generative artificial intelligence on the training of researchers in higher education contexts, especially in countries where structural challenges coexist with the urgent need for educational innovation. This article, therefore, aims to analyze the effects of the use of generative artificial intelligence on the productivity and acquisition of skills by doctoral students at a university in Mozambique. The research adopted an exploratory and descriptive approach, using a questionnaire applied to students of two postgraduate programs at a higher education institution.

The article is organized into five parts. After this introduction, a brief review of the literature on the relationship between AI and academic practices is presented. Next, the methodology used to conduct the study is described. The fourth section is dedicated to the presentation and discussion of the main results obtained. Finally, the conclusions and final reflections are presented, highlighting the challenges, potentialities, and implications of the use of AI in the training path of doctoral students.

2. PRODUCTIVITY AND SKILLS DEVELOPMENT OF STUDENTS

The use of AI in education is expanding rapidly(Roos, 2018) and has been integrated into administration and teaching-learning processes (Chassignol *et al.*, 2018). This evolution, particularly of generative AI, has significantly transformed the higher education landscape, creating possibilities for research, scientific writing, and learning processes. Nalbant (2021) Identifies a series of advantages of AI for education, namely: access to information, distance learning, personalization in education, global knowledge, student attendance detection, digitalization and environmental protection, removal of barriers, reduction of errors, carrying out challenging explorations, among others.

The growing adoption of AI in higher education, especially among PhD students, has led to profound changes in the way research activities are planned, organized, and executed.

Recent studies show that the use of generative AI tools, such as ChatGPT, can significantly increase the productivity of postgraduate students. For example, one study found that, with proper instruction, the use of AI reduced writing time by 64.5% and improved the quality of texts (from a B+ to an A), especially benefiting students for whom English is not their first language (Usdan et al. 2024). This suggests a leveling effect and a facilitation of the academic production process.

Another example is a study published by MIT Technology Review that showed that writers who used ChatGPT took 40% less time to complete their tasks, and the texts produced received 18% higher grades compared to participants who did not use the tool (Camara, 2023; MIT Technology Review, 2025).

These results are corroborated by another study by MIT and Stanford, which indicates an increase of up to 14% in productivity and a reduction of 35% in task execution time among workers who used generative AI, with even more significant gains for less experienced workers (Pergentino, 2023). Such evidence reinforces the idea that AI facilitates the academic production process and contributes to improving the quality of work (MIT Technology Review, 2025).

Beyond these examples, the literature highlights that the impact of AI is not limited to productivity: there are also significant effects on skills development. A recent study introduced the concept of "critical literacy in generative AI," which combines technical understanding, ethical awareness, and self-directed learning. PhD students who participated in micro curricula focused on this approach reported a significant transformation in the way they use AI: they moved from a skeptical stance to a more critical and strategic understanding, learning to create effective prompts, critically evaluate AI responses, and reflect on the writing process as knowledge creation (Palm, 2025).

Systematic reviews have identified six key areas in which AI offers significant contributions to academic research: idea generation and research project design; content structuring and development; synthesis of existing literature; data management; review and editing of texts; and support for compliance with ethical guidelines. These contributions highlight the potential of AI to offer comprehensive support throughout all stages of the scientific process, promoting gains in efficiency, rigor, and innovation (Khalifa and Albadawy, 2024).

However, despite the benefits, there are ongoing concerns that over-reliance on AI tools can compromise the development of essential skills such as critical thinking, creativity and intellectual autonomy (Aldulaijan and Almalky, 2025; Univeristy of London, 2025) Educational institutions such as City St George's recommend the responsible and ethical use of AI, emphasizing that it should support — not replace — the development of a researcher's capabilities. Use should be discussed with supervisors, and, where AI contributes to the final product, it should be appropriately acknowledged.

In short, recent literature points to a significant expansion in the use of AI in higher education, with applications ranging from administration to teaching-learning processes and scientific production. Generative AI, in particular, has transformed the way researchers, especially doctoral students, plan, organize, and execute their academic activities. Studies show that tools such as ChatGPT increase productivity, reduce writing time, and improve text quality, especially for students whose first language is not English. In addition to operational gains, AI also contributes to the development of skills such as critical literacy, combining technical, ethical, and reflective aspects. However, there are warnings about the excessive use of these tools, which can compromise essential cognitive skills.

3. METHODOLOGY

To answer the research objective initially formulated, we carried out exploratory and descriptive research. We analyzed the impact of using generative AI on productivity and the acquisition of skills by doctoral students at a university in Mozambique.

As a data collection tool, we applied a questionnaire survey to all the students in the two doctoral courses at the respective institution. The questionnaire has closed and open questions, which shows that it is a mixed instrument (quantitative and qualitative). The survey was administered using Google Forms, and the link was sent to all PhD students.

The sample consisted of 117 students from both courses. Of these, 62 responded to the survey with valid answers, which corresponds to a response rate of 52.99%.

The following table shows the relationship between the total number of individuals surveyed and the socio-economic variables (gender, age group, doctoral course, and year attended).

Table 1. Relationship between the total number of individuals surveyed and the socio-economic variables

Variables	n	%	Total n	Total %
Gender				
Male	53	85,50	62	100,00
Female	9	14,50		
Age group				
Under 25	0	0,00	62	100,00
25 to 29 years old	0	0,00		
30 to 34 years old	5	8,1		
35 to 39 years old	17	27,4		
40 years or older	40	64,5		
Doctoral course				
Doctoral course 1	28	45,2	62	100,00
Doctoral course 2	34	54,8		
Year attended				
1st year	21	33,9	62	100,00
2nd year	16	25,8		
3rd year	25	40,30		

Data collection took place between January 15 and February 15, 2025. Quantitative data was analyzed using descriptive statistics (percentages) to identify trends in the use of AI and its impact on the productivity of doctoral students. As for the qualitative data, a content analysis was carried out to recognize patterns in the answers that would show how AI contributes to the development of skills in the field of academic research.

About ethical issues, the study ensured that informed consent was obtained from the participants, guaranteeing the anonymity and confidentiality of the information collected, by the institution's ethical guidelines.

4. DISCUSSION OF RESULTS

The data collected provides a clear picture of the impact of AI on the research practices of doctoral students. Analysis of the results shows significant changes in the way these students plan and conduct their research activities, reflecting the growing incorporation of AI into their academic processes. The main findings are presented below, organized around three central dimensions: Organizing research with AI, ethics, and AI addiction and impact on skills development.

4.1 Organizing Research with AI

This category aims to understand how these technologies are being integrated into academic work routines, influencing efficiency, autonomy, and time management. It aims to analyze whether AI acts as a complement to traditional methodologies or whether it has caused a significant restructuring in students' research approach.

Changes in the planning and organization of tasks

The data shows that the majority of doctoral students (48,4%) consider that the use of AI has significantly changed the way they plan and organize their research tasks. This result indicates a clear and direct impact of AI on the structuring of academic work, suggesting that these tools are being actively and transformatively integrated into research routines.

In addition, 38,7% of respondents recognize changes, albeit to a moderate degree, which reinforces the idea that even among those who have not yet adopted AI in depth, there is already a perception of a positive or adaptive influence. Only 12,9% indicated that the use of AI had not caused any changes, which represents a negligible minority in the universe studied.

These data lead us to conclude that AI is indeed redefining research practices among doctoral students, promoting a reorganization of strategies for planning and carrying out tasks. The prevalence of positive responses, both significant and moderate, shows a trend of growing adoption and a potential for structural transformation in the way postgraduate students manage their academic careers.

Role of AI as a complement or substitute for the researcher's skills

The survey reveals that the majority of students believe that AI complements their skills, with 58,1% of respondents saying that AI helps to improve research work, without completely replacing the researcher's skills. This indicates that students recognize the value of AI as a useful tool that increases productivity and facilitates tasks, but without eliminating the need for essential human skills.

On the other hand, a smaller proportion of students (19,4%) believe that AI partially replaces researchers' skills, suggesting that, although it won't replace everything, AI can take over some functions previously done manually, such as data analysis or proofreading. Only 1,5% of participants see AI as capable of completely replacing the skills of the researcher.

On the other hand, 21,0% of students say that AI won't replace their skills, indicating that they still value the cognitive and creative skills that AI can't replicate.

Balance between increased productivity and skills development

Of the 62 doctoral students, 50,0% consider the use of AI to be "very balanced", stating that it increases productivity without compromising their skills. Another 29,0% find the balance "reasonable", recognizing efficiency gains but identifying gaps in the development of some skills. For 8,1%, the use of AI is "poorly balanced" because, although it helps productivity, it hinders the learning of essential skills. Only 3,2% rate the balance as "unbalanced", feeling that AI prioritizes productivity to the detriment of their academic and personal development. Finally, 9,7% still have no opinion on this impact.

These results show that half of the students enjoy the benefits of AI without seeing significant losses in their skills, while almost one in four (29,0%) perceive a trade-off between productivity and skills development. A minority (11,3%) express more severe concerns about the imbalance, indicating the need for usage practices that guarantee both efficiency and continuous learning.

4.2 Ethics and AI Addiction

This category explores the ethical dilemmas faced by doctoral students when using these tools and the risks associated with possible over-reliance. At the same time, it seeks to reflect on responsible and sustainable practices for integrating AI into advanced training, ensuring that the technology is used critically and consciously.

Ethical issues related to the use of AI in research

The results indicate that a considerable proportion of doctoral students (24,2%) faced significant ethical doubts when using AI in their research, while the majority (53,2%) reported having had these doubts to a moderate degree. These figures show that almost 78,0% of participants recognize some level of ethical questioning related to the use of AI, which demonstrates a growing awareness of the limits and responsibilities associated with the application of these tools in the academic context.

On the other hand, 21,0% of respondents said that the use of AI had no relevant ethical impact, which may indicate different levels of awareness of these issues, or even contexts of use where ethical dilemmas are less evident. A residual number (1,60%) mentioned that AI reduced their productivity, which, although a minority, suggests that, for some, the effort required to deal with ethical issues could compromise the fluidity of research work.

In terms of conclusions, the data points to the need to include the ethical dimension in the training of doctoral students, providing spaces for reflection and guidance on the responsible use of AI. The existence of doubts, albeit moderate, should not be undervalued, as it shows that students are questioning the impacts and limits of these technologies - an essential step towards their conscious and critical integration into the scientific process.

Possible over-reliance on AI tools

Of the 62 doctoral students, 27,4% say that AI often leads to excessive dependence, 54,8% see this risk in specific cases, and 17,7% consider it only a complement. These figures indicate that almost all students (82,2%) recognize some degree of dependence in the use of AI, although for the majority, this only occurs in specific situations. This suggests the need for clear guidelines for the conscious use of these tools, combined with training that reinforces the autonomy of the researcher and the development of critical skills without resorting solely to AI.

Strategies and recommendations for the responsible integration of AI in academic training

The growing adoption of AI tools in doctoral research poses the challenge of balancing efficiency gains with the preservation and development of critical skills. From the respondents' answers, a set of recommendations emerged that can guide the responsible integration of AI into advanced training:

Continuing education and practical training:

- Insert compulsory modules on AI into curricula, with workshops that teach everything from basic functionalities to advanced applications in research, writing, and data analysis;
- Offer regular refresher training on new AI tools and extensions.

Ethical policies and guidelines:

- Create an institutional code of conduct that defines limits on use, rules for citing automated tools, and penalties for inappropriate practices.
- Guarantee regular supervision and auditing, ensuring transparency and accountability in the research process.

Balanced use and support for autonomy:

- Encourage AI as an aid, not a substitute, highlighting the importance of maintaining critical thinking and independent reflective processes;
- Develop protocols that identify specific tasks where AI is recommended and points where manual or collaborative work is preferable.

Integration of specialized tools:

- Guide doctoral students in the selection and use of reference management software with AI plugins, assisted writing systems, and libraries for data analysis;
- Promote practical sessions to apply these tools to real research cases.

Encouraging debate and awareness:

- Organize seminars, study groups, and interdisciplinary forums dedicated to the ethical dilemmas, methodological possibilities, and impacts of AI in science;
- Include reflections on plagiarism, authorship, and validation of results produced with the help of AI.

The conscious adoption of AI in doctoral programs requires a multifaceted approach: continuous education, clear regulation, respect for intellectual autonomy, and constant dialogue on ethics and methodologies. With these practices, it is possible to maximize the transformative potential of AI while cultivating the critical and creative skills indispensable to the academic researcher.

4.3 Impact on Skills Development

This category investigates which skills have been most influenced, positively or negatively, by the use of AI, and how doctoral students perceive this influence on their educational and scientific path.

Skills most impacted by the use of AI

To understand the impact of AI on the development of academic and research skills, a Likert scale with three response options - "Improved", "Had no impact", and "Harmed" - was used to assess perceptions of different skills. Below are the main results in percentage terms, based on the responses collected.

- The majority of participants believe that AI has had a positive impact on scientific writing, with around 69,0% reporting improvements, while around 31,0% indicated that they had felt no impact at all.
- About critical thinking, around 63,0% recognized improvements, but around 34,0% saw no change, and a small minority (around 3,0%) pointed to negative impacts. This balance indicates that, although there is a positive trend, there is also some caution about the depth of this influence.
- In data analysis skills, both quantitative and qualitative, the results reveal a split: around 47,0% perceived improvements, while the remaining 53,0% identified no impact. This suggests that the use of AI in these areas may not yet be fully disseminated or exploited.
- Planning and organizing research projects was identified as a skill that benefited, with around 60,0% of respondents noting improvements, although around 39,0% saw no change, and around 2,0% saw a loss;
- Searching for and organizing bibliographic references stood out as one of the most positively impacted areas, with around 77,0% reporting improvements, and only around 23,0% noting no impact - a clear indicator of the usefulness of AI in this process;
- In solving complex problems, around 53,0% saw benefits, 45,0% felt no change, and 2,0% mentioned negative effects;
- As for creativity in generating ideas and hypotheses, AI was widely valued, with around 71,0% indicating improvements, compared to 29,0% who felt no impact;
- In the critical interpretation of results, around 57,0% reported improvements, while around 42,0% noted no difference, and 2,0% indicated harm;
- Time management and personal productivity were the skills with the most positive impact, according to around 79,0% of respondents, as opposed to 21,0% who felt no change, highlighting AI's strong influence on personal organization and efficiency;
- Scientific communication also benefited from the use of AI for around 60,0% of participants, while the remaining 40,0% saw no impact;
- Team collaboration or interdisciplinarity, on the other hand, was the skill least positively impacted, with only 37,0% recognizing improvements, compared to 61,0% who felt no change, and 2,0% who pointed out losses. This result indicates that AI still plays a limited role in collaborative dynamics.
- Finally, in data-driven decision-making, around 52,0% reported improvements, while 47,0% saw no impact, and 2,0% reported a negative effect.

The data reveals a largely positive perception of the impact of AI, especially in the areas of organization, productivity, creativity, and support for scientific writing. However, in areas such as data analysis and interdisciplinary collaboration, the impact is not yet as evident, which may indicate a transitional phase in the use of AI or less maturity in its application in these domains.

Influence of AI on critical thinking skills

The results on the impact of AI on the development of critical thinking reveal diverse opinions. A small proportion of students, 9,7%, believe that AI significantly reduces their ability to think critically, while 22,6% point to a moderate reduction. On the other hand, 22,6% say that AI does not affect their critical thinking skills. A larger proportion, 29,0%, believe that AI moderately increases their ability to think critically, and 16,1% consider that it significantly increases this ability. Thus, the majority of students believe that AI has a positive or neutral impact on the development of critical thinking, with a considerable proportion stating that it moderately or significantly increases this ability. However, some still see risks that AI could reduce this ability, especially if it is used excessively.

5. CONCLUSION

This study allowed us to understand how AI is transforming the research practices of PhD students. The data show that AI is increasingly being integrated into academic routines, changing the organization of work, influencing the development of skills, and raising important ethical reflections. Most students demonstrate openness to using AI as a support tool, recognizing substantial benefits in productivity and the quality of scientific work, without, however, losing sight of the risks associated with its excessive or uncritical use.

It was also found that the perception of the impacts of AI varies according to the areas and individual experiences, which indicates a transition phase in the integration of these technologies in the academic context. The conscious and critical adoption of AI, therefore, requires adequate institutional support, which combines technical, ethical, and methodological training.

Organizing research with AI

IA is already significantly changing the way students organize their research. A significant proportion of PhD students feel that AI has changed the way they plan tasks, structure projects, and manage their time. The perception that AI complements, rather than replaces, human skills predominates, serving as support in the execution of repetitive or technical tasks. Even so, it is essential to ensure that this support does not reduce students' critical involvement in the more complex research processes. In general, the data point to a structural transformation in academic work practices, mediated by these new tools.

Ethics and dependence on AI

Ethical issues emerge as a relevant concern among students. The majority report having faced doubts, albeit moderate, about the limits and responsibilities of using AI in scientific production. The perceived risk of excessive dependence is also highlighted, although in many cases this only manifests itself in specific tasks. These concerns demonstrate the need for clear regulation and targeted training that helps students make informed decisions about the use of AI, promoting responsible and transparent research practices. It is urgent to foster a culture of conscious use that balances innovation with academic integrity.

Impact on skills development

The perception of the impact of AI on skills development is generally positive. Students report particularly clear improvements in scientific writing, creativity, organisation of references, and time management. However, some more analytical and collaborative skills, such as data analysis or interdisciplinary work, do not show such significant impacts, suggesting that AI is not yet fully explored or integrated in these contexts. The influence of AI on critical thinking is mixed: although the majority perceive improvements or neutrality, some students identify risks of cognitive impairment, especially in cases of excessive or dependent use.

This study shows that AI is a profoundly transformative element in the research practices of doctoral students. However, this integration is not without its challenges. Ethical issues, risks of dependency, and possible impacts on the development of fundamental skills, such as critical thinking, continue to raise legitimate concerns among students.

In this scenario, the need for an institutional response that goes beyond the mere provision of technology becomes clear. Higher education institutions must take an active role in the ongoing training of their students, promoting critical and up-to-date digital literacy on the applications of AI in science. This training should include not only the technical mastery of the tools but also ethical and methodological reflection on their use, ensuring that students are prepared to make responsible and informed decisions.

In addition, it is urgent to define clear regulatory policies and good practices that guide the use of AI in academic contexts. The existence of codes of conduct, rules for citing AI-generated content, and scientific oversight mechanisms is are fundamental element to guarantee the integrity of research.

In short, AI has the potential to significantly enrich the experience and results of doctoral research, provided it is integrated in a conscious, critical, and targeted manner. This study contributes to this debate by offering empirical data that illustrates both the benefits and risks of using AI, and by highlighting the importance of educational and regulatory strategies that ensure a balance between technological innovation and full academic development. The responsible adoption of AI will therefore be one of the decisive factors for the future of scientific research in an academic context.

ACKNOWLEDGEMENT

This work has been supported by FCT – Fundação para a Ciência e Tecnologia within the Project Scope: UIDB/05777/2020 (https://doi.org/10.54499/UIDB/05777/2020).

REFERENCES

Aldulaijan, A. T. and Almalky, S. M. (2025). 'The Impact of Generative AI Tools on Postgraduate Students' Learning Experiences: New Insights Into Usage Patterns', Journal of Information Technology Education: Research, 24, p. 3.

Camara, L. (2023). Estudo do MIT: IA Generativa Aumenta Produtividade em Tarefas de Redação.

Chassignol, M. et al. (2018). 'Artificial Intelligence trends in education: a narrative overview', Procedia Computer Science, 136, pp. 16–24.

Khalifa, M. and Albadawy, M. (2024). 'Using artificial intelligence in academic writing and research: An essential productivity tool', Computer Methods and Programs in Biomedicine Update, p. 100145.

MIT Technology Review (2025). 'O ChatGPT ajuda as pessoas a escrever melhor', MIT Technology Review. Available at: https://www.mittechreview.pt/2023/humanos-e-tecnologia/o-chatgpt-ajuda-as-pessoas-a-escrever-melhor/.

Nalbant, K. G. (2021). 'The importance of artificial intelligence in education: a short review', Journal of Review in science and engineering, 2021, pp. 1–15.

Palm, J. (2025). 'Critical AI literacy enhances doctoral students' writing skills'. Available at: https://www.chalmers.se/en/current/news/cls_critical-ai-literacy-enhances-doctoral-students-writing-skills/.

Pergentino, C. (2023). 'Inteligência artificial pode aumentar em 14% a produtividade de trabalhadores, aponta estudo', Globo. Available at: https://epocanegocios.globo.com/futuro-do-trabalho/noticia/2023/04/inteligencia-artificial-aumenta-em-14percent-a-produtividade-de-trabalhadores-aponta-estudo.ghtml.

Roos, S. (2018).'Chatbots in education: A passing trend or a valuable pedagogical tool?'

Univeristy of London (2025). 'Use of Generative AI Tools in Doctoral Research'. Available at: https://www.citystgeorges.ac.uk/research/city-doctoral-college/doctoral-training-resources/use-of-generative-ai-tools-in-doctoral-research.

Usdan, J., Connell Pensky, A. and Chang, H. (2024). 'Generative AI's Impact on Graduate Student Writing Productivity and Quality', Available at SSRN.

INFORMAL LEARNING AND AFFINITY SPACES: LESSONS LEARNED FROM THE ANALYSIS OF THE CURIOSITY, INTEREST AND ENGAGEMENT CONSTRUCT OBSERVED IN PROJECT NOOA

Juliana C. C. Monteiro[1], Carla Morais[2] and Miguel Carvalhais[3]
[1]CIQUP / Faculty of Engineering, University of Porto
Rua Dr. Roberto Frias, s/n, 4200-465 Porto, Portugal
[2]CIQUP / Faculty of Sciences, University of Porto
Rua do Campo Alegre s/n, 4169-007 Porto, Portugal
[3]i2ADS, Faculty of Fine Arts, University of Porto
Av. Rodrigues de Freitas, 265, 4049-021 Porto, Portugal

ABSTRACT

Mentoring within affinity spaces holds an invaluable potential for informal learning. It promotes continuous learning through reflective practices and ongoing feedback, which are vital for developing expertise and for fostering lifelong learning. However, ensuring active engagement in these spaces can be challenging. Understanding the mechanisms of curiosity, interest and engagement appears as key to comprehending the conscious and unconscious negotiations that occur in participation, collaboration, and their subsequent outcomes in this setting. This paper examines the curiosity, interest and engagement (CIE) development throughout the informal learning processes that underpin the development of digital and cultural literacy through intergenerational digital storytelling within the project *Narratives of Our Age (NOOA)*. The project facilitated intergenerational digital storytelling, focusing on local myths, crafts, and traditions, based on the creation of a hybrid affinity space. Drawing from the collaborative and participatory potential of web 2.0 applications, the project fostered in-person intergenerational exchanges to develop digital stories that raise acknowledgment and have the potential to preserve and to disseminate regional cultural identity, while contributing to participants' cultural and digital literacy. In parallel, the project also developed a cross media online space to disseminate the digital stories that were produced by the intergenerational groups, and to expand the discussions around the topic of cultural identity. With an action-research methodology and drawing from qualitative data obtained through interviews and participative observation, we investigated how the CIE construct manifested within both in-person and digital affinity spaces of the project. Findings indicate that in-person interactions were particularly effective in fostering all levels of the CIE construct, promoting deeper engagement and sustained interest. Conversely, online spaces primarily stimulated initial curiosity and interest but encountered challenges in maintaining prolonged engagement. These results underscore the value of an integrated, hybrid approach that combines face-to-face and digital interactions to effectively support informal learning and cultural and digital literacy development in an intergenerational setting. The study contributes to the understanding of the opportunities and challenges of hybrid affinity spaces for enhancing community-based digital storytelling initiatives, and of how innovative informal educational approaches can leverage technology and cultural practices to create engaging, culturally responsive learning environments that bridge generational gaps and promote lifelong learning.

KEYWORDS

Informal Learning, Affinity Spaces, Intergenerational Digital Storytelling, Curiosity, Interest, and Engagement (CIE), Digital Literacy, Cultural Literacy

1. INTRODUCTION

Informal learning represents a departure from conventional, institutionally driven educational models, characterized instead by its intrinsic, self-directed, and often tacit acquisition of knowledge and skills (Ang et al. 2018). Unlike formal educational settings, with predetermined curricula and structured assessments, informal learning thrives in environments where learning is embedded within meaningful activities, driven by the learner's own volition, curiosity, and intrinsic motivation (Rogoff et al. 2016). Affinity spaces, on the other hand, are physical and virtual environments where individuals converge based on shared interests, passions, or goals, fostering a sense of community and collective identity (Gee 2005; Rogoff et al. 2016). These spaces serve as crucibles for informal learning, as participants collaboratively build knowledge, exchange expertise, and engage in peer-to-peer mentorship, blurring the lines between teacher and learner (Prestridge et al. 2021).

The mentoring dynamics within affinity spaces are invaluable given its capacity to foster sustainable learning through reflective practices and iterative feedback loops, essential components for cultivating expertise and promoting lifelong learning (Bennetts 2001). However, creating and sustaining engagement within them can be a challenge. Literature suggests that the consolidation of affinity spaces is deeply dependent on a dimension of participatory culture that requires the observation of an interest construct (Arnone et al. 2011, p. 184). Arnone and colleagues (2011) argue that curiosity is intrinsically linked to related constructs, such as interest and engagement. They propose a theoretical framework for understanding curiosity, interest, and engagement within technology-rich learning environments, underscoring the need to examine how individuals approach research and information-seeking tasks, as well as the factors that sustain these efforts (Arnone et al. 2011, p.181). The Curiosity, Interest, and Engagement (CIE) model suggests that curiosity and interest are the main forces driving engagement in technology-rich learning settings, with new media technologies acting as effective tools to spark curiosity and interest, thereby enhancing the learning experience. However, the model also recognizes that CIE is shaped by various personal, situational, and contextual factors, including individual personality traits, prior knowledge, learning styles, personalized feedback, cultural influences, expectations, and the resources available to learners.

The connection between informal learning and affinity spaces is particularly salient in the contemporary landscape, marked by the proliferation of digital technologies and the increasing importance of lifelong learning in a rapidly evolving global economy. Digital platforms and online communities have created unprecedented opportunities for individuals to connect with like-minded peers, access vast repositories of information, and engage in collaborative projects that transcend geographical boundaries (Shonfeld et al. 2021; Haythornthwaite 2019). In parallel, the increasing availability of digital tools and participatory media has transformed how societies preserve and transmit cultural knowledge. This is particularly relevant in aging communities, where intergenerational gaps threaten the continuity of local traditions and regional identity, and informal learning plays a crucial role in bridging these gaps through authentic, community-based processes.

This work is framed within the project *Narratives of Our Age (NOOA)*, that was conceived to explore how intergenerational digital storytelling could serve as a mechanism to support cultural identity maintenance in the Vale do Sousa region, in Portugal (Monteiro 2024). This paper focuses on how affinity spaces, both online and in-person, created opportunities for informal learning, analyzed through the Curiosity, Interest, and Engagement (CIE) construct. In this work, we intersect academic discourse with salient findings in project NOOA, beginning with scrutinizing the entourages of in-person affinity spaces in the project, and proceeding with discussing the processes of involving the community in online affinity spaces. We conclude by reflecting about moving from interest to engagement and the components that we identified as core to this transition.

2. METHOD

This study employed an action research methodology, resorting to data derived from semi-structured interviews and participatory observation, to critically examine the manifestation of the Curiosity, Interest, and Engagement (CIE) construct across both in-person and digital affinity spaces within the context of the project NOOA (Monteiro 2024). From 2018 to 2021, this project brought together three intergenerational groups, consisting of 18 participants ranging from 16 to 85 years old, in activities aimed at enhancing intergenerational interactions towards cultural literacy. We devised a five-step framework for intergenerational storytelling to promote the sharing of cultural identity knowledge, which included digital documentation and dissemination, both in-person and online. This framework underwent cyclical evaluation and was executed in three phases: 1) planning group dynamics and stories; 2) creating digital stories in in-person intergenerational storytelling workshops; and 3) disseminating and discussing stories in the project's online platforms. These steps were carried out through eight sessions of in-person intergenerational digital storytelling activities, interspersed with periods of autonomous work, that was moderated online through a private Facebook group. This resulted in the participatory and collaborative creation of 35 digital stories resorting to the use of web 2.0 tools. These stories were subsequently shared and discussed on the project's online platforms.

The in-person intergenerational digital storytelling workshops streamlined the project's in-person affinity space around the topic of cultural identity of the region, over eight in-person sessions. Additionally, the project's digital spaces streamlined the basis of a digital affinity space (Gee 2005), designed to be the stage for discussion of the artifacts produced during the in-person activities, and ultimately to aggregate both artifacts and discussion around cultural identity of the region of Vale do Sousa. The project's digital affinity space was designed with a cross media strategy, consisting of ensuring the social media presence for the project through a YouTube channel and a Facebook page, in parallel with the use of the NOOA website. Subsequently, we also prepared a final presentation event and a set of news in regional media so that the broader community could be aware of the project and consider to contribute to it with their stories and discussion.

Figure 1 synthetizes how the dynamization of the project's affinity spaces was prepared.

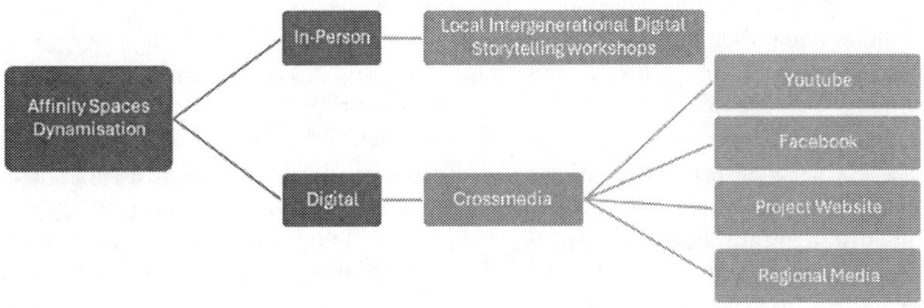

Figure 1. Affinity Spaces dynamization in project NOOA

The consolidation of affinity spaces is deeply dependent on a dimension of participatory culture that requires the observation of an interest construct (Arnone et al. 2011, p. 184). Aligned with this idea, we endorsed multiple activities in pursue of such consolidation. We strategically started our participatory approach by building the basis of an in-person affinity space. This choice was based on two main premises, namely both our perceptions of the digital literacy and skills of the groups, and our conception of the potential of the online contact as an extension of the face-to-face contact in this context, as a stage potentially capable of allowing the conversations initiated during the in-person interactions to go further, in reach and in depth.

The CIE framework by Arnone and colleagues (2011) provided a structured lens to analyze participants' involvement in the project NOOA in both spheres, as understanding its dynamics is crucial to comprehend the conscious and unconscious negotiations that occur in terms of participation, collaboration and its consequent outcomes. This model highlights the dynamic interaction among curiosity, interest and engagement and their impact on the learning process in environments dominated by technology.

At the end of the empirical work, nine interviews were conducted (May 2021) with a sample composed of three seniors and six juniors. We resorted to a semi-structured interview guide prepared to collect qualitative insights of the participants regarding topics such as intergenerational storytelling, cultural identity and the impact of technology. After finishing the transcription of the interviews, we submitted the resultant data set to a thematic analysis process (Braun and Clarke 2012).

Identifying curiosity, interest, and engagement relied on a careful analysis of both verbal and non-verbal cues, that manifested over the in-person activities' observation and during the thematic analysis of the final interviews. As indicators of *Curiosity*, we gathered insights from explicit questions and inquiries of the participants. Direct questions, such as "How did you...?" or "What was it like when...?", consistently signaled a desire for specific information or contextual understanding. This was further supported by follow-up questions that delved deeper into prior topics, thereby revealing a commitment to comprehensive understanding and often highlighting pre-existing knowledge gaps. Key indicators also included the exploration of novelty and difference, wherein participants expressed surprise or wonder when encountering divergent perspectives or historical contexts. Probing into generational distinctions in trends, technologies, or social norms also exemplified a drive to grasp unique era-specific knowledge. Finally, the presence of hypothetical thinking, through "what if" scenarios or imagined experiences, denoted an active cognitive process of internalizing and connecting with recounted narratives.

On another hand, the presence of *Interest* was discernible through voluntary topic initiation and active listening behaviors, often complemented by emotional expressions of enthusiasm and sustained attention, alongside the articulation of personal relevance drawn from the conversation.

Finally, we looked for *Engagement* on indicators of active and reciprocal participation, evidenced through collaborative dialogue and involving co-construction of meaning, highlighting deep interaction. Also, we could perceive that the use of expressions of collectivity such as "we" often underscored shared investment, and it was considered a strong indicator of engagement in our analysis, complemented by demonstrations of learning or application, where previously discussed insights were referenced in new contexts.

3. OVERALL CIE DEVELOPMENT OVER THE COURSE OF NOOA

According to Arnone and colleagues (2011), curiosity is defined as a "state of wanting to know or learn about something," often sparked by new stimuli, uncertainty, or gaps in knowledge. Curiosity drives individuals to seek information and explore new ideas, leading to a more profound understanding of their surroundings. In contrast, the authors describe interest as a more lasting form of curiosity, marked by a positive emotional connection to a specific topic or activity. It emerges from a sense of personal relevance and the belief that the topic is valuable or enjoyable, thereby fostering persistence and motivation in learning, as individuals are drawn to activities that stimulate their curiosity and align with their interests. Lastly, Arnone and colleagues (2011) define engagement as active and focused participation in a learning activity, characterized by immersion in the learning experience, a sense of flow, and a strong connection to the learning material. Engagement is essential for effective learning, as it encourages deeper processing of information and knowledge acquisition.

The authors argue that three key dynamics guide the interactions and flow within the CIE model. The first is the curiosity dynamic, which proposes that when curiosity is triggered, it leads to interest, which can then result in engagement, provided the trigger or triggers are resolved. This poses that, essentially, curiosity must be satisfied for this process to unfold. If curiosity remains unresolved due to perceived incompetence or other unfavorable factors, the outcome may be negative, leading to withdrawal, anxiety, frustration, disinterest, or changes in participatory collaboration. In such cases, interest and engagement may not develop.

The second dynamic, the interest dynamic, starts with a triggered situational interest. This model suggests that interest can re-trigger curiosity, depending on environmental conditions. If sustained, interest may evolve into a maintained situational interest, potentially leading to emerging individual and group interest, which could further develop into a well-established individual and group interest. This concept underscores the connection between interest and varying levels of engagement, as well as the significance of time in fostering the development and implied negotiations of these levels.

Finally, the engagement dynamic can occur in three forms: participative, affective, and cognitive. Participative engagement happens when learning is driven by an external factor, such as a parent, teacher, or affinity space, with minimal intrinsic motivation. Affective engagement arises when the experience is enjoyable and may motivate interest. Cognitive engagement occurs when an individual is fully and intrinsically committed to learning more about a subject.

In light of this, we resorted to the aforementioned qualitative indicators to operationalize our analysis of the development of CIE in the project's affinity spaces.

Following, we summarize the main findings concerning participants and audiences' engagement, and we analyze the development of CIE model, by associating each model's stage with the phases of the empirical work that it was observed in during the project (see Figure 2).

We noticed that curiosity evolved into interest and then into engagement during phase 1 of the in-person activities. This phase involved brainstorming and connecting, where participants engaged in activities of collective remembering and planning the creation of stories.

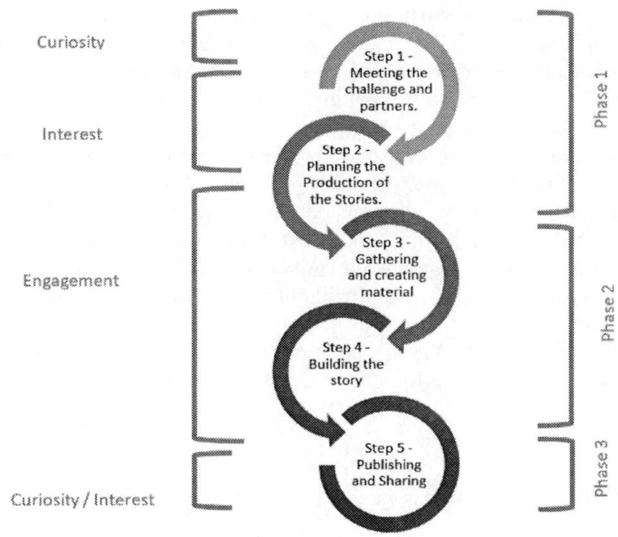

Figure 2. Stage of Arnone et al. (2011) CIE model observed in each phase of project NOOA 's empirical work

Engagement persisted throughout phase two, with the participatory intergenerational production of stories, but reverted to curiosity and interest in phase 3, which focused on sharing online and continuing the conversation. When activities transitioned to online, curiosity was evident but did not transition into interest. This was mainly monitored through the interactions in the project's Facebook page, where the main online interactions were observed.

3.1 Participants and Audiences' Engagement: Between Interest and Agency

In the course of developing our in-person affinity space, the project's activities effectively sparked curiosity, interest, and engagement among both participant groups. Indeed, participants typically showed a keen eagerness to learn, often driven by the novelty or unpredictability of the tasks' outcomes and methods, which led to a state of *Curiosity* (Arnone et al. 2011). This curiosity was maintained and enhanced by positive emotions and a belief in the value or enjoyment of the subject, resulting in a state of *Interest* (Arnone et al. 2011). Ultimately, participants exhibited active involvement, immersion, and a profound connection to the topic, reaching a state of *Engagement* (Arnone et al. 2011). However, this final state did not develop uniformly across all groups of participants, that diverged in terms of relationship development and length, highlighting the importance of allowing sufficient time for group members to become familiar with each other, as well as the need for appropriate moderation due to two specific factors: a) the authority of one of the intergenerational sides, which discouraged the other side from taking a more active and critical role at a certain point, and b) the generational gap and underdeveloped intergenerational understanding. Moderation emerged as crucial for overcoming engagement barriers and bridging the intergenerational gap, in parallel with consistency and time to establish a sense of community in this context.

In NOOA, we observed that curiosity evolved into interest and engagement during phase 1 of the in-person activities (see figure 2), which involved brainstorming and connecting. Engagement and a state of flow were frequently linked to activities of collective intergenerational remembering and planning the creation of stories. Engagement persisted through phase two, with the participatory intergenerational production of stories, but reverted to curiosity and interest in phase 3, which focused on online sharing and conversation. Indeed, when the project transitioned to the online realm, curiosity was evident but did not naturally transform into interest. In fact, right after the project's presentation event on Facebook, there was a noticeable organic increase in pageviews, shares, and likes. Although the page managed to maintain and gradually increase its followers over time, indicating curiosity about the project and its stories, it appeared that the community was not yet prepared for deeper forms of involvement. In response, we implemented a coordinated communication strategy to enhance interest and, consequently, streamline engagement. We began by seeking partnerships with similar

Facebook pages and groups. This led to a bidirectional expansion of reach and a temporary boost in engagement through page views, post sharing, liking, and commenting. Nevertheless, the impact was modest within the analysis timeframe, which proved to be somewhat short for consolidation.

Overall, our results indicate that engagement in the project's online spaces experienced temporary increases but struggled to sustain themselves independently over time, highlighting the need for structured and consistent moderation until the community can consolidate. This finding emphasizes the significant challenges in engaging our target demographics online with the topic of cultural identity.

With the experimental approach that we applied, rather than implementing a pre-established plan, we aimed to explore and evaluate the dynamics that could naturally emerge in these spaces, using an ethnographic approach. This helped us assess what dynamics might occur after the project's implementation period and assess the potential impact of these spaces without moderation of the research team. Nonetheless, we acknowledge that this approach required a delicate balance between providing necessary triggers for curiosity and interest to ignite and convert into engagement and allowing the online affinity space to develop independently. This was an adaptive effort that needed to be continuously refined and adjusted, in order to progress positively.

4. IMPLICATIONS: MOVING FROM INTEREST TO ENGAGEMENT

In examining the dynamics that emerged in the project's affinity spaces, we identified three key areas for reflection: our audience, the context, and our engagement expectations. Firstly, concerning our audience, we observed that two groups showed a pronounced interest in these stories and discussions, both in-person and online: seniors, who were keen to pass on their stories and knowledge, and juniors in their late 20s or 30s, who had enough background to appreciate these narratives. However, during the project , it became evident that these groups are less inclined to engage online, preferring face-to-face interactions, which underscores the need for mixed strategies. This also prompted us to consider our audiences and how interests are formed. We identified two primary audience types: a) Agents: participants, including seniors and some engaged juniors, who took an active and critical role in the dynamics; b) Listeners: participants, such as younger juniors or those with less developed interest, who adopted a more passive stance. We also noticed that juniors sometimes switched roles based on their interest level: if their interest was undeveloped, they acted as listeners, but when interest grew, they became active participants. Online, however, these engaged juniors often reverted to listeners, likely due to a lack of new stimuli to engage them in online interactions. Although we anticipated younger juniors to be the main online agents in project NOOA, given their propensity for online interaction compared to other participants, their interest primarily manifested in observation rather than interaction. This highlights the implications for audience engagement in online dynamics.

Contextual and situational factors are crucial in these dynamics, as highlighted in the literature. According to Arnone and colleagues (2011), contextual factors help convey meaning, while situational factors include "in the moment" elements like interactions, attention competition, unexpected events, etc., which, along with personal factors like traits or predispositions, can affect how curiosity is acted upon. The context in which these dynamics occurred significantly influenced their development. Indeed, part of the project's dynamics were still developing when the COVID-19 pandemic emerged, right in the transition between phase 2 and phase 3 of the project. People worldwide began to realize the uncertainty surrounding the duration of social isolation, and despite initially we speculated that there could be an increased demand for virtual interactions, while this may have been true for other aspects of social life during the pandemic, it posed a challenge within the context of this project, as some participants lacked internet access or the skills to navigate it. Additionally, this situation made it difficult to assess how online interactions could contribute to maintaining cultural identity, as this issue was not priority during such uncertain times.

Likewise, when discussing online interactions with project NOOA's spaces, we must consider the niche nature of our project and its impact on establishing online audiences, content specificity, and the organic growth of engagement. Expanding our analysis beyond in-person participation to online spaces, we must recognize the complexities of organically growing a Facebook page focused on the cultural identity of a specific region. This requires a multifaceted approach that emphasizes community building and discoverability, leading to a more gradual path toward sustained engagement. To explore this, we examined the nuanced factors contributing to the slower development of engagement within hyper-focused niche Facebook pages like ours. First, we considered the specificity of audience demographics. Niche Facebook pages inherently target specific audiences, resulting in a smaller potential audience compared to mainstream topics. This limits the page's organic reach and initial engagement potential, further challenged by the specific demographics we successfully attracted to the project's digital spaces through our dissemination efforts. Second, we note the importance of thematic relevance concerning content specificity within this niche, as the alignment between a page's content and its audience's interests is crucial. However, content focused on a niche cultural identity isn't prone to resonate as widely as content with broader appeal. For instance, references or traditions specific to the featured culture might not be easily understood by a wider audience or those less familiar with it, leading to slower engagement growth on a diverse platform like Facebook and explaining the gradual development observed in NOOA. Third, we also recognized the potential algorithm bias resulting from Facebook's algorithm prioritizing content that generates high engagement metrics such as likes, shares, and comments. A hyper-focused niche page may struggle to achieve these metrics due to its limited audience size, creating a feedback loop that hinders organic growth. We tested diverse strategies to address the aforementioned challenges, such as exploring diverse content while maintaining thematic consistency, fostering discussions and interactions, and collaborating with complementary pages or influencers. However, persistence appears to be essential, with regular posting, authentic interactions, and a long-term perspective necessary to allow engagement to solidify over time.

Considering the experimental nature of our initial approach to online spaces and the specificity of our audiences, we structured this initial dynamization in a semi-organized manner, enabling us to gather indicators of spontaneous interactions and the primary challenges that emerged from this first approach. Consequently, further research is required to tackle these challenges and to identify and assess effective solutions for fostering sustained meaningful engagement in this context.

Nevertheless, the findings of this approach highlight practical implications for educators and policymakers to optimally leverage this educational potential. The first points to the importance of strategic implementation of moderation and design approaches that cultivate robust peer-to-peer interactions and facilitate expert-novice collaborations. To encourage peer-to-peer interactions, moderators can initiate discussion prompts, create dedicated channels for specific sub-topics, and highlight contributions from various members. This also opens space for facilitating expert-novice collaborations, that can significantly enhance knowledge transfer and ensure sustained involvement, with fluidity of roles in both in-person and online spaces, taking advantage of the complementarity of strengths that can be fostered in intergenerational settings.

The development and tailored design of hybrid activity proposals, resorting to effective triggers, is also crucial to bridge online and offline experiences. The success of the latter, however, is dependent on digital inclusion conditions. Hence, promoting digital inclusion is paramount for ensuring equitable access and broad participation within these hybrid learning ecosystems. In this sphere, policy initiatives should observe the provision of access to technology and reliable internet connectivity. Concurrently, comprehensive training programs designed to cultivate digital literacy skills are essential, for both seniors and juniors, to enable all participants to effectively engage with the digital facets of these spaces.

By systematically integrating these practical recommendations, educators and policymakers can effectively harness the inherent power of informal learning within affinity spaces, thereby cultivating profound curiosity and sustaining long-term interest and intellectual engagement across a diverse spectrum of subjects.

5. CONCLUSIONS

This paper explored the development of curiosity, interest, and engagement fostered through informal learning activities that support the development of digital and cultural literacy, within the context of intergenerational digital storytelling in NOOA. Grounded in an action-research methodology and relying on qualitative data, the project demonstrated that sustained engagement was most effectively cultivated through in-person intergenerational digital storytelling interactions. Conversely, digital platforms showed strength in sparking initial curiosity and enabling wider participation, but faced limitations in maintaining prolonged engagement, particularly among older participants. These findings contribute to a growing body of work on informal learning by expounding how hybrid spaces—where digital and face-to-face interactions coalesce—can support culturally responsive and generationally inclusive learning environments.

Evaluating informal learning in affinity spaces requires a mix of qualitative insights, focusing on social interaction, participant engagement, and the influence of space or platform design to provide a comprehensive understanding of how learning occurs in these dynamic, interest-driven communities. This work adds nuance to our understanding of the CIE construct by illustrating how each component may be differentially activated depending on the different modes of participation. In doing so, it offers practical insights for educators and community practitioners designing interventions that aim to promote cultural literacy and heritage while enhancing digital competencies. It also highlights both the opportunities and potential challenges that underscore the need for intentional scaffolding in hybrid settings, as well as the importance of tailoring effective interest triggers, underlining the iterative and facilitative role of moderation in sustaining participation—particularly in hybrid and intergenerational settings.

Despite the valuable insights offered, this study is not without limitations, beginning with the methodological choices that shape both the research process and its outcomes. While the action-research approach was appropriate to enable iterative reflection and adaptation, it also introduced the potential for researcher bias. This positionality of the researcher, as both facilitator and analyst, was critically reflected upon throughout the process, with reflexive practices employed to mitigate its influence. Moreover, interpreting the findings requires careful attention to the specificities of the audience, context, and topic. Looking ahead, future research should not only address these methodological and contextual limitations but also explore strategies to strengthen engagement in online affinity spaces, particularly for underrepresented or digitally excluded groups. Further investigation is also needed into how hybrid affinity spaces can be scaled and sustained in varied sociocultural contexts, all of which influence the dynamics of engagement, in-person and online. The lessons learned here hold strong potential for informing community education programs, cultural heritage initiatives, and policies aimed at digital inclusion—ultimately reinforcing the value of culturally grounded, participatory media practices in bridging generational divides and fostering lifelong learning in an increasingly hybrid world.

ACKNOWLEDGEMENT

The authors gratefully acknowledge the support of the Portuguese Foundation for Science and Technology (FCT–Portugal), through the Ph.D. Grant PD/BD/114139/2015 and COVID/BD/151585/2020, with the DOI https://doi.org/10.54499/COVID/BD/151585/2020.The authors also acknowledge the support of CIQUP, Faculty of Science, University of Porto (Project UIDB/00081/2025), and of IMS- Institute of Molecular Sciences (Project LA/P/0056/2020).

REFERENCES

Ang, S.S. et al., (2018). Learning in the Context of Work in a Digital Age: The Use of Digital Media in Informal and Formal Learning Contexts. *Professional and Practice-based Learning*, 21, 87–101. 10.1007/978-3-319-63257-5_7.

Arnone, M.P. et al., (2011). Curiosity, interest and engagement in technology-pervasive learning environments: A new research agenda. *Educational Technology Research and Development*, 59 (2). 10.1007/s11423-011-9190-9.

Bennetts, C., (2001). Lifelong learners: in their own words. *International Journal of Lifelong Education*, 20 (4), 272–288. 10.1080/02601370110048782.

Braun, V. and Clarke, V., (2012). Thematic analysis. *APA handbook of research methods in psychology, Vol 2: Research designs: Quantitative, qualitative, neuropsychological, and biological*, 57–71. 10.1037/13620-004.

Gee, J.P., (2005). Semiotic social spaces and affinity spaces: From the age of mythology to today's schools. *Beyond Communities of Practice: Language, Power and Social Context*. 10.1017/CBO9780511610554.012.

Haythornthwaite, C., (2019). Learning, connectivity and networks. *Information and Learning Science*, 120 (1–2), 19–38. 10.1108/ILS-06-2018-0052.

Monteiro, J.C.C., (2024). *Narratives of Our Age: Intergenerational Digital Storytelling and Cultural Identity*. University of Porto. Available at: https://sciproj.ptcris.pt/83786DFA.

Prestridge, S. et al., (2021). New alignments for the digital age: insights into connected learning. *Educational Technology Research and Development*, 69 (4), 2171–2186. 10.1007/s11423-021-09968-5.

Rogoff, B. et al., (2016). The Organization of Informal Learning. *Review of Research in Education*, 40 (1), 356–401. 10.3102/0091732X16680994.

Shonfeld, M. et al., (2021). Learning in digital environments: a model for cross-cultural alignment. *Educational Technology Research and Development*, 69 (4), 2151–2170. 10.1007/s11423-021-09967-6.

DESIGN OF A SYNTAX–ANALOGY-BASED LEARNING TOOL FOR ENJOYABLE HARMONIC EDUCATION USING NATURAL LANGUAGE STRUCTURES

Yoshitsugu Kohjiba[1], Hinata Nakamura[1], Soma Arai[2] and Takuro Shibayama[1, 2]
[1]*Tokyo Denki University*
Ishizuka, Hatoyama, Hiki, Saitama 362–0074, Japan
[2]*Tokyo College of Music*
1–9–1 Kamimeguro, Meguro, Tokyo 153–0051, Japan

ABSTRACT

Harmony, which is necessary for understanding and creating tonal music, is a widely known and useful discipline that is not limited to music learners. In existing learning methods, it is necessary to learn basic music theory, harmony theory, the environment as well as performance techniques for practicing the theory and matching and coordinating the theory with sonic images. Therefore, the learning cost is high and beginner students are often hesitant to commence studying. The software under study enables students who have not yet learned harmony theory to learn a sense of harmonic progression from a sound using English grammar, which is a skill they have already acquired. The purpose of this software is to expand the entry points for learning harmony.

The software is intended to create a learning effect through learners' self-reciprocal and proactive structuring. This is not the case with conventional games, where the player intervenes in the game to find a specific goal or strategy; rather, the individual player intervenes in a self-exploratory manner; in other words, the learning effect is created by exploiting their own senses. We believe that this method has the potential to broaden the concept of learning through games and to be linked to the essence of learning skills that are difficult to articulate, such as the arts.

KEYWORDS

Expansion of Entertainment, Gamification, Learning of Harmony

1. BACKGROUND: THE SOCIAL CONTEXT OF THE RESEARCH

Many people intuitively sense the changes in chords when listening to music, even if they do not explicitly know the term "harmonic progression". However, understanding and composing tonal music require explicit knowledge of how the chords are connected. Thus, mastery of harmony theory requires foundational skills in reading music, perceiving harmonies aurally, and accessing suitable performance environments. The high cost and complexity associated with acquiring these skills frequently become barriers, leading to resistance to structurally comprehend tonal music.

Understanding "harmonic progression" is not easy. Nevertheless, learning harmony is necessary because it is recognised as a necessary part of understanding and composing tonal music. In practice, learners may intuitively grasp harmony without explicit structural knowledge, guided by sensory perception and experiential familiarity. The motivation for learning harmony relies on aspiring learners' sensory convictions.

Although learners recognise the practical value and rationale for studying harmony, their ultimate objective typically focuses on composing music. Personal digital audio workstation (DAW) environments have become widely accessible, substantially lowering the barriers to music production. Consequently, more learners feel the need to acquire harmony theory and related concepts to support their compositional activities.

Despite reduced environmental barriers, learners frequently become frustrated, indicating that the overall effort required still outweighs the perceived benefits. High-level music notation skills and associated audio production technologies remain essential, imposing additional learning requirements. In addition, there is the possibility of a disconnect between the learner's sensory perception of an image and the need to construct it logically based on harmony theory and the image created by thinking based on harmony theory.

Thus, learning harmony theory helps with composition and other logic-based activities, but it simultaneously leads to the contradictory result of standardising the thinking process. This paradoxical outcome generates internal conflicts that complicate learners' experiences. Therefore, it is necessary for the learner to discriminate between the goals of learning, but it is arduous for beginners to do so, and learning with these things in mind is demanding, so it is difficult to find step-by-step results. Based on the above, the authors believe that the difficulty of learning harmony stems from two points: the external problem of learning structural logic and the need to match the internal logic established through learning, and the internal problem of learners' difficulty in setting their own goals. This discrepancy between external requirements and internal objectives characterizes the central challenge of traditional harmony education methods and remains difficult to resolve using conventional theory-based approaches.

2. DESIGN CONCEPT OF A NEW LEARNING METHOD

2.1 Concept of a Learning Model with the Aid of Linguistic Knowledge

As indicated in the Problem Statement, the difficulty in learning harmony stems from the high cost of learning until the acquisition of theory is fed back into the learner's senses. The difficulty in setting the stages of learning progress arises from the need for comprehensive learning of theory. By reducing this cognitive load and clearly defining the learning steps, it becomes possible to create accessible entry points for harmony education. Leveraging the structural knowledge already possessed by learners could significantly lower these barriers.

In music education, precedents exist for enhancing learning by linking concepts across different disciplines. Examples include Lerdahl and Jackendoff's structural-linguistic approach in "GTTM" (1983), Sumiko Tanaka's "Iroompu" (1957), Masaki Matsubara et al.'s "Score Illuminator" (2012), and Herbert Grüger's "Liederfibel" (1927), which integrate visual and musical information. These examples highlight both the potential and limitations of structural musical analysis. Notably, while quantitative structural analyses often encounter methodological limitations, alternative approaches such as "Iroompu" and "Score Illuminator" successfully facilitate learning despite cognitive differences between instructors and learners. This asymmetrical learning illustrates that learners can independently construct meaningful connections across various information domains.

Considering these insights, we selected English grammar and vocabulary as concrete sources of structural knowledge to facilitate the learning of harmonic progressions, an abstract target. To achieve this, English language elements must be structurally reinterpreted beyond their conventional linguistic functions to guide learners toward abstract cognitive processing. Consequently, we developed a method that allows learners to form abstract inferences independently and recursively apply them, enabling the structural recognition of harmonic progressions without traditional harmony training.

2.2 Design of a Learning Model Utilizing Linguistic Knowledge

When designing a learning model that leverages linguistic knowledge, it is critical to define clear guidelines for using natural language structures and harmonic progressions. The following common characteristics of spoken language and harmonic progression informed our integrated model.

A) Both assumptions are that the listener will definitely finish within a limited time.
B) Both possess a structural regularity.
C) Both are irreversible and cannot be corrected retrospectively.

Based on these shared traits, our approach involves extracting structural analogies between natural language and harmonic progressions from symbolic and phonological perspectives.

2.2.1 Linguistic and Acoustic Conditions

We established the following specific conditions.

Natural language and related acoustic characteristics:

A) Adoption of grammatical structures with strict word order dependency, prohibiting scrambling.

B) The vocabulary used is easily pronounced by Japanese speakers and is structured in units of one foot and two morae.

C) Vocabulary that is estimated to fall within a speech time range of 340–430 milliseconds (to reduce the load on short-term memory).

Harmonic progression and related acoustic characteristics a) use rules that do not conflict with English grammar, and b) the length of the sound source of the chords used, is uniform.

In phonology, a mora is a unit that represents a quantitative segmentation of phonemes, essentially indicating where phonemes may be divided. (this is, admittedly, a rather simplified explanation.) In Japanese, morae are conceptualised differently than they are in Indo-European languages. Specifically, each mora typically corresponds to a single character in written Japanese, excluding small kana and other modifiers. A foot, in the context of native Japanese prosody, is generally composed of two morae and serves as an intuitively familiar rhythmic unit in spoken Japanese.

2.2.2 Cognitive Integration

This model strategically utilises the similarities and differences between linguistic and musical auditory information. Language and music share linear temporal properties essential for information transfer, but differ significantly in their semantic specificity. By intentionally employing these acoustic similarities and differences, our method allows learners to structurally infer harmonic progressions without requiring traditional systematic harmony training.

2.3 Building an Integrated Learning Model without Structural Exceptions

However, it is necessary to clarify how structural musical analysis differs from linguistic structures, particularly in terms of acoustic perception. Previous studies have shown that while certain linguistic or visual elements can be linked to musical acoustics, fully integrated structural decoding remains challenging, as evidenced by the limitations observed in “GTTM” and related approaches. This difficulty arises primarily from the lack of unified methods for handling exceptions within structural theories. Additionally, as Peretz (2003) and Satoh (2018) indicated, distinctions between language and music based on auditory perception are inherently flexible, hinging upon individual perception: learners constantly address the subjective question, “How do we perceive it?” This is accomplished by subjectively constructing structures, similar to how people formulate meaning from the acoustics of natural language or how they construct harmonic theory from the acoustics of music. In other words, although using concrete non-musical information does not always yield an integrated understanding of musical structures – subjectively meaningful information can still positively influence the learning process. Importantly, effective learning outcomes can be achieved even without explicit answers from teachers. Such subjective learning strategies contrast with approaches like the “GTTM”, which aim for strict structural consistency (Jackendoff, 1983). Nevertheless, methods such as Matsubara et al.’s “Score Illuminator” (2012), which links abstract performance images with visual elements (colours), demonstrate the value of abstract associative strategies, as suggested by Kyoya (2014). Similarly, “Iroompu” (1957) linked pitch to colours to facilitate intuitive comprehension, demonstrating its effectiveness even without a precise structural representation.

Based on the above, providing learners unfamiliar with formal harmony theory with a step-by-step learning experience based on auditory information is considered effective in facilitating self-awareness of progress through recursive reflection without requiring prior knowledge.

3. DESIGN AND IMPLEMENTATION OF SOFTWARE *ENGLISCHTOWASEI*

3.1 Implementation Environment and Design

The *EnglischToWasei* software was developed to accomplish this task. It comprised the following:

(1) To be able to acquire the acoustic body of a harmonic progression without the use of instruments.

(2) The ability to use English with only knowledge of English grammar, except for listening to the acoustic form of English.
(3) Learners should be able to define and assess the effects of their learning both spontaneously and recursively.
(4) The design concept is based on the following principles.

The English sentences, English words, harmonic progressions and chords used were in accordance with those defined in the "Basic Theory of the Learning Method" section, but some English words that deviated from the definitions were used. This was performed to verify the effectiveness of the proposed method.

The execution environment was assumed to be a screen touch device running Android OS (8.1), and Unity was used for creation.

3.2 Operation Screen and Interface

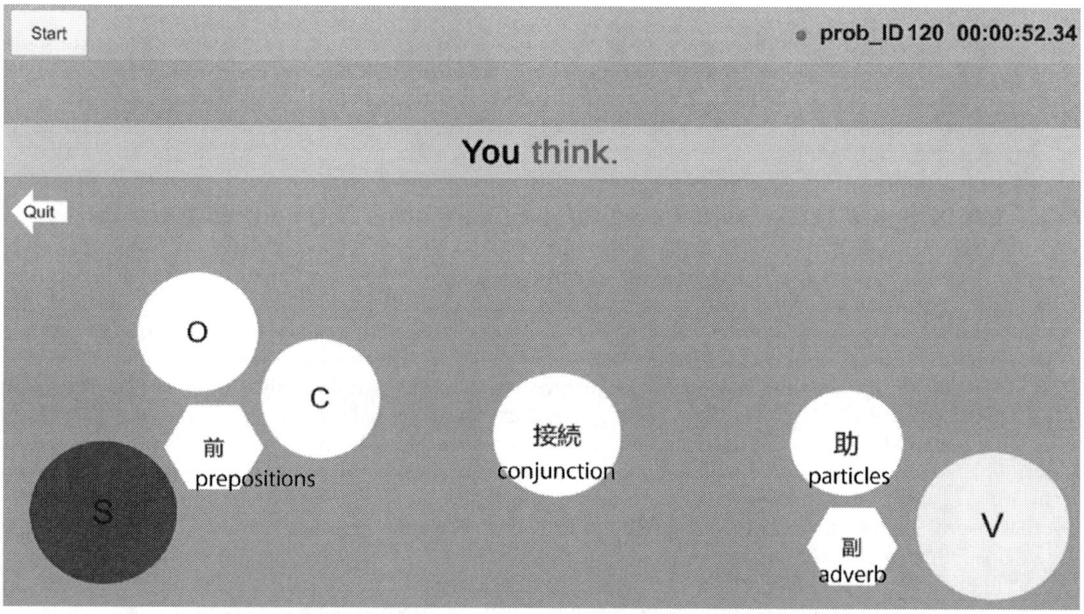

Figure 1. Graphical user interface (GUI) of *EnglischToWasei*

Note: The four buttons on the left are operated with the left hand, and the three buttons on the right are operated with the right hand. The Conjunction button is positioned in the centre so that it could be pressed with either hand. This is to accommodate future method expansions that may require the reversal of the order of the left- and right-hand operations.

The screen consists of buttons for Subject+Verb+Object+Complement (SVOC) + part-of-speech + pronouns in conjunctive sentences (selection buttons below), English text, question ID, elapsed time, and start and end buttons. Pressing the Start button initiates the recording, and after a countdown, the text is displayed at the upper centre of the grey ribbon. The text simultaneously turns orange in word order, and when the Select button is pressed, the colour returns to black as soon as a chord is played, following which the next word turns orange. Pressing a button sounds like a chord, but if one presses a button that is not the correct answer, the correct button will glow red and the chord will sound different from the originally intended chord progression. The program ends after completion of a specified number of repetitions. A CSV file contains the recorded audio file, ID of the button pressed, time from the start to the point when the button is pressed, and the question ID.

The sound source is a recording of isochronous chords on a piano, created with the assistance of a music specialist.

The user interface (UI) consists of the following elements:

A) English text (displayed in units of button operation requests)

B) Selection buttons (SVOC structure and part-of-speech)
C) Problem ID, elapsed time display
D) Start button
E) Quit button.

3.3 Instructions for Use (Manual)

The learner presses a button that matches the text in the orange region, which produces a chord. The orange area is simultaneously read at the pitch that is subjectively considered appropriate.

3.4 Design Philosophy and Learning Method Characteristics

As this software prioritises logical consistency and effective learning methods, it has the following characteristics:

(1) Exclusive use of English grammar information
The software provides learners exclusively with English grammar-based information, deliberately omitting explicit musical details such as notation or harmony theory. The learner can only obtain this information by listening to the sonorous chords and harmonic progressions. This is aimed at acquiring the acoustics of harmonic progressions as abstract information, as explained in the basic theory of this method.

(2) No corrections or mistakes
When an incorrect selection occurs, the correct choice is visually indicated, but the learner cannot redo or correct it. This approach aligns with Saussure's principle of linearity, which reflects the temporal and irreversible nature of musical perception. Because auditory images differ with each listening session, correcting mistakes would not help novice learners effectively recognise harmonic progressions. Instead, learners must learn to differentiate between correct and incorrect progressions purely through auditory experience.

(3) Optimization of selection button placement
If an incorrect button is selected, the system is designed to play a chord that is not in the same key as the originally intended chord. As mentioned above, there is visual information that the correct button turns red, but there is nothing that the learner can recognise as a mistake in the acoustic body only. This is to suggest (guide) the learner that the chord progression, including the chord caused by the wrong button operation and the chord progression before and after it, is "a chord progression that does not need to be memorized". This is to accomplish the seemingly contradictory task of having the learner listen to the harmonic progression in a holistic manner by not allowing correction, etc. At the same time, the learner perceives the harmonic progression discriminatively through pressing buttons, taking into consideration how the presence or absence of learning harmony theory functions when listening tonally. The purpose of this study was to examine the role of harmonic theory in its development of the harmonic theory. As indicated in the Background section and problem statement, it is suggested that even those who have not learned harmony theory perceive harmonic progression as structural information through the senses; however, it is true that harmony theory makes this inference even clearer at the same time. This difference in accuracy, with no conceptual difference, is similar to the contrast between those with and without an absolute pitch. As shown by Mikumo (1990), absolute pitch holders reproduce melodies and their memories better than non-absolute pitch holders, suggesting that they recognise melodies discriminatively. However, it has been suggested that those who do not have absolute pitch perceive melodies holistically. As mentioned, this is similar to the difference in the reproduction of heard music or images of music, depending on whether they have learned harmony theory. In other words, the way in which we perceive the sound of a harmonic progression changes progressively as we learn it, and we can expect to move from a holistic to a discriminative way of perceiving it as we become more proficient. It should be noted that the "correct harmonic progression" and "incorrect harmonic progression" here do not refer to harmonic progressions per se, but rather to harmonic progressions that are assumed by the method.

(4) Learning support through ergonomic device design

Anticipating reduced operation times as proficiency increases, the software employs an ergonomic design that encourages the alternating use of both hands, enhancing ease of use. Buttons associated with similar grammatical functions are placed close to each other to facilitate intuitive selection. This arrangement simultaneously promotes efficient operation and reinforces learners' awareness of accuracy during the learning process (Figure 2).

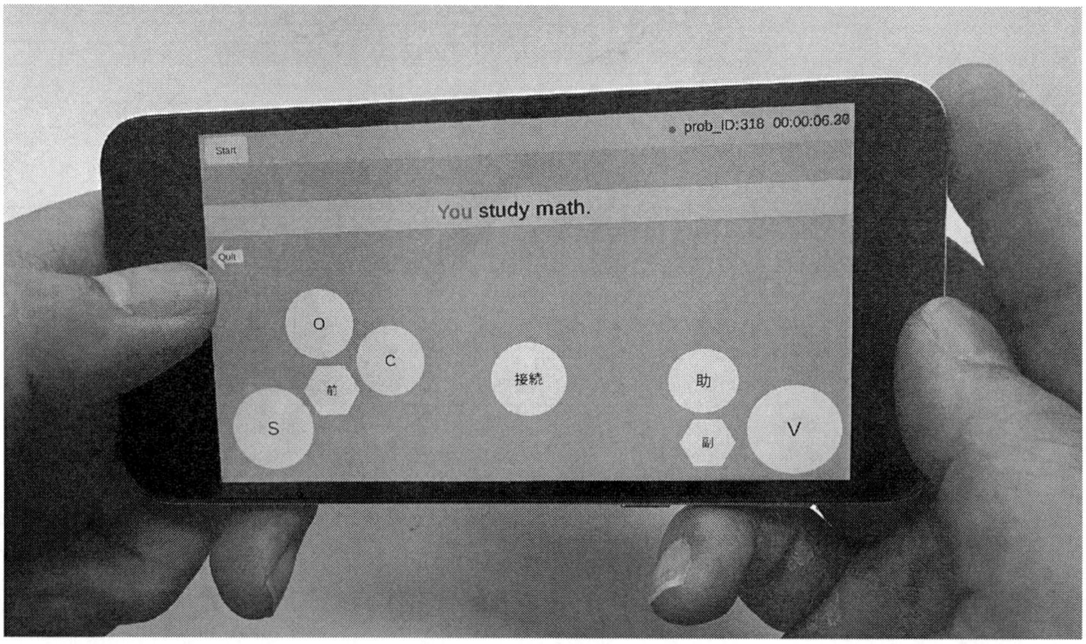

Figure 2. The design allows learners to hold their smartphones with both hands and proceed with learning using both thumbs

4. EXPERIMENTAL RESULTS AND OBSERVATIONS

To measure the effectiveness of this software, the authors selected participants who met the requirements of the target learners of this method, and conducted data acquisition on two occasions during 2020–2023 with the support of the Shibayama Laboratory of Tokyo Denki University. Participants were provided with the device, a manual showing the operating procedures, and pre- and post-questionnaires to assess learning effectiveness. Four participants took part in the first session, and 13 in the second.

The first and second subjects were asked to perform Steps 1–11 and 1–12, respectively. The number of questions for each step was 30 for Steps 1–8, 15 for Step 9, 18 for Step 10, 10 for Step 11, and 50 for Step 12, which were randomly selected from Steps 2–11. The number of operations required per question for each step was two for Step 1, three for Steps 2, 3, 5, and 7, four for Steps 4, 6, 8, and 9, six for Step 10, and seven for Step 11. Therefore, if the user proceeds to Step 11, 528 button operations are performed ($30 \times 2 + 30 \times 3 + 30 \times 4 + 15 \times 4 + 18 \times 7 + 12 \times 6 = 528$).

A music specialist listened to the sound sources obtained from the two experiments, and the pitches uttered by the participants were recorded. Any special notes on pronunciation and pitch were recorded separately. When compared to the CSV file in which the input information was recorded, several phenomena (characteristics) were observed.

A) Confirmation of structural recursion in harmonic cognition

Although there were individual differences in the "subjective pitch considered appropriate" for each participant in combination with chords, there was individual recency in the pitches associated with words used in the same sentence pattern and grammar. This result suggests that participants were able to use their previous learning from another step. Additionally, all participants used one of the chordal component tones for the accompanying pitch. Until the pitch of the accompanying notes was determined, the player was committed to only listen to the chords by pressing the buttons, and the operation time was irregular; however, once the pitch was determined, it became regular, and the average time required for the operation converged. No mistakes (except for misreading the text) were found in the selection of the tonalities. When a selection or operation error occurs, there is no continuity between the selection and error before and after the selection or operation error.

B) Relationship between proficiency in linguistic knowledge and learning effectiveness
A few participants made mistakes in the choice of complement, object, auxiliary, and adverb, but the corresponding participants vocalised at a pitch that matched the chord heard based on the wrong choice, and the mistakes were recurrent (they regularly made mistakes and vocalised at the wrong pitch).

C) Formation of intermediate acoustic perception
The recorded speech showed that all participants spoke English isochronously, although there were individual differences. In the case of the question in which the number of moras was intentionally greater than that of other words, the average time required to pronounce the word varied, although it was still isochronous. This is a characteristic of both natural language speech and pronunciation accompanied by an acoustic body in a normal harmonic progression, and is spontaneous because the participants were not instructed to vocalise in this way in this experiment. This suggests that participants performed the method according to an intermediate concept that was neither a natural language nor a conventional harmonic progression.

Table 1. An overview of the problems used in this method

Step	Model melody	Harmonic progression	English grammar	Tonality	Number of questions
1	i→v	I→V	S + V	[A]	30
2	i→v→i	I→V→I	S + V + C	[A]	30
3	i→v→iii	I→V→I	S + V + O	[A]	30
4	i→ii→v→iii	I→II→V→I	S + Av + V + O	[A]	30
5	i→iv→v	I→IV→V	S + Ad + V	[A]	30
6	i→iv→v→iii	I→IV→V→I	S + Ad + V + O	[A]	30
7	i→iv→v	I→IV→V	S + Ad + V	[S]	30
8	i→iv→v→iii	I→IV→V→I	S + Ad + V + O	[S]	30
9	i→iv→v→iii	I7→II7→V7→I7	S + Ad + V + O	[S]	15
10	i→v→iii→vi →i→v→i	I→V→I→VI→IV→V→I	S + V + O + Rlp + S+V+C	[S]	18
11	i→v→iii→vi (modulation) →v→i	I→V→I→VI (modulation) →V→I	S + V + O + Rlp +V+C	[A]	12
12	Random from Step 2 to 11				50

Note: Step: problem category; Model melody: the ideal example melody line (although other sounds may also be used); Harmonic progression: harmonic progression; English grammar: the grammatical structure of the problem; Tonality: type of tonality; Number of questions: the number of questions per step; [A]: affirmative sentence (major key); [N]: negative sentence (minor key); [S]: singular; [P]: multiple.

5. DISCUSSION: SIGNIFICANCE OF LEARNING EFFECT IN UNLEARNERS OF HARMONY

Based on the above, we believe that the learning effect of this method and the software is that even those who have not yet learned harmony can learn to read the tonal information of harmonic progressions structurally, make choices based on discrimination, and learn recursively using the abilities they already possess.

Some of these results may overturn conventional wisdom in music education, such as "it is difficult for students who have not yet learned harmony to discriminate tonality" and "it is difficult for students to articulate the pitch of the fourth degree increment" (although this was omitted in this paper).

6. FUTURE ISSUES AND PROSPECTS

As the software uses the device's built-in microphone for recording, it can only create sound sources according to the performance of the hardware and does not separate the audio source from the software source. The experimental environment will be reconstructed for future research. At the current stage, the design prioritises research rather than gameplay to motivate players, as the objective of measuring learning effects is given priority, but once the research is completed, puzzle-like games that utilise the symbolic aspect of the learning method (e.g. Mojipittan), or games in which the sound elements of the harmonic progression are matched with the game progression, will be developed. After the research is complete, we plan to move on to the stage of creating games such as games that structurally utilise sound elements in the gameplay to create a sense of immersion (e.g. Rez), by matching the sound elements of the harmonic progression with the game progression. The challenge is how to improve the game while maintaining the learning effect. In parallel, we would like to improve the game so that more complex harmonic progressions can be learnt, contributing to both the learning effect and the gameplay.

ACKNOWLEDGEMENTS

This study was supported by the Research Institute for Science and Technology of Tokyo Denki University (Grant Number Q25D-12) and JSPS KAKENHI (23K00243).

REFERENCES

Grüger, H., (1927), *Die Liederfibel*, Ostdeutsche Verlangsansalt, Germany.

Kawano, M., (1998), Mora, Syllable, and Rhythm: A Psycholinguistic Study. *Journal of the Phonetic Society of Japan 2*, No. 1, pp.16-24.

Kyoya, I., (2014), The Relation Between Abstract Concepts and Concrete Information: A Comparison of Concrete Words and Abstract Words. *Journal of Ritsumeikan University Humanities Society*, 2014, pp. 156-163.

Lerdahl, F. and Jackendoff, R., (1983), *A Generative Theory of Tonal Music*, Cambridge, MA: MIT Press.

Matsubara, M., Suwa M. and Saitoh, H., (2012), An Interactive Learning-aid System for Analytical Comprehension of Music by Highlighting Orchestral Score in Colors. *The Journal of Japanese Society for Artificial Intelligence*, Vol. 27, No. 5, pp. 281-295.

Mikumo M., (1990), Coding Strategies and Recognition of Melodies. *The Japanese Journal of Psychology*, Vol. 61, No. 5, pp. 291-298.

Nagano, S. and Uesawa, K., (1961), The Effect of Color Notes on Organ Learning in Early Childhood. *The Journal of The Japanese Association of Educational Psychology*, Vol. 9, No. 2, pp. 92-100.

Peretz, I., (2003), Modularity of Music Processing. *Nature Neuroscience*, No. 25, pp. 686-691.

Satoh, M., (2018), Cognitive and Emotional Processing in the Brain of Music. *The Japanese Journal of Neuropsychology*, Vol. 34, No. 4, pp. 274-288.

Shirai, N. and Takami, H., (2018), A Study of "E-fu" in Germany: Liederfibel, a Songbook with a Fusion of Music and Illustration, *Bulletin of the Faculty of Education of Bukkyo University* 2018, Vol. 17, pp. 63-74.

Tanaka, S., (1957), Iroompu, https://www.ongakunotomo.co.jp/catalog/detail.php?id=451411 (in Japanese), accessed Apr. 20, 2015.

ASPECTS OF COMPUTATIONAL THINKING DEVELOPMENT IN FIRST-YEAR HEALTH SCIENCE STUDENTS

Edita Butrime[1] and Virginija Tuomaite[2]
[1]*Vytautas Magnus University, Lithuanian University of Health Sciences*
K. Donelaicio g. 58, Kaunas, A. Mickeviciaus 9, Kaunas, Lithuania
[2]*Kaunas University of Technology*
K. Donelaičio g. 73, Kaunas, Lithuania

ABSTRACT

In recent decades, there has been a worldwide focus on the integration of information technology into educational curricula to adapt to the changing technological world. Schools are being equipped with computers and software to engage students and facilitate the completion of assigned tasks. In the mid- to late 2000s, the focus of computational thinking (CT) began to shift away from the computer science field and toward thinking that it was a core skill for everyone. The paper aims to reveal some aspects of CT development of first-year health science students.

KEYWORDS

Computational Thinking, Virtual Learning Environment of University, e-Learning of Gen Z

1. INTRODUCTION

In recent decades, there has been a worldwide focus on the integration of information technology into educational curricula to adapt to the changing technological world. Schools are being equipped with computers and software to engage students and facilitate the completion of assigned tasks. However, the use of technology is not enough to ensure relevant and quality teaching. The 21st century is unthinkable without media and information flow, and students who are currently learning have different attitudes toward learning and learning tools than their teachers or parents do (Bakonis, 2014). Some of the most important cognitive skills of the 21st century, according to Ng (Bureau of Education, U. & Ng, S., 2019): information processing - data interpretation and data analysis; problem solving and engineering thinking; scientific investigation; computational thinking and ICT; manipulative and technological skills; collaboration and communication skills. This set of skills is needed in the present, as well as in future life and professional activities. According to Lockwood & Mooney (2017) "although there is a lot of work currently being done around the world in many different educational contexts, the work relating to computational thinking (CT) is still in its infancy. Along with the need to create an agreed-upon definition of CT lots of countries are still in the process of, or have not yet started, introducing CT into curriculums in all levels of education" (p. 1).

The paper aims to reveal some aspects of computational thinking (CT) development of first-year health science students.

In this paper e-learning is analysed as a socio-cultural system (Scedrovitskij, 1995; Mamardasvili, 1958; Kvedaravicius, 2006; Butrime, Zuzeviciute, 2014). Such an approach towards e-learning enabled the presentation of 'a multi-dimensionality' in the concept. The analysis of e-learning as a socio-cultural system enabled the formulation of an interdisciplinary problem, for the solutions of which it is necessary to invoke theories and outcomes of computer science, culture, and education. The analysis allowed the enumeration of the forms and contents related to educational support for the participants (students) of the system.

The article is based on the generational theory. This approach allows us to reveal the significance of computational thinking in contemporary studies. At the moment in educational institutions You can meet late Baby Boomers, Gen X and Millennials professors who are teaching representatives of late Millennials and early Gen Z. Thus, a generation is considered a group of people born in similar years with very similar tastes and preferences (Stanisauskiene, 2015; Martín-Critikián, & Medina-Núñez, 2021; Espejo et al., 2025).

2. THE CONCEPT OF COMPUTATIONAL THINKING

ICT's influence on the learning process is not only significant but its significance has evolved over the last few decades. Since 1990, computational thinking has been taught primarily at universities. Such thinking was linked to the study of exact and computer sciences. At the turn of the century, educators and policymakers began to recognize an understanding of the mechanisms of digitization as an important 21st-century skill (Denning, Tedre, 2019). 21st Century Skills Partnership is visualised in Figure 1. Key subjects and themes for the 21st century are mathematics, economics, science, English, reading, language arts, and foreign languages. Also art, geography, history, citizenship, and political science. These topics must also include new content areas that will impact the workplace and community, i.e., learning and innovation skills, information, media, and technology skills, and life and career skills (Christopoulos, 2023).

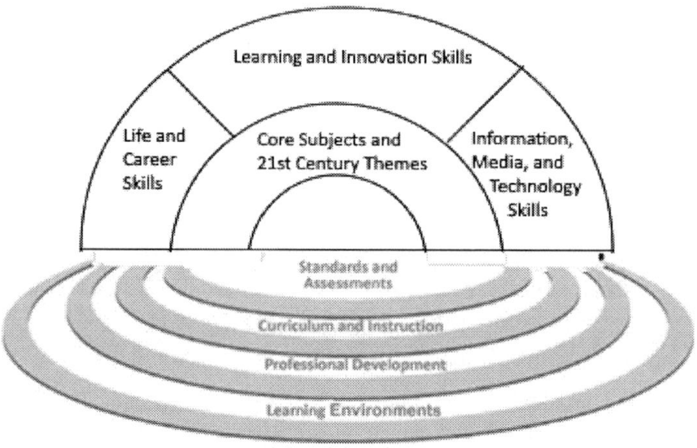

Figure 1. 21st Century Skills Partnership (Christopoulos, 2023; Fadel, Velaga; 2019)

Authors note that European societies are facing a positive challenge to create a new, dynamic education system that prepares citizens for creative work and non-linear career paths, as well as for the new phase of digital globalisation (Tsekeris, 2019; Schwab, 2018; Baldwin, 2019). We are now in the Web 4.0 phase. According to Wadhwa (2016), technology is rewriting the basic rules of society.

In the latter stage, there is a lot of interdisciplinary testing. One such is the integration of STEM into STEAM, where the letter A stands for Art (Christopoulos, 2023). The STEAM concept emphasises creativity as a factor in creating an environment conducive to learning, as a way to develop creative and critical thinking, and as a prerequisite for the development of communication and collaboration skills (Christopoulos et al., 2023; Hunter-Doniger et al., 2018; Maeda, 2013; Peppler and Wohlwend, 2018).

Recognising that the arts (arts education) improve achievement in maths and language arts, the arts have been included in STEM curricula. On the other hand, arts, and creativity education has been linked to higher PISA scores, problem-solving and critical thinking skills, teamwork, change management, collaboration, and other personal qualities that are highly valued in today's labour market. Learning and innovation competencies aim to ensure that students learn in a productive environment. Learning and innovation competencies include critical thinking and problem-solving; communication; collaboration; creativity and innovation, i.e., the ability to generate a broad range of concepts to create new and useful concepts. Knowledge, media, and technology competencies (1) individuals in the 21st century should be able to use a wide range of information; 2) adapt to rapid changes in technological tools; and 3) contribute by working both individually and collaboratively). Knowledge Literacy; Media Literacy; Information and Communication Technology (ICT) Literacy: ICT knowledge and skills (to be able to use technical tools to conduct research, process, evaluate, and communicate data); Life and Career Skills (Christopoulos, 2023).

Such an approach corrected the concept of computational thinking. Computational thinking is the ability to recognize and formulate various environmental problems (tasks), organize and analyze data logically, represent them using diagrams and models, assess the solvability of a problem, and attempt to automate a solution using information technology (Sudeikiene, Gaucyte, 2020). According to Payne et al. (2022) "the definition is subject to debate, many experts agree that CT involves a set of thinking skills, and based on an analysis of former studies". According to Denning & Tedre (2019) "computational thinking is sometimes portrayed as a universal approach to problem solving" (xiii p.). "Computational thinking is the mental skills and practices for designing computations that get computers to do jobs for us, and explaining and interpreting the world as a complex of information processes" (Denning & Tedre, 2019; p. 4). In the mid- to late 2000s, the focus of CT began to shift away from the computer science field and toward thinking that it was "a core skill for everyone, not just computer scientists" (Payne et al., 2022; Wing, 2006, p. 33). Korkmaz et al. (2017) suggested that this concept (set of skills) should include creativity and cooperation, in addition to algorithmic thinking, critical thinking, and problem-solving. According to Dagiene at al. (2024), computational thinking is the ability to solve problems (tasks) in any domain in a series of steps, including refining the problem formulation, analysing the data, decomposing, abstracting, automating, and testing.

According to Dagiene at al. (2022), the components of informational thinking are data processing (gathering and selecting information, processing data, recognizing patterns, drawing conclusions); data representation (representing data in diagrams, text, figures, tables); breaking a problem down into smaller parts and combining them; and algorithmization (planning and organizing sequences of actions to solve a problem); abstraction (reducing complexity by focusing on key features and forming models); modelling (applying models and simulations, for example, to experiments); automation (recognising how technology can help solve new problems); parallelism (organising resources to solve a problem using synchronous transfer and collaboration tools); rationale; summarisation. According to CSTA (Computer Science Teacher Association) and ISTE (International Society for Technology Education), the components of informational thinking are data collection (gathering and selecting appropriate information); data analysis (processing data, discovering patterns, concluding); data representation (organising and presenting data in graphs, charts, text, figures, tables, etc.); data analysis (using data as a basis for analysis).); problem decomposition (breaking a problem into smaller components and combining the components); algorithmization (planning and organizing sequences of steps to solve a problem); abstraction (reducing complexity by defining the basic idea, building characteristics and creating models); simulation (using modelling or simulation, e.g. for experiments); automation (recognising how technology can help with new tasks that would otherwise be repetitive, infeasible or overly burdensome); parallelism (organising the organisation of resources around a problem to solve it and to achieve an objective, through the use of tools for simultaneous transfer and collaboration).

Given today's trends in education (building a knowledge-based society, changing learning environments, competency-based learning), the role of education can be seen as unique in bringing about change towards higher-quality education. It can also be seen as an experiment based on the very rapid penetration of modern information and communication technologies.

3. INVESTIGATING ASPECTS OF COMPUTATIONAL THINKING IN FIRST-YEAR HEALTH SCIENCE STUDENTS

This is a pilot study, i.e. an initial review of the data obtained. Only a small part of the data is presented in the paper, which is related to the students' abilities related to computational thinking.

The aim: to review the characteristics of first-year students' communication and information search skills in the conditions of Intergenerational and Cultural Diversity. Study participants: Veterinary medicine, Odontology, Medical and Veterinary Biochemistry (135 first year students studying in Lithuanian), Odontology, Veterinary medicine (98 first year students studying in English). Data collection - 2024 fall semester. The participants of the research are the representatives of the Z generation.

Students work in a computer lab during the Introduction to Studies course and test (i.e. communicate on) the university intranet Microsoft 365, Teams, OneDrive, etc. Independent work task content for first-year students is simulation during which students become familiar with the following:

- Study administration system;
- Student portal;

- Technical support;
- University library automated catalog, subscription databases;
- Virtual learning environment Moodle.

The text of the task is presented step by step - an algorithm (Figure 2.). At the end of the task description, it is explained what file must be uploaded to the virtual learning environment (VMA) moodle. A file name template is provided. At VMA, students find a task description and a word file - a cover page template. Before starting work, students are verbally explained how to complete the work. Students' questions are answered. The deadlines for completing the work are discussed.

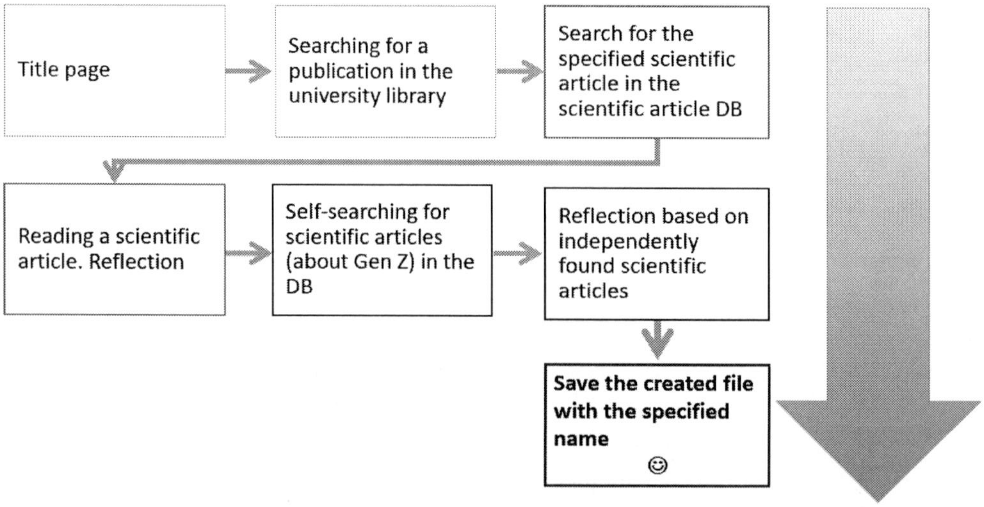

Figure 2. The algorithm of the task given to students

Simulation tasks are prepared in order of increasing difficulty (Figure 2). Students must use information search and critical thinking skills. All tasks are formulated in such a way that the methodology of the previous task can be used.

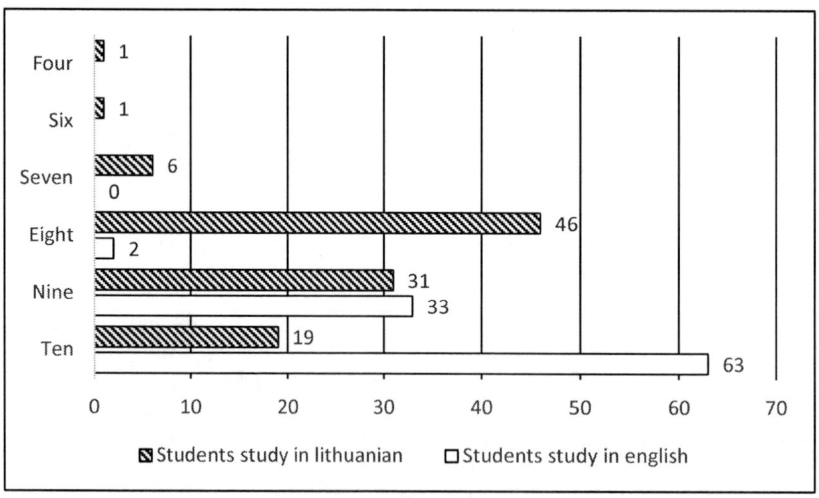

Figure 3. Results of the assessment of students' independent work

Students from different study programs performed differently. The results (Figure 3) allow us to formulate the following hypothesis: subcultures develop differently, and, therefore, students' abilities differ. Figure 3 shows the results of the assessment of the work of only those students who have already settled their independent work: Veterinary Medicine (104 students studying in Lithuanian), Odontology, and Veterinary Medicine (98 students studying in English). No assumptions can be made yet, as these are only initial data because not all students have submitted their work.

Having analyzed the mistakes made by students in individual assignments, we can conclude that they are varied and no trend can be observed. Communication problems include incorrect group code on the title page, the title of the work is not in the place where it was indicated, and the text of the work is written on the title page directly below the title of the work. Technical errors are the errors in the layout of the text and problems with uploading images to a Word file. Information retrieval and processing include citation errors, and errors in the list of references used. The analysis of working in a virtual learning environment revealed that not everyone was able to upload their work to Moodle, they either emailed or were late in submitting their work. Some students submit a file in a format other than the one specified in the assignment.

There are some limitations of data analysis. It is an analysis of the tasks completed by students who studied the subject Introduction to Studies in the fall semester of 2024. The method employed was qualitative research. The analysis was carried out on the reflections of students in Veterinary Medicine (students studying in Lithuanian), and in Odontology, and Veterinary Medicine (students studying in English). Mind maps were created during work groups (Odontology, Medical and Veterinary Biochemistry (students studying in Lithuanian). The qualitative study highlighted students' perceptions of the use of ICT in learning environments (Table 1).

Table 1. Analysis of students' mind maps. Topic: University studies and information and communication technologies

Categories	Subcategories	Contextual content
Students notice a change (9)	A frightening change (5)	„...some students are stressed by the need to work across multiple platforms and tools...“ „...students often feel anxious about technical issues that may disrupt their learning...“
	The need to adapt to constant change(4)	„...I noticed that the number of people reading books in my immediate environment has almost disappeared, while the use of more accessible information technologies and the amount of time spent on them has increased...“ „...there is a need to adapt to the constantly changing information environment...“
Communication in virtual environments (1)	Interactive IT solutions (2)	"...interactive <...> elements <...> increase interest and help memorize the material better..."
	Augmented reality (1)	"...virtual and augmented reality can be used in laboratories, architectural simulations, or medical practice..."
	Remote communication(2)	"Digital platforms help share material and assignments faster and make feedback easier." "...You can study from any location in the world..."
	Data security (2)	"...insecure platforms can damage student data..."
	Gamification (1)	"...gamified elements (Quizlet, Kahoot) increase interest and improve retention..."

Categories	Subcategories	Contextual content
Requirements for improving the competences of teachers (2)		"...It is necessary to invest in teacher training to use the tools effectively and ensure the quality of learning..."
Need to learn(3)		"...there is the need to adapt to a constantly changing information environment..."
Generation Z is in a hurry (1)		"...They often opt for quick and convenient solutions because they are used to quick access to information..."
Attitude towards studies (3)		"...I was pleasantly surprised and curious about the section on the aims of higher education institutions, which are not only to provide information and instruction, but also, as the author puts it, to prepare for lifelong learning...", "...to develop <...> the ability to seek, find, critically evaluate and use the information for problem-solving, decision making, continuous updating of knowledge, professional and personal development...>"
Critical thinking (3)		"...the concept of critical thinking is now more important than ever..."
Information competence (1)		"...I would not agree that the library plays an essential role in the development of information literacy. In my opinion, this should be done in lessons and lectures..."
Using diagrams and models (1)		"...I had never heard of mind-mapping apps before, so I learned how to use them..."

The students were given a lecture on visual learning techniques before drawing the mind maps. During the exercise, the students were divided into groups and the stages and methodology of brainstorming and mind mapping were explained to them. During this activity, the students not only drew mind maps but also tried out a remote group work tool. The analysis of the mind maps created by the students concluded that this was a challenging exercise. Only one map was accurate, i.e. in line with the topic. During the discussion, it was asked whether the students had drawn mind maps in secondary school. It turned out that the authors of the exact mind map had experience from secondary school. Despite the difficulties, the analysis of the mind maps allowed us to identify aspects related to computational thinking. Students were able to identify and formulate different environmental problems (tasks), organize and analyze data logically, represent them using diagrams and models (if the task is difficult you need to learn), assess the solvability of the problem (give advice to lecturers), try to automate the solution using information technology (expressed satisfaction with the use of interactive tasks, remote group work tools).

4. CONCLUSION

The pilot study is based on primary data processing. The number of participants is small, but the participants come from different high schools and different countries, which allows us to observe some trends in the different subcultures of Generation Z.

Students (representatives of Generation Z) from different subcultures have different experiences of using ICT-based tools in their study communication and practice, which influences their abilities. Thus, it is necessary to share the experience of using ICT-based tools in study practice for teachers working in different universities and in different countries.

Understanding students' strengths and weaknesses helps to understand their specific needs, teacher-directed expectations, and learning opportunities, which helps to create an effective learning environment.

Students (representatives of generation Z) notice changes in their study communication influenced by the penetration of ICT-based tools.

Students (representatives of Generation Z) evaluate the use of ICT-based tools in study communication and practice both positively and negatively.

Creativity and problem-solving need to be learned. New tools - new challenges. This study shows that this is important in the current phase of change in education, i.e. working with Generation Z students and working in an environment of intense and rapid ICT penetration.

REFERENCES

Bakonis, E., (2014). Mediju ir informacinis rastingumas Lietuvoje: laikas keisti požiurį. *Svietimo problemos analize*, 9(114), 1-8.

Baldwin, R., (2019). *The Globotics Upheaval: Globalization, Robotics, and the Future of Work.* New York, NY: Oxford University Press.

Bureau of Education, U. & Ng, S., (2019). *Exploring STEM competences for the 21st century*, UNESCO: United Nations Educational, Scientific and Cultural Organisation. Switzerland. Retrieved from https://coilink.org/20.500.12592/518tc8 on 15 Mar 2025. COI: 20.500.12592/518tc8.

Butrime, E., Zuzeviciute, V., (2014). *E-Learning as a socio-cultural system: towards a balanced development.* Saarbrücken: LAP LAMBERT Academic Publishing.

Christopoulos, A., (2023). *Praktinis STEAM ugdymo mokykloje įgyvendinimo ir tobulinimo vadovas (A Practical Guide to Implementing and Improving STEAM Education in Schools).* Europos Sąjungos finansuojamo strateginių partnerysčių projekto „STEAM ugdymo tobulinimas" (sutarties Nr.: 2020–1–LT01–KA201–077935) intelektinis produktas

Dagienė, V., at al., (2022). Teaching computational thinking in primary schools: Worldwide trends and teachers' attitudes. *Computer Science and Information Systems, 19*(1), 1-24.

Dagienė, V., at al., (2024). Fostering computational thinking in compulsory education in Europe: a multiple case study. *Baltic journal of modern computing.*, 12(2), 189-221.

Denning, P. J., & Tedre M., (2019). *Computational Thinking.* The MIT Press.

Espejo, L. et al, (2025). A Systematic Review on Unique Characteristics of Generation Z and Their Impact on Purchasing Decisions. *Journal of Ecohumanism.* 4(1), 12-25.

Fadel, C., & Velaga, K. C., (2019). File:Framework for 21st Century Learning.jpg. 19 March 2019. https://commons.wikimedia.org/wiki/File:Framework_for_21st_Century_Learning.svg.

Hunter–Doniger, T. et al, (2018). STEAM through culturally relevant teaching and storytelling. *Art Education*, 71(1), 46–51. https://doi.org/10.1080/00043125.2018.1389593

Korkmaz, O. et al, (2017). A validity and reliability study of the computational thinking scales (CTS). *Computers in Human Behavior*, 72, 558–568. https://doi.org/10.1016/j.chb.

Kvedaravicius. J., (2006). *Organizaciju vystymosi vadyba* vadovelis / Kaunas.

Lockwood, J., & Mooney, A., (2017). Computational thinking in education: Where does it fit? A systematic literary review. arXiv preprint arXiv:1703.07659.

Maeda, J., (2013). STEM +. Art = STEAM. *The STEAM Journal*, 1(1), 1–3. https://pdfs.semanticscholar.org/5d30/c4d0ffd67a4cd4d8ffbfa0eab32bf6ba8806.pdf

Mamardasvili M., K., (1958). Procesy analiza i sinteza. *Voprosy filosofii*, No 2. Moskva.

Martín-Critikián, D., & Medina-Núñez, M., (2021). Redes sociales y la adicción al like de la generación z. *Revista de Comunicación y Salud.* 11, 55-76. https://doi.org/10.35669/rcys.2021.11.e281

Payne, L., (2022). Tawfik, A., & Olney, A. M. Computational thinking in education: Past and present. *TechTrends*, 66(5), 745-747. https://doi.org/10.1007/s11528-022-00766-1

Peppler, K., & Wohlwend, K., (2018). Theorizing the nexus of STEAM practice. *Arts Education Policy Review*, 119(2), 88–99. https://doi.org/10.1080/10632913.2017.1316331

Scedrovitskij, G., P., (1995). *Izbrannye trudy. Moskva: Skola kulturnoj politiki.*

Schwab, K., (2018). Globalization 4.0 – what does it mean?. *World Economic Forum. In World Economic Forum Annual Meeting.* Available at https://www.weforum.org/agenda/2018/11/globalization-4-what-does-it-mean-how-it-will-benefit-everyone/

Stanisauskiene, V., (2015). Karjeros sprendimus lemianciu veiksniu dinamika kartų kaitos kontekste. (The Dynamics of Factors Determining Career Decisions in The Context of Generation Change). *Tiltai* (2): 1-20.

Sudeikiene, I., Gaucyte, D. (2020). *Ismaniosios technologijos ir informatinis mastymas* [viewed 2025-03-15]. https://www.nsa.smm.lt/wp-content/uploads/2021/10/ismaniosios-technologijos-ir-informatinis-mastymas-pradinukai.pdf

Tsekeris, C., (2019). Surviving and thriving in the Fourth Industrial Revolution: Digital skills for education and society. *Homo Virtualis*, 2(1), 34-42. doi:http://dx.doi.org/10.12681/homvir.20192.

Wadhwa, V., (2016). *6 Big Ways Tech Is Rewriting Society's Rules.* [viewed 2025-03-15]. https://singularityhub.com/2016/11/16/6-big-ways-tech-is-rewriting-societys-rules/

Wing, J. M., (2006). Computational thinking. *Communications of the ACM.* 49(3), 33–35. https:// dl. acm. org/ doi/ fullHtml/ 10. 1145/1118178.1118215

FRAMEWORK FOR NET ZERO PATHWAY FOR INDIAN CHEMICAL INDUSTRY

Mukundan Ramakrishnan, Wilson Kia Onn Wong and Suchismita Das
SP Jain School of Global Management, Mumbai, India

ABSTRACT

There is an increasing urgency to achieve net-zero carbon emissions to combat climate change. While all companies in a sector or industry have similar external environment, regulations etc, the key differentiator could be the engagement with stakeholders of the company i.e. customers, suppliers and employees in addition to management's efforts green investments and strategy and green financing to support this. This study provides a firm-level framework on the impact of these elements in the chemical industry by analysing the independent variables of customers, suppliers, employees, strategy & investment decisions and green finance on the dependent variable of financial outcome (i.e., revenue growth) to highlight the relationships between them. Based on the hypothesis, the questionnaire developed was administered online to obtain the current view of the independent and dependent variables from managements across 6 Indian chemical companies, with 40 respondents. Personal privacy was ensured during the data collection. The data was analysed using a multivariable regression model (deploying fit model by least squares), as the key aim was to understand the relationship between the dependent variable and independent variables. This study highlights that in this transition, while the dependent variable of revenue growth is impacted by independent variables of customers, employees, investment & strategy and green financing, the key differentiator is investment & strategy decisions by implementing decarbonisation projects and supporting these investments with green financing. Thus, industry leaders and policymakers need to focus on green project execution and green finance to accelerate the delivery of net zero. Currently, there is no specific study on an Indian perspective on the framework for a pathway to deliver a net zero in an energy-intensive sector such as the chemical industry. Similar studies could be extended using this framework for net-zero pathways in other sectors and regions.

KEYWORDS

Net Zero, Employee Engagement, Customer Engagement, Supplier Engagement, Strategy, Financing

1. INTRODUCTION

Globally, energy-intensive industries (EIIs) comprising steel, chemicals, cement, aluminium, and glass contribute more than 24% of total GHG emissions. This research focuses on the chemicals sector, which accounts for approximately 8.5% of industrial emissions worldwide (Bashmakov *et al.*,2022). While R&D and technology are significant accelerator in the net-zero pathway, there are several business enablers, such as leadership awareness of decarbonisation issues, motivating and empowering employees, development of strategy for transition from current to net-zero state, communicating and committing to customers and suppliers, and access to green finance. These non-technology enablers are the focus of this microeconomic research.

The IEA (International Energy Agency) in 2020, presented an ambitious net zero by 2050 plan for the energy sector, with an investment of US $1 trillion per year for each of the next three years or about 1% of the US $100 trillion annual global GDP, increasing to US $ 5 trillion per year by 2030. Such a transition would involve accelerated investment in renewable generation capacity, deploying storage systems, adding power transmission lines, and demand-side management. (Deutch, 2020).

In the case of wind and solar power systems, their declining cost structures have made them highly competitive renewable energy sources. From 2010 to 2020, the cost of electricity generated by utility-scale solar photovoltaics (PV) declined by 85%, from USD 0.381/kWh to USD 0.057/kWh (International Renewable Energy Agency, 2020). Further, technological advances like in "Small Modular Reactors" (SMRs) designed to generate up to 300 MW (megawatts) could also provide the medium- and long-term solution. The construction cost of an SMR is approximately US$2.6bn as opposed to the US$26bn needed for a large reactor, making this a highly scalable. SMRs could easily be built with decommissioned coal plants in brownfield sites, which offer substantial value in terms of their existing power grid and other infrastructure, ready access to water sources and a local skilled workforce (Wong, 2023).

Net-zero transition requires industry to green its processes by decarbonising. Several drivers for this at the firm level include green industrial policy, market, natural resources, human capital, and innovation. (Nyangchak, 2022). The facilitators to overcome barriers, include organisational leadership (top management commitment and high ambition), organisational capabilities (knowledge and learning, shared vision, R&D and innovation capability, long term focus, stakeholder engagement), and external support (public policies, synergistic relationships, and public grants and resources) (Johnson *et al.*, 2023). While all firms in a sector or industry segment face similar external environment including public policy, public grants and resources, the key differentiator is how the firm engages with customers, suppliers, employees, its investment & strategy decisions and green financing and the impact of these on revenue growth of the firm.

Thus, this microeconomic study at the firm level focuses on the relationship and impact of the independent variables 'customer engagement', 'supplier engagement', 'employee engagement', 'strategy and investment decisions', and 'green financing' on the dependent variable of revenue increase arising out of the transition to net zero. This leads to a competitive advantage for the firm, resulting in both environmental and financial performance and outcomes.

2. LITERATURE REVIEW

The chemical industry in India faces several challenges in its ambition to deliver net-zero emissions. These include sustainable feedstock based on biomass (which could be potentially limited by the local availability of water and land), reducing energy intensity as chemical manufacturing processes are often energy-intensive, relying heavily on fossil fuels (Gabrielle *et al.,* 2024; Trambouze, 1993). Ensuring sustainability across the entire chemical supply chain, including raw material sourcing, transportation, and waste management, is essential for achieving net-zero emissions. Collaboration with suppliers and customers to adopt sustainable practices can help address this challenge.

Transitioning to net zero emissions necessitates substantial financial resources, which can present a challenge for smaller companies or those operating with narrow profit margins. Access to financing options and incentives for green investments can aid this transition. Overcoming these challenges will require coordinated efforts from industry stakeholders, including companies, government agencies, research institutions, and non-governmental organisations, to foster innovation, investment, and policy support towards a sustainable, net zero future for the Indian chemical industry (Rajendran *et al.,* 2022).

Business models can serve as crucial mechanisms to facilitate sustainability transitions regarding how companies innovate their business models for system-level transformation towards sustainability (Hernandez-Chea *et al.,* 2021). Additionally, a systematic approach to managing environmental strategy can enhance financial performance and may help manufacturing managers strengthen their internal resources to improve their competitive advantage and outcomes (Tan *et al.*, 2022).

There are six internal company barriers (risk aversion, short-term orientation, economically driven thinking, reluctance to engage in trade-offs, resource shortages, and a lack of knowledge) and four external company barriers (technology, market, legislative, and societal and consumer factors). Furthermore, their interrelationships and integration into a holistic sustainable strategic management framework influence outcome. (Takacs *et al.*, 2022). The company's environmental strategy significantly impacts the transition to net zero (Hart and Ahuja, 1996). Firms with strong organisational capacities are more likely to adopt a strategy that supports the transition to net zero through emission reductions, product stewardship, and clean technology. Additionally, company boundaries encompass relationships between firms (Dyer and Singh, 1998). All companies are embedded within a network of relationships, and the critical resources of firms may extend across boundaries in inter-firm routines and processes. These connections between companies represent sources of relational rents and competitive advantages.

Customer engagement is an interactive process that fosters deep and meaningful relationships and interactions between a company and its customers (Hollebeek *et al.*, 2014). Only a limited number of existing studies have examined customer engagement in relation to sustainability or carbon management (Jarvis, 2017), and these typically adopted consumer behavioural perspectives to analyse customers' willingness to participate in sustainability initiatives (O'Brien *et al.,* 2020). Some studies highlighted the social attributes and platforms through which customer engagement can be pursued and measured (Jarvis, 2017). Suppliers' high-level, pro-environmental practices boost sales if their customers exhibit similar levels of environmental management

(Chen and Ho, 2019). Therefore, investing in environmental practices may not benefit suppliers if their customers do not engage similarly, and benefits could be realised only if these customers appreciate their suppliers' efforts. It has also been discovered that suppliers could engage their customers and heighten their dependency on business through instrumental motives and international environmental policies (Peng *et al.*, 2022).

The importance of engaging suppliers to enhance companies' sustainability performance has been highlighted in several previous studies (Sancha *et al.*, 2016; Tidy *et al.*, 2016). Although it is challenging for companies to control suppliers, powerful buyers can act as role models and establish standards, thereby promoting their suppliers' involvement in sustainability efforts (Amaeshi *et al.*, 2008). By employing sustainable supply management practices, firms can influence and engage their suppliers (Zahn, 2021), and by fostering collaborative relational practices, improved outcomes can be anticipated in terms of relational rents and competitive advantages (Dyer, 1998; Sancha *et al.*, 2016). Recently, the relational view has been broadened to encompass not only economic but also environmental and social benefits (Villena *et al.*, 2021). Inter-firm supplier-partner collaboration is one of the elements contributing to a firm's ability to achieve decarbonisation.

Further, a study on green finance and carbon markets (Zhang and Umair, 2023) found that the importance of green finance is growing for funding sustainable projects and reducing carbon emissions. The results reveal significant dynamic spillover effects between green bonds and renewable energy stocks, as well as between carbon markets and renewable energy stocks, highlighting a complementary relationship between green bonds and carbon markets. Thus, they play a crucial role in enabling net zero by de-risking sustainable projects, unlocking capital, and shifting capital allocation. Overall, green finance provides tools and funding to make sustainable projects more appealing to investors (Azoulay *et al.,* 2023). This assists in overcoming the initial hurdle of potentially higher upfront costs compared to traditional options. By mitigating risks and offering competitive returns, green finance encourages private sector investment in clean technologies and solutions that are vital for achieving net zero. Additionally, green finance urges financial institutions to move away from investments in high-carbon industries toward sustainable alternatives. This redirects capital towards a net-zero future (Bertram *et al.,* 2021).

Hence, a firm level of research could provide a framework and identify key drivers for pathways to net zero and decarbonisation by enabling managers and policymakers to focus on the critical levers for accelerating towards net zero.

3. HYPOTHESIS AND CONCEPTUAL FRAMEWORK

The proposed conceptual framework, as detailed below, seeks to examine the interplay of firm level independent variables of 'customer engagement', 'supplier engagement', 'employee engagement', 'strategy and investment decisions' and 'green finance' on the dependent variable of financial performance arising out of net zero pathways and decarbonisation adopted by a firm. This framework is examined in the context of the Indian chemical industry.

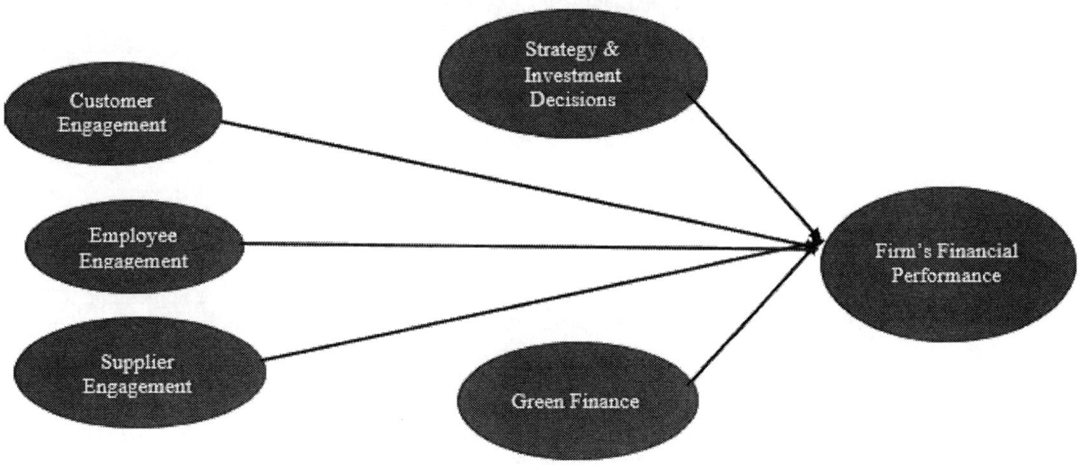

Figure 1. Proposed conceptual framework

3.1 Hypothesis Development

In our hypothesis, a holistic value chain approach could provide a clearer understanding of the variables that enable and drive outcomes. In this framework, the outcome (i.e., dependent variable) is the revenue growth of companies. This single financial indicator drives almost all other financial outcomes and is easy to communicate and validate. The independent variables include customers, suppliers, employees, strategy and investment decisions, and green finance. The net zero transition is also enabled and driven by customers, suppliers, and employees. Companies are pledging to decarbonise and deliver net zero in increasing numbers each year (Buettner *et al.*, 2022), with the key objective of being able to sell their products (i.e., generate and increase revenues, enhance attractiveness to employees, and appeal to providers of finance/capital).

The five independent variables identified for the study include employee engagement, customer engagement, supplier engagement, strategy and investment decisions, and green finance. The developed questionnaire relied on the Gallup and Aon models to measure employee engagement (Harter *et al.*, 2002). Employee engagement was emphasised while examining the effects of employees' environmental awareness as an intermediate variable (Khadair *et al.,* 2022). Determinants of supply chain engagement in carbon management (Lintukangas *et al.*, 2023) underscored the need to commit, communicate, and collaborate, which was further elucidated in the detailed study of the relationship between customer concentration and corporate carbon emissions (Deng *et al*., 2022). Businesses are incentivised to adopt a proactive environmental strategy as it reduces costs by enhancing operations, minimising waste and energy consumption, and utilising recyclable resources. At the firm level, identifying and implementing green and carbon projects is critical, as highlighted in a study titled "Value relevance of corporate environmental performance" (Choi *et al*., 2020). Regarding capital and access to green finance, it is argued that knowledge, assessment, and access to green finance are crucial, as detailed in the role of green finance in environmental protection (Wang and Zhi, 2016). However, most studies provide limited evidence in the emerging market context, where growth pressures could challenge the net-zero transition. Hence, focusing on the Indian industry could yield valuable insights.

The hypothesis tested in this study include:

Customer engagement:

a. H_0– 'Customer engagement' does not create revenue growth from net zero.
b. H_1– 'Customer engagement' creates revenue growth from net zero.

Supplier engagement:

a. H_0 – 'Supplier engagement' does not create revenue growth from net zero.
b. H_1 – 'Supplier engagement' creates revenue growth from net zero.

Employee engagement:

a. H_0 – 'Employee engagement' does not create revenue growth from net zero.
b. H_1 – 'Employee engagement' creates revenue growth from net zero.

Strategy and investment decisions:

a. H_0 – 'Strategy and investment decisions' does not create revenue growth from net zero.
b. H_1 – 'Strategy and investment' decisions creates revenue growth from net zero.

Green Finance:

a. H_0 – 'Green finance' access does not create revenue growth from net zero.
b. H_1 – 'Green finance' access creates revenue growth from net zero.

4. METHODOLOGY

For this research, we studied leading Indian chemical firms through an online survey of the firms' management. The study includes five independent variables: customer engagement, supplier engagement, employee engagement, strategy and investment decisions, and green finance. Its aim is to identify the relationship between these independent variables and financial performance. The data was tested for normality and homogeneity of variance to determine the necessity for data transformation or removal of outliers. We subsequently analysed the data using a multivariable regression model, applying the least squares method. The primary objective was to understand the nature of the relationship (i.e., not prediction) between the dependent variable of revenue growth and the independent variables of engagement with customers, suppliers, employees, strategy, and green financing. The data performed well regarding normality and homoscedasticity and exhibited relatively low multicollinearity.

While Partial Least Squares Structural Equation Modelling (PLS-SEM) is an option, it is more suitable for preliminary studies where researchers aim to develop a theoretical model and identify significant dependencies between concepts, often working with smaller sample sizes. It is also better suited to studies that require the estimation of individual construct scores for prediction or subsequent analyses. This approach is widely utilised, especially for non-normal data and some degree of multicollinearity and is primarily recommended for examining models with latent variables, i.e., underlying constructs that are not directly measured (Riou *et al.,* 2016). Although multivariate regression models are capable of prediction, they are relatively weaker than PLS-SEM models in this respect.

Partial least squares (PLS) is a method for constructing predictive models when the factors are many and highly collinear with an emphasis is on predicting the responses and not necessarily on trying to understand the underlying relationship between the variables. (Tobias, 2016)

Based on the hypothesis, the questionnaire developed was administered to obtain the current view of the independent and dependent variables from management teams across six chemical companies, with 40 respondents.

4.1 Questionnaire Development

The questionnaire measured, the perception score of a firm's approaches on a 7-point Likert scale. In terms of parameters, the questionnaire included key elements of organisational engagement with the entire value chain, (i.e., with customers, suppliers, and employees), strategy & investment decisions and green finance including green bonds.

The questionnaire developed leaned on the Gallup, Aon model to measure employee engagement (Harter *et al.*, 2002) and the study of the effects of employees' environmental awareness as an intermediate variable (Khadair *et al.*, 2022). Determinants of supply chain engagement in carbon management (Lintukangas *et al.*, 2023) highlighted the need to commit, communicate, and collaborate, which was further clarified in the details of customer concentration and corporate carbon emissions (Deng *et al.*, 2022). At the firm level, identifying and implementing green and carbon projects is critical, as highlighted in Value Relevance of Corporate Environmental Performance (Choi *et al.*, 2020). Finally, in terms of capital and access to green finance, it is argued that knowledge, evaluation, and access to green finance are critical, as detailed in the role of green finance in environmental protection (Wang and Zhi, 2016).

4.2 Data Analysis

The data were assessed for normality and homogeneity of variance to determine the necessity for data transformation or the elimination of outliers. Various tests were performed for normality (i.e., Kolmogorov–Smirnov test, Wilcoxon test, Kruskal–Wallis test, and Van der Waerden test) and homoscedasticity (i.e., Brown-Forsythe, Levene, and Welch tests).

The test for normality is a crucial first step. The goodness-of-fit test yielded a Shapiro–Wilk value with Prob $< W$ at 0.9338 and an Anderson–Darling A^2 of 0.19722, accompanied by a simulated P value of 0.8852. These results confirm that the data originate from a normal distribution, as all P values are significantly above 0.05. To assess homoscedasticity and heteroscedasticity, a test of variances was conducted using the residuals of each parameter, including the dependent variable (revenue increase) and the median split for predicted revenue increase. The table below presents the P values for the Brown-Forsythe, Levene, and Welch tests for the entire model.

Table 1

	Brown Forsythe	**Levene**	**Welch**
Whole model	0.1372	0.1269	0.5644

Consequently, the P values in both the Brown-Forsythe and Levene tests exceed 0.05, while in the Welch test they are significantly higher than 0.05. This suggests that there is no difference in the variances of the two groups (i.e., those above and below the median of the split predicted revenue increase). Therefore, the data demonstrate homogeneity of variances.

5. RESULTS

5.1 Bivariate Fit ANOVA Yielded the Following Results:

One key finding of bivariate ANOVA is that none of the supplier-related parameters have a Prob > F above 0.05. This highlights that supplier-related independent parameters do not explain the variance in the dependent variable of revenue increase due to NZ. Therefore, firms that concentrate their efforts on employees, customers, strategy, and financing for NZ benefit most from revenue increases.

5.2 Univariate Regression Analysis:

In the fitted model by least squares for each variable, the key finding was that no single variable has a strong relationship with the dependent variable of increased revenue due to NZ. However, strategy and investment decisions (M), customers (C), and finance (F) have maximum R^2 values of approximately 0.29, 0.29, and 0.26, respectively. In all cases, the Durbin-Watson score is well above 1.79, thus exhibiting negligible autocorrelation.

5.3 Multivariate Regression Analysis:

In the Fit Model by Least Squares, the R^2 value was 0.54 and the p-value was 0.0104. However, to achieve this, we excluded responses to four questions.
a) The firm provides opportunities in NZ to employees (E3)
b) Net zero communication with customers (C2)
c) Net zero collaboration with suppliers (S3)
d) Assess green finance (F2)

None of the parameters had a VIF greater than 5. This resulted in a Durbin–Watson score of 1.40 and an autocorrelation value of 0.2675, with a probability less than DW of 0.082. The Durbin–Watson score of 1.40 indicates the presence of autocorrelation. We conducted the test for non-randomness of the data despite obtaining data around the same time, as we lack information on respondents' temporal references during their responses. Based on the univariate analysis, it appears that the supplier parameters do not adequately explain the dependent variable of net zero increasing revenue, leading us to eliminate all supplier parameters from the model. This adjustment reduced the model's R^2 to 0.51 but improved the D–W score to 1.68 and decreased AICc and BIC to 143 and 152, respectively. Furthermore, the removal of parameter C3 (collaboration with customers) lowers the Durbin–Watson score and increases AICc and BIC. Consequently, the optimal model is S No 3, which is the initial model with all supplier parameters eliminated. This model has the highest VIF score of 3.86, an R^2 of 0.51, and a Durbin–Watson score of 1.68.

Given the key parameters outlined above, it is evident that independent of supplier is not as a key differentiator as customers, employees, strategy, and finance that play a vital role in the pursuit of net zero. Amongst these, the two parameters are of greatest significance, with a P Value of less than 0.05, are:

1. The company implements decarbonisation projects and green investments.
2. The company actively updates itself on green finance products and options.

Therefore, implementing green projects and actively updating green finance options to fund them enhances the outcomes of revenue growth in the net-zero transition.

6. CONCLUSION

Most net-zero strategies are long-term and entail short-term cost impacts and investment demands. The findings indicate that companies are beginning to recognise the positive influence of net-zero initiatives on financial performance, as demonstrated by revenue growth. Organisations have made substantial progress in evaluating net zero as part of their risks and opportunities, followed by engaging employees in the net-zero conversation. They have also experienced considerable success in involving customers and suppliers in discussions about net zero. Nonetheless, knowledge of and access to green finance is an area that requires improvement, and companies should dedicate greater effort. Furthermore, strategy and investment decisions related to the transition to net zero need more emphasis.

The analysis indicates that 'strategy and investment decisions' (including the implementation of decarbonisation and net-zero projects) have the greatest impact on revenue growth. Implementing net-zero projects builds credibility with employees, customers, and suppliers as stakeholders witness tangible actions. The results also emphasise that 'green finance' is a crucial tool in the journey towards net-zero transition. By staying informed about green finance, management can take timely actions to facilitate the net-zero transition.

Therefore, industry leaders and policymakers must concentrate on executing green projects and facilitating green finance to expedite the journey towards achieving net zero for the chemical industry in India. This study offers an Indian perspective on the framework for accelerating decarbonisation and establishing a pathway to attain net zero in an energy-intensive sector such as the chemical industry. At the firm level, it is essential to identify the enablers and barriers, and to develop strategies, internal processes, and reward and recognition mechanisms to drive net zero projects, bolstered by green finance. At the industry level, collaboration, disclosure, and the sharing of best practices are necessary to enable the implementation of green projects and green finance, fostering the development of a comprehensive ecosystem for the transition to net zero.

While this research establishes a framework for Indian chemical companies, similar studies could be conducted in other regions and industrial sectors to develop a comprehensive view and gain further insights.

REFERENCES

Azoulay, M. *et al.* (2023). *Financing the net-zero transition: From planning to practice, McKinsey & Company* https://www.mckinsey.com/capabilities/risk-and-resilience/our-insights/financing-the-net-zero-transition-from-planning-to-practice

Amaeshi K., Osuji O., Nnodim, P. (2008). Corporate Social Responsibility in Supply Chains of Global Brands: A Boundaryless Responsibility? *Clarifications, Exceptions and Implications. Journal of Business Ethics 81, 223–234.*

Bashmakov, I.A., Nilsson, L.J, Acquaye, A., Bataille, C., Cullen, J.M., de la Rue du Can S., Fischedick, M., Geng, Y., Tanaka, K. (2022). Industry. In IPCC, 2022: Climate Change 2022: Mitigation of Climate Change. *Contribution of Working Group III to the Sixth Assessment Report of the Intergovernmental Panel on Climate Change; Cambridge University Press, Cambridge, UK and New York, NY, USA.*

Bertram, C. et al. (2021). A path to Zero: How to reach net Zero https://www.imf.org/en/Publications/fandd/issues/2021/09/how-to-reach-net-zero-emissions-bertram

Buettner, S.M., König, W., Vierhub-Lorenz, F., Gilles, M. (2022). What Motivates Companies to Take the Decision to Decarbonise? *Preprints*, 2022100395.

Chen, C-M, Ho, H. (2019). Who pays you to be green? How customers' environmental practices affect the sales benefits of suppliers' environmental practices. *Journal of Operations Management*; 65: 333–352.

Chen, L., Msigwa, G., Yang, M., Yap, P. S. (2022). Strategies to achieve a carbon neutral society: a review. *Environmental Chemistry Letters 20:2277–2310.*

Choi, H., Han, I., Lee, J. (2020). Value Relevance of Corporate Environmental Performance: A Comprehensive Analysis of Performance Indicators Using Korean Data. *Sustainability*; 12(17):7209.

Deng, S., Duan, T., Li, F. W., & Pu, X. (2022). Major Customers and Carbon Footprints along the Supply Chain. *Journal of Corporate Finance*, Volume 92, 2025, 102752, ISSN 0929-1199, https://doi.org/10.1016/j.jcorpfin.2025.102752.

Deutch, J. (2020). Is Net Zero Carbon 2050 Possible? *Joule 4, 2237–2243, November 18, Elsevier Inc.*

Dyer J.H. and Singh, H. (1998). The Relational View: Cooperative Strategy and Sources of Interorganizational Competitive Advantage. *Academy of Management Review, 23.*

Gabrielle, B, Loyce, C. (2024). How to Reconcile Biomass Supply for Green Chemistry and Food Resources? *Green Chemistry and Agro-food Industry: Towards a Sustainable Bioeconomy. Springer, Cham.*

Hart, S.L. and Ahuja, G. (1996). Does it pay to be green? An empirical examination of the relationship between emission reduction and firm. *Business Strategy & Environment.*

Harter, J. K., Schmidt, F. L., Corey, L., Keyes, M. (2002). Wellbeing in Workplace and its relationship to business outcomes *A review of Gallup studies.*

Hernández-Chea, R., Jain, A., Bocken, N.M.P., Gurtoo, A. (2021). The Business Model in Sustainability Transitions: A Conceptualization. *Sustainability*. 13(11):5763.

Hollebeek, L., Glynn, M., & Brodie, R. (2014). Consumer Brand Engagement in Social Media: Conceptualization, Scale Development and Validation. *Journal of Interactive Marketing. 28. 10.1016/j.intmar.2013.12.002.*

Jarvis, W., Ouschan, R., Burton, H.J., Soutar, G. and O'Brien, I.M. (2017). Customer engagement in CSR: a utility theory model with moderating variables, *Journal of Service Theory and Practice, Vol. 27 No 4, pp. 833.*

Johnson, M. P., Rötzel, T. S., Frank, B. (2023). Beyond conventional corporate responses to climate change towards deep decarbonisation: a systematic literature review *Management Review Quarterly, Jan 2023.*

Khadair, W. A., Jabbar, A. D. (2022). Studying the Effect of Employees' Environment Awareness as an Intermediate Variable for the Relationship Between Institutional Pressures and Intentions of Human Resources for Adoption of Green Information Techniques: An Exploratory Study of Micro-Companies *International Journal of Professional Business Review 7 (2).*

Lintukangas, K., Armine, H., Khknen, A. K., Karttunen, E. (2023). Determinants of Supply Chain Engagement in Carbon Management. *Journal of Business Ethics 186:87–104.*

Lovelock, J. (2007). *The revenge of Gaia: Earth's climate in crisis and the fate of humanity.* Basic Books.

Nyangchak, N. (2022). Emerging green industry toward net-zero economy: A systematic review. *Journal of Cleaner Production 378 134622.*

O'Brien, I. M., Ouschan, R., Jarvis, W., Soutar, G. N. (2020). Drivers and relationship benefits of customer willingness to engage in CSR initiatives. *Journal of Service Theory and Practice, Vol. 30 No. 1, pp. 5-29.*

O'Brien, I. M., Jarvis, W., Soutar, G. N. (2015). Integrating social issues and customer engagement to drive loyalty in a service organisation. *Journal of Services Marketing, Vol. 29 No. 6/7, pp. 547-559.*

Peng, Y., Wang, W., Li, S., Veglianti, E. (2022). Competition and cooperation in the dual-channel green supply chain with customer satisfaction. *Economic Analysis and Policy, Volume 76, Pages 95-113, ISSN 0313-5926.*

Rajendran, V., Maheshwari, S., Nayak, A., Aggarwal, U., Dhev, P. (2022). Mission net-zero - A roadmap for the Indian Chemical Industry - Kearney. https://www.kearney.com/industry/energy/article/-/insights/mission-net-zero-a-roadmap-for-the-indian-chemical-industry

Riou, J., Guyon, H., Falissard, B. (2016). An introduction to the partial least squares approach to structural equation modelling: a method for exploratory psychiatric research. *International Journal of Methods Psychiatric Research. 2016 Sep;25(3):220-31.*

Sancha, C., Gimenez, C., Sierra, V. (2016). Achieving a socially responsible supply chain through assessment and collaboration. *Journal of Cleaner Production, Volume 112, Part 3, Pages 1934-1947, ISSN 0959-6526.*

Smil, V. (2021). Grand Transitions. *Oxford University Press.*

Takacs, F., Brunner, D., Frankenberger, K. (2022). Barriers to a circular economy in small- and medium-sized enterprises and their integration in a sustainable strategic management framework. *Journal of Cleaner Production, 362 132227.*

Tan, K., Siddik, A. B., Sobhani, F. A., Hamayun, M., Masukujjaman, M. (2022). Do Environmental Strategy and Awareness Improve Firms' Environmental and Financial Performance? The Role of Competitive Advantage. *Sustainability, 14, 10600.*

Trambouze, P. (1993). Energy saving in the petrochemical industry. SpringerLink. Available at: *https://link.springer.com/chapter/10.1007/978-94-011-1454-7_8*

Tidy, M., Wang, X., Hall, M. (2016). The role of Supplier Relationship Management in reducing Greenhouse Gas emissions from food supply chains: supplier engagement in the UK supermarket sector. *Journal of Cleaner Production, Volume 112, Part 4, 2016.*

Tobias, R. D. (2016). An introduction to partial least squares regression .

Villena, V.H, Choi, T.Y, Revilla, E. (2021). Mitigating Mechanisms for the Dark Side of Collaborative Buyer–Supplier Relationships: A Mixed-Method Study. *Journal of Supply Chain Management, 57: 86-116.*

Wang, Y., Zhi, Q. (2016). The Role of Green Finance in Environmental Protection: Two Aspects of Market Mechanism and Policies, *Energy Procedia, Volume 104, Pages 311-316, ISSN 1876-6102.*

Wang, Y., Zhia, Q. (2016). The role of green finance in environmental protection: Two aspects of market mechanism and policies *Energy Procedia 104 311.*

Wong, W. K. O. (2023). Creating Artificial Suns: the Sino Western race to master limitless clean energy through nuclear fusion *Asian Education and Development Studies Vol.12 No1.*

Zhan, Y., Chung, L., Lim. M., Ye, F., Kumar, A., Tan, K.H. (2021). 'The Impact of Sustainability on Supplier Selection: A behavioural study', *International Journal of Production Economics, vol. 236, 108118.*

Zhang, Y., Umair, M. (2023). Examining the interconnectedness of green finance: an analysis of dynamic spillover effects among green bonds, renewable energy, and carbon markets. *Environmental Science and Pollution Research vol. 30 pp. 77605-77621.*

INTEGRATED INTRUSION DETECTION AND PREVENTION MODEL FOR MOODLE LEARNING MANAGEMENT SYSTEM

Olayinka Anthony Ojo, Celestine Iwendi and Adedapo Paul Aderemi
Department of Computing, School of Creative Technologies, University of Greater Manchester
Deane Road, Bolton, BL3 5AB, United Kingdom

ABSTRACT

This study developed and evaluated an integrated intrusion detection and prevention (IDP) model for Moodle Learning Management System (LMS), utilizing Snort 3, Open-Source Security (OSSEC), ModSecurity, and Moodle's security settings. The increasing security threats facing LMS platforms was addressed in the study by leveraging the strengths of each tool: Snort 3 for network-level detection, OSSEC for host-based monitoring, ModSecurity for web application protection, and Moodle's native security features for enhanced control. An experimental approach was adopted, beginning with a literature review to identify vulnerabilities, followed by system design, tool configuration, and integration. The model was tested against simulated attacks, with performance measured by detection accuracy. The results demonstrated the model's effectiveness in identifying and mitigating common security threats within Moodle LMS such as distributed denial of service, brute force attack, SQL injection and aggressive scan. The study concludes by recommending the deployment of the IDP model in a live environment for both private/individual owned and public owned Moodle platforms, for the provision of a robust framework for enhancing security. This work contributes to the broader field of LMS security through the provision of a comprehensive, multi-layered approach to protecting educational platforms from cyber threats.

KEYWORDS

Cyber Threats, Integrated IDP, Intrusion Detection, Intrusion Prevention, Moodle LMS, Snort 3

1. INTRODUCTION

Moodle, an acronym for *modular object oriented dynamic leaning environment*, is a learning management system (LMS) ranked as the most widely adopted eLearning platforms among institutions and organisations across the world (Abdelsalam, Idrees, & Shokry, 2023; Ally, 2022). This claim is backed up by the statistics on the Moodle website (Table 1) which reveals an enormous figure in terms of number of users, sites, enrolment, and resources among others. Moodle LMS is pivotal in modern education due to its versatility and accessibility. As an open-source platform, it supports a wide range of pedagogical methods (including video lecture, demonstration and discussion), enabling educators to create personalised learning environments that cater to diverse student needs (Ukhov, Borshchenko, Kabanov, Bergen, & Ryapukhin, 2021). The availability of plugins and tools (such as forums and quizzes) in Moodle fosters interactive and collaborative learning. It facilitates continuous assessment and feedback, essential for student development. Furthermore, Moodle is scalable, making it suitable for both large and small educational institutions. The platform's ongoing relevance in offering learning solutions both now and in the future is guaranteed by its capacity to work with various educational technologies and adherence to global standards. Moreover, by prioritizing accessibility and inclusivity, this resource guarantees that every student, including those with disabilities, has an equal chance to succeed, thus proving to be a crucial tool in achieving fair education. As a result, Moodle was considered the most targeted eLearning platform for attackers taking advantage of its common vulnerabilities, despite having more vulnerabilities than other learning management systems (Djeki, Degila, Bondiombouy, & Alhassan, 2021).

Table 1. Moodle statistics

Variables	Figure
Countries	240
Sites	161,666
Courses	47,750,401
Users	419,493,082
Resources	426,890,896
Enrolments	2,390,868,228

(Data source: https://stats.moodle.org/ June 2024)

Common vulnerabilities in Moodle LMS pose significant risks to educational institutions that rely on this platform for delivering and managing online learning (Perera, Sirisuriya, & Gunathilake, 2022). Figure 1 show a ten-year total security vulnerability of Moodle LMS as indicated by data sourced from common vulnerabilities and exposures (CVE Details, 2024). It was revealed that cross-site scripting (XSS), cross-site request forgery (CSRF), Input validation and SQL Injection respectively took the lead among the most prominent Moodle vulnerabilities over the last ten years.

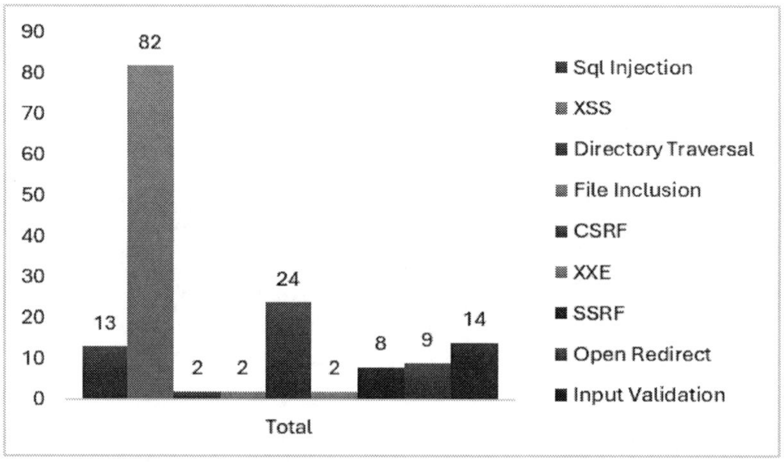

Figure 1. Moodle LMS vulnerability from 2014-2024

Educational institutions face considerable dangers due to prevalent weaknesses in the Moodle LMS, which they depend on to administer online education (Perera et al., 2022). Insufficient input validation is a common concern that can result in different types of injection attacks like SQL injection and cross-site scripting (XSS) (Ally, 2022; Perera et al., 2022; Djeki et al., 2021; Bhatia & Maitra, 2018). These security flaws enable hackers to alter database queries or insert harmful scripts, which could jeopardize user data and system stability. Another important vulnerability is the lack of proper access control, which could allow unauthorized individuals to obtain higher levels of permissions, resulting in access to restricted sections and confidential data (Bhatia & Maitra, 2018). Moreover, the presence of obsolete or inadequately set-up third-party plugins may create security vulnerabilities, given that these plugins may not follow the same strict security protocols as the main Moodle codebase (Milošević, Kuk, Popović, & Čisar, 2022).

Moreover, unsecure data transmission without adequate encryption can potentially leak sensitive information like login details and personal student data to malicious third parties while in transit (Bhatia & Maitra, 2018). Insufficiently secure sessions can be taken over, posing another worry about poor session management (Perera et al., 2022). Several solutions have been suggested to mitigate these risks. These are: essential regular updates and patches which will address known vulnerabilities and improve overall security; implementing robust security practices, such as comprehensive input validation, strict access controls, and secure communication protocols; and regular security audits and vulnerability assessments to identify and remediate potential threats proactively (Zabala et al., 2022; Perera et al., 2022; Djeki et al., 2021). Addressing these common vulnerabilities will ensure the availability of a more secure and reliable learning environment in Moodle LMS, with protection of sensitive educational data and the privacy of its users (Zabala et al., 2022). Vulnerability through database violation or compromise was also identified (Milošević et al., 2022). This dangerous unauthorised access to Moodle user's data privacy is enhanced by careless installation of third-party plugins not recommended by the official organisation.

Observations by authors suggest detecting vulnerabilities in Moodle LMS involves a multi-faceted approach that combines automated tools, manual testing, and regular audits to ensure the platform's security and integrity (Mudiyanselage & Pan, 2020; Ukhov et al., 2021; Yaseen, 2023). Automated vulnerability scanners like Nessus and OWASP ZAP are crucial for detecting typical problems such as SQL injection, cross-site scripting (XSS), and obsolete software components. These tools scan the Moodle environment in a systematic manner, identifying possible vulnerabilities that need to be looked into further. Manual penetration testing complements automated scans by simulating scenarios of real-world attacks. Security professionals manually check the system for vulnerabilities that automated tools may overlook, such as logic errors and intricate injection spots (Yaseen, 2023). This method offers a more thorough insight into the security status of the system and aids in pinpointing weaknesses in both bespoke code and third-party plugins. Regular security assessments are necessary to uphold Moodle's security throughout the duration. The audits need to encompass code evaluations, configuration evaluations, and compliance checks for security standards like ISO/IEC 27001.

Continuous monitoring of the Moodle environment is necessary to secure the Moodle MLS platform (Ukhov et al., 2021). Utilizing tools like Snort to deploy intrusion detection systems (IDS) and intrusion prevention systems (IPS) can aid in promptly identifying suspicious activities and potential breaches (Abdulrezzak & Sabir, 2023). Additionally, involving the Moodle community in responsible disclosure initiatives promotes users and developers to report newly identified vulnerabilities. These programs make it easier to provide regular updates and patches, dealing with security issues before they are maliciously taken advantage of. Figure 2 illustrates the various factors that contribute to improving Moodle security.

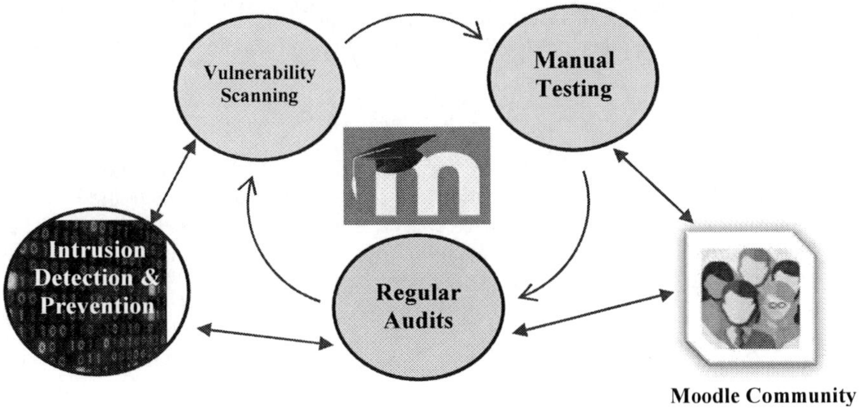

Figure 2. Mitigating Moodle Vulnerability

In Moodle LMS, intrusion detection and prevention systems (IDPS) are essential for identifying vulnerabilities through real-time monitoring, detection, and response functions. Intrusion detection and prevention systems (IDPS) aim to detect and prevent possible threats from taking advantage of weaknesses, thus improving the security of the Moodle platform. IDS mainly concentrate on observing network and system actions to detect indications of malicious behaviour. They employ signature-based detection to identify anomalies in normal behaviour. IDS generates alerts when potential threats are identified, allowing administrators to promptly investigate and address the threat. This proactive strategy assists in detecting and resolving vulnerabilities like unauthorized access attempts, data exfiltration, and suspicious network traffic (Zabala et al., 2022).

IPS go beyond IDS by not just identifying, but also proactively preventing threats. Automated actions can be implemented to prevent potential attacks from compromising the Moodle LMS, including dropping malicious packets, blocking IP addresses, and terminating suspicious sessions. This is highly successful in preventing vulnerabilities such as SQL injection, cross-site scripting (XSS), and brute force attacks, which can greatly impact the security and performance of the LMS. Combining IDPS with Moodle LMS enables ongoing system monitoring, quick detection of new threats, and fast response to risks, resulting in a multi-layered security strategy. This will guarantee a successful handling of weaknesses, leading to a more secure and durable online learning setting. Additionally, IDPS can produce extensive logs and reports, assisting in conducting forensic investigations and adhering to security standards and regulations.

An IDPS that is fully integrated will greatly improve Moodle LMS security through offering thorough, flexible, and immediate threat management capabilities. This comprehensive strategy will be more effective in responding to new threats and changing attack methods, which is essential for safeguarding the security of dynamic educational platforms such as Moodle. Utilizing advanced IDP tools like Snort 3, OSSEC, and ModSecurity along with Moodle's built-in security guarantees that the integrated IDPS can effectively detect and react to possible vulnerabilities and attacks. Contrary to conventional signature-based detection systems, this model integrates network-level, host-based, and web application security measures to enable analysis and detection of multilevel security threats. This ability is crucial for identifying vulnerabilities and complex attacks that could go undetected.

1.1 Statement of the Problem

Educational institutions and mentors around the globe extensively utilize the Moodle Learning Management System (LMS) to support online learning and oversee educational content. Nevertheless, as digital platforms become more essential, Moodle LMS is encountering a rising number of cybersecurity risks emanating from network intrusions, data breaches, and unauthorized access attempts (Akacha & Awad, 2023). Current Moodle's security architecture cannot sufficiently safeguard against the daily myriads of threats that could emerge from varying and multiple sources. Additionally, previous studies have not explored the capability of a network intrusion detection and prevention system (especially, Snort 3) combined with a host-based intrusion detection system for an enhanced, cost-effective Moodle security. This deficiency in existing security models for Moodle LMS, therefore, necessitates a study that would harness both network and host based IDPS, combined with the potency of web application firewall and Moodle's inbuilt security settings.

1.2 Objectives

The specific objectives that will help achieve the purpose of this research are as follows.

(a) To design an integrated intrusion detection and prevention system for Moodle LMS.
(b) To implement an integrated intrusion detection and prevention system designed for Moodle LMS.
(c) To test the intrusion detection and prevention model's ability to identify and stop different types of cyber-attacks and intrusions on Moodle LMS to assess its performance and accuracy.

1.3 Research Questions

(a) How is the structure of an integrated IDPS framework designed for Moodle LMS?
(b) How can integrated IDPS framework for Moodle LMS be implemented?
(c) How effective is the integrated IDPS in detecting and preventing different types of cyber-attacks on Moodle LMS?

2. MATERIALS AND METHODS

2.1 Research Design

This experimental research project concentrates on creating, applying, and evaluating a unified intrusion detection and prevention system for Moodle LMS that utilizes network and host-based tools alongside Moodle security configurations. The research employs a systematic method, starting with an examination of relevant literature to pinpoint weaknesses and deficiencies in current IDP solutions for Moodle. The design involves setting up specific tools to perform effectively in Moodle, then testing the integrated model against simulated attacks after implementation. The performance of the model was assessed by metrics such as detection accuracy and false positives. Ultimately, the model will be managed with an administrative interface when deployed in a live environment, guaranteeing constant monitoring, feedback, and updates on Moodle's security status.

Integrated IDP design (Figure 3) reveals how the IDPs are integrated into the system to protect Moodle from an emanating attack by filtering through every packet originating from every kind of users. This multilayered protection provides a lowered tension on the Moodle's inbuilt security settings which must also be tuned for any unplanned attack on its resources. The system administrator has both direct and indirect access to the IDPs (Snort/OSSEC) and WAF (ModSecurity) via configuration and rules files, and the SIEM.

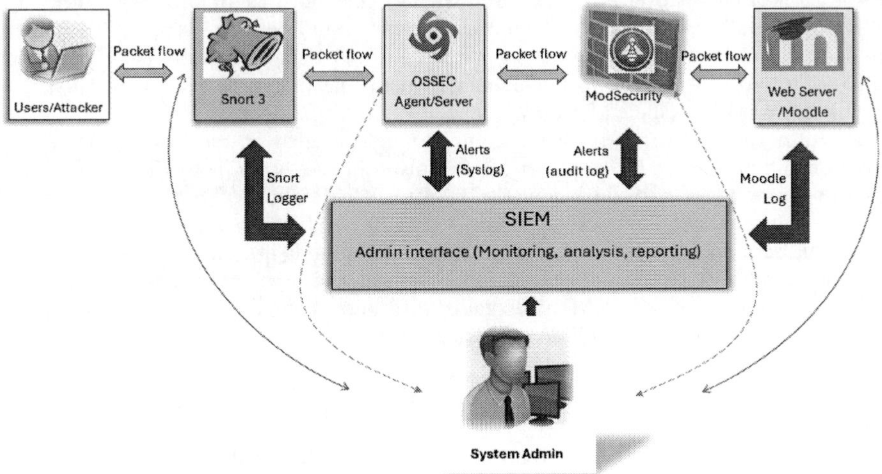

Figure 3. Integrated IDP Design

2.2 System Configuration

Table 2. Host and VMs Configuration

Machine	Operating System	CPU	Hard disk	RAM
Host	Windows 11 Pro, 64-bit	Intel(R) Core(TM) i5-7200U CPU @ 2.50GHz 2.71 GHz	SSD 250 GB	16 GB
VM1	Kali Linux 2024.2, Debian-64 bit	2	80 GB	4 GB
VM2	Ubuntu 22.04, 64-bit	2	45 GB	4 GB
VM3	Metasploitable2	1	8 GB	512 MB
VM4	Windows 7, 32-bit	1	25 GB	2 GB

The system configuration for the experiment as shown in Table 2 reveals the hardware resources of the host system and the hardware resources allocation for the installed virtual machines that were used for the experiment.

2.3 Moodle and Security Installation

Moodle 4.4.1 was downloaded and setup on the Ubuntu 22.04 VM as the host system. Moodle security was configured such that added users can only use secure passwords and could be temporary denied access after five failed login attempts. This is to prevent the platform against password guessing or account hacking. A security provision in Moodle which allows blocking a known security threat on the network was also employed to block the only known attack system in the simulated experiment.

ModSecurity, a free web application firewall (WAF) that defends against cyber threats by examining and controlling HTTP traffic (Alagoz, Tok, & Bicakci, 2021; Sobola, Zavarsky, & Butakov, 2020), was installed on the Ubuntu system to function as a component for the Apache 2.4.52 web servers for the Moodle LMS. Its role is to identify and prevent attacks through customizable rules or the OWASP Core Rule Set (CRS) that protects against vulnerabilities such as SQL injection, XSS, and remote code execution.

OSSEC is an acronym for Open-Source Security Host-based Intrusion Detection System (HIDS). This security tool was installed as hybrid (both server and agent) on the Ubuntu system which hosts Moodle in order to observe and analyse the events on it. OSSEC offers immediate insight into various security incidents by scanning log files, file integrity, rootkits, and registry keys on the systems (Tanner, 2019). This function helps to identify unapproved actions, system irregularities, infringement of policies, and other issues, playing a vital role in a security-oriented framework which the model is meant to provide. The web user interface for this security software was also installed for easy management of alerts.

Snort 3.1.43 (also known as Snort ++) was compiled and installed on the Ubuntu machine to provide intrusion detection and prevention (IDP) security mechanism for the virtual network environment on which the Moodle host belongs. Snort 3 provides upgraded performance through multi-threading support, improved syntax for easier rule creation, seamless dynamic reloading, and enhanced compatibility with modern network setups. Lua scripting, thorough protocol analysis, and a modular structure make customization and scalability easier for advanced threat detection. Both built-in and local rules were activated whilst also creating additional local rules to secure Moodle against distributed denial of service (DDoS), SQL injection and brute force attacks emanating from within or outside the network.

2.4 Research Testbed

The control environment designed for the experimentation and testing of the integrated intrusion detection and prevention model for Moodle LMS is depicted in Figure 4. The Windows 11 system which hosts the virtual machines (VMs) has VirtualBox 7 installed on it and its resources are shared among the VMs (Table 2). A Technicolor network modem provided the network infrastructure which enables network access virtually connecting the machines. The Ubuntu VM acts as the administrator's system for managing the Moodle platform while Windows 11 and Windows 7 VMs serves as the users' systems from which Moodle users' accounts (instructors and students) were accessed. Kali Linux VM is the attack system whilst the Metasploitable 2 VM was used as the 'hacker's playground' for pre-testing dangerous attacks.

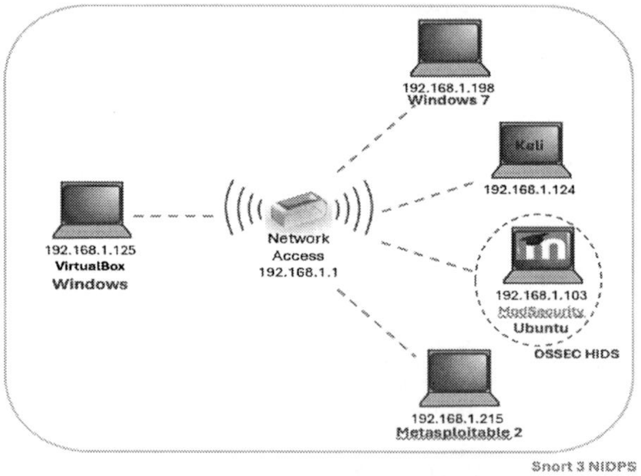

Figure 4. Research Testbed Network

2.5 Data Collection

This included gathering logs and monitoring information from SIEM and security software while conducting controlled simulations of different cyber-attacks. Snort 3 monitored network traffic, OSSEC observed host-level activities, and ModSecurity evaluated web application requests. The IDP tools were meticulously set up to identify and document possible breaches originating from both internal and external sources in the network. Moreover, Moodle's included logging capabilities were activated in order to monitor user actions and system occurrences. Intrusion detection log data from Snort 3 were saved in comma separated value (CSV) formats and analysed to evaluate how accurate and efficient the integrated IDP model is.

Ethical declaration

Privacy and confidentiality of user data were safeguarded during testing to address possible ethical concerns in the study. The simulated attacks in this research were carried out in a controlled setting with the use of virtual machines to avoid unauthorized access to real user data.

3. RESULTS AND DISCUSSION

All the simulated attacks were unsuccessful. Moodle security configuration successfully blocked the attack system from gaining access to the platform while also preventing unauthorised access after five unsuccessful attempts at login. DDoS attack was blocked by Snort 3 as revealed in Figure 5 which shows a combine screen shot of the test bed. Also, the brute force attack conducted using the hydra tool and SQL injection using SQL map failed as the malicious packets were blocked by integrated IDPS jointly provided by the security mechanism of the model. Intrusion detection and prevention activities ongoing within the Moodle host system are captured and analysed by the OSSEC web interface to alerts administrator to take necessary actions.

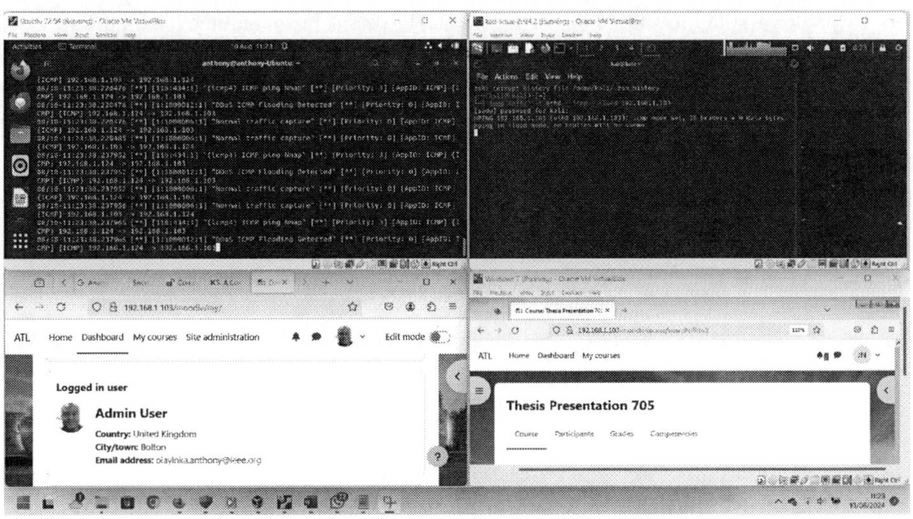

Figure 5. Combine screen shot of testbed

The IDP model created for Moodle LMS effectively reduced cyber threats by utilizing a strategic, multi-layered security approach. The model makes use of essential tools: Snort for strong network-level detection, OSSEC for constant host-based monitoring, and ModSecurity for web application protection. This integration targeted multiple levels of vulnerabilities to provide thorough security protection. Snort's detection at the network level highlights the significance of network-based intrusion detection systems in thwarting large-scale attacks (Abdulrezzak & Sabir, 2023; Lai et al., 2023; Boukebous, Fettache, Bendiab, & Shiaeles, 2023; Kaur & Sharma, 2023; Kaushik, Athmaraman, John, & Sridhar Raj, 2024). OSSEC strengthened the framework by overseeing system integrity and reducing vulnerabilities caused by rootkits and ransomware (Venkatesan et al., 2018; Tanner, 2019; Djeki et al., 2021; Lee, Shim, Lee, Park, Park, & Lee, 2021; Lee et al., 2022a; Lee et al., 2022b). ModSecurity was instrumental in protecting Moodle's online platforms by preventing typical cyber-attacks such as SQL injection and cross-site scripting, hence effective in web application security (Muzaki et al., 2020; Mukhtar & Azer, 2020; Sobola et al., 2020; Alagoz et al., 2021; Lokesh et al., 2021). The integrated model showed effectiveness in threat detection which could facilitates quick responses in dealing with security breaches, thus illustrating the benefits of a layered defence approach. However, difficulties could emerge when implementing this model in practical situations, reflecting worries about the disparity between theoretical and real outcomes. Continuous improvement and integration of real-time threat intelligence are crucial to staying flexible in the face of ever-changing cyber threats. In general, the model provides a strong structure that improves the security of Moodle and can be customized for different educational settings.

4. CONCLUSION

This research effectively created and assessed an integrated intrusion detection and prevention (IDP) system for Moodle LMS, which integrates Snort 3, OSSEC, ModSecurity, and Moodle's built-in security features. The experimental method showed how the model effectively identifies and reduces different cyber threats, improving Moodle LMS's overall security. The combination of network, host, and web security tools created a thorough, multi-layered defence strategy which effectively secured the Moodle platform against various cyber threats. The findings indicated that the integrated model effectively enhanced accuracy in detecting threats and reducing false alarms, making it a strong solution for educational institutions utilising Moodle. The study therefore recommended as follows:

a) Individuals or organisations using Moodle LMS should consider deploying the integrated IDP model to enhance their security posture against increasingly sophisticated cyber threats at little or no cost.
b) Moodle installation should be secured with HTTPS. This guarantees that all information exchanged between users and the Moodle server is protected through encryption, thereby enhancing security levels.
c) The security tools and Moodle should be regularly updated to address new vulnerabilities and maintain the effectiveness of the IDP model.

Future research should explore customizing the model for different LMS environments and scaling it for larger institutions with diverse user bases.

REFERENCES

Abdelsalam, M., Idrees, A. M., & Shokry, M. (2023). A proposed model for improving the reliability of online exam results using blockchain. *IEEE Access*.

Abdulrezzak, S., & Sabir, F. A. (2023). Enhancing intrusion prevention in Snort system. In *15th International Conference on Developments in eSystems Engineering (DeSE)* (pp. 88–93). IEEE. https://doi.org/10.1109/DeSE58274.2023.10099757

Akacha, S. A. L., & Awad, A. I. (2023). Enhancing security and sustainability of e-learning software systems: A comprehensive vulnerability analysis and recommendations for stakeholders. *Sustainability, 15*(19), 14132.

Alagoz, M., Tok, M. S., & Bicakci, K. (2021). Exploring and improving the usability of ModSecurity web application firewall. In *International Conference on Information Security and Cryptology (ISCTURKEY)* (pp. 51–56). IEEE. https://doi.org/10.1109/ISCTURKEY53027.2021.9654294

Ally, S. (2022). Review of online examination security for the Moodle learning management system. *International Journal of Education and Development using Information and Communication Technology, 18*(1), 107–124.

Bhatia, M., & Maitra, J. K. (2018). E-learning platforms security issues and vulnerability analysis. In *2018 International Conference on Computational and Characterization Techniques in Engineering & Sciences (CCTES)* (pp. 276–285). IEEE.

Boukebous, A. A. E., Fettache, M. I., Bendiab, G., & Shiaeles, S. (2023). A comparative analysis of Snort 3 and Suricata. In *IEEE IAS Global Conference on Emerging Technologies (GlobConET)* (pp. 1–6). IEEE. https://doi.org/10.1109/GlobConET56651.2023.10150141

CVE Details. (2024). *Moodle vulnerability statistics*. https://www.cvedetails.com/vulnerability-list/vendor_id-2105/product_id-3590/Moodle-Moodle.html

Djeki, E., Degila, J., Bondiombouy, C., & Alhassan, M. H. (2021). Security issues in digital learning spaces. In *2021 IEEE International Conference on Computing (ICOCO)* (pp. 71–77). IEEE.

Kaur, G., & Sharma, K. (2023). Contemplate and investigate a network-based intrusion detection system. In *Second International Conference on Electronics and Renewable Systems (ICEARS)* (pp. 812–816). IEEE. https://doi.org/10.1109/ICEARS56392.2023.10085479

Kaushik, Y., Athmaraman, A., John, A. M., & Sridhar Raj, S. (2024). Signature-based intrusion prevention system for software defined networks using SNORT. In *3rd International Conference on Artificial Intelligence for Internet of Things (AIIoT)* (pp. 1–6). IEEE. https://doi.org/10.1109/AIIoT58432.2024.10574719

Lai, Y.-C., Yu, C.-L., Liao, M.-L., Lin, Y.-S., Chang, Y.-C., & Chen, J.-L. (2023). An intelligence defense system with SNORT rules. In *25th International Conference on Advanced Communication Technology (ICACT)* (pp. 249–254). IEEE. https://doi.org/10.23919/ICACT56868.2023.10079506

Lee, S.-J., Shim, H.-Y., Lee, Y.-R., Park, T.-R., & Lee, I.-G. (2022). Ransomware detection using open-source tools. In *24th International Conference on Advanced Communication Technology (ICACT)* (pp. 1385–1391). IEEE. https://doi.org/10.23919/ICACT53585.2022.9728873

Lee, S.-J., Shim, H.-Y., Lee, Y.-R., Park, T.-R., & Park, S.-H., Lee, I.-G. (2021). Study on systematic ransomware detection techniques. In *23rd International Conference on Advanced Communication Technology (ICACT)* (pp. 297–301). IEEE. https://doi.org/10.23919/ICACT51234.2021.9370472

Lee, S.-J., Shim, H.-Y., Lee, Y.-R., Park, T.-R., Park, S.-H., & Lee, I.-G. (2022). Study on systematic ransomware detection techniques. In *24th International Conference on Advanced Communication Technology (ICACT)* (pp. 297–301). IEEE. https://doi.org/10.23919/ICACT53585.2022.9728909

Lokesh Raju, S., Sheshware, S., & Patel, R. R. (2021). Use ModSecurity web application firewall to mitigate OWASP's top 10 web application vulnerabilities. In *Implementing Enterprise Cybersecurity with Open-Source Software and Standard Architecture* (pp. 193–235). River Publishers.

Milošević, Đ., Kuk, K., Popović, B., & Čisar, P. (2022). Endangered data in Moodle platform with malicious plugins. In *2022 21st International Symposium INFOTEH-JAHORINA (INFOTEH)* (pp. 1–5). IEEE.

Mudiyanselage, A. K., & Pan, L. (2020). Security test Moodle: A penetration testing case study. *International Journal of Computers and Applications, 42*(4), 372–382. https://doi.org/10.1080/1206212X.2017.1396413

Mukhtar, B. I., & Azer, M. A. (2020). Evaluating the ModSecurity web application firewall against SQL injection attacks. In *15th International Conference on Computer Engineering and Systems (ICCES)* (pp. 1–6). IEEE. https://doi.org/10.1109/ICCES51560.2020.9334626

Muzaki, R. A., Briliyant, O. C., Hasditama, M. A., & Ritchi, H. (2020). Improving security of web-based application using ModSecurity and reverse proxy in web application firewall. In *International Workshop on Big Data and Information Security (IWBIS)* (pp. 85–90). IEEE. https://doi.org/10.1109/IWBIS50925.2020.9255601

Perera, P. H., Sirisuriya, S. C. M., & Gunathilake, H. R. W. P. (2022). Security vulnerabilities and security elements of frequently used e-learning platforms: A review.

Sobola, T. D., Zavarsky, P., & Butakov, S. (2020). Experimental study of ModSecurity web application firewalls. In *IEEE 6th International Conference on Big Data Security on Cloud (BigDataSecurity), IEEE International Conference on High Performance and Smart Computing (HPSC), and IEEE International Conference on Intelligent Data and Security (IDS)* (pp. 209–213). IEEE. https://doi.org/10.1109/BigDataSecurity-HPSC-IDS49724.2020.00045

Tanner, N. H. (2019). Monitoring with OSSEC. In *Cybersecurity Blue Team Toolkit* (pp. 57–66). Wiley. https://doi.org/10.1002/9781119552963.ch5

Ukhov, P. A., Borshchenko, D. A., Kabanov, D. D., Bergen, M. E., & Ryapukhin, A. V. (2021). Customization of open-source solutions on the example of the LMS Moodle distance learning platform. *Journal of Physics: Conference Series, 1889*(2), 022002. IOP Publishing.

Venkatesan, R., Devi, D. R., Keerthana, R., & Kumar, A. A. (2018). A novel approach for detecting DDoS attack in H-IDS using association rule. In *IEEE International Conference on System, Computation, Automation and Networking (ICSCA)* (pp. 1–5). IEEE. https://doi.org/10.1109/ICSCAN.2018.8541174

Yaseen, K. A. Y. (2023). Enhance Moodle platform security against denial-of-service attack (DoS). *Journal of Critical Reviews, 10*(2), 140–147.

Zabala, L. N. B., Velasco, C. S. R., & Parada, H. D. J. (2022). Security scheme for Moodle platforms based on a multi-layered model. In *2022 Congreso Internacional de Innovación y Tendencias en Ingeniería (CONIITI)* (pp. 1–5). IEEE.

TOWARDS SUSTAINABLE LEATHER GOODS. INNOVATION IN MATERIALS

Dalia Gallico
Università San Raffaele Roma, Italy

ABSTRACT

This paper explores the environmental and ethical challenges associated with traditional leather goods production and investigates sustainable alternatives through material innovation and circular design. The study begins with a critical assessment of the environmental footprint of conventional leather—highlighting issues such as greenhouse gas emissions, water consumption, chemical pollution, and animal welfare concerns. It then presents a comparative analysis of next-generation biomaterials derived from plants and agricultural waste, including cactus leather (Desserto®), pineapple fiber (Piñatex®), mushroom mycelium (Mylo™), and apple waste (AppleSkin®). These alternatives demonstrate significant potential in reducing carbon emissions, water usage, and chemical toxicity. The core of the paper is EcoModular project: a modular, customizable bag designed with circular principles and constructed from eco-innovative materials. The study evaluates EcoModular's sustainability performance through key indicators, highlighting its potential as a replicable model for ethical luxury fashion. This research contributes to the broader discourse on sustainable fashion by offering concrete design strategies and business insights for reducing the industry's ecological impact.

KEYWORDS

Sustainable Fashion, Eco-Leather Alternatives, Circular Design, Biobased Materials, Ethical Luxury, Modular Bags

1. INTRODUCTION

The traditional leather goods sector is today at the center of a debate regarding the environmental and social impact of fashion. The production of leather is linked to intensive cattle farming, one of the major sources of greenhouse gases (up to 14.5% of anthropogenic emissions according to the FAO) and to deforestation to make space for pastures. Furthermore, tanning processes use large quantities of water and polluting chemical substances, contributing to water and soil pollution (FAO, 2013).

From an ethical point of view, the use of animal skins raises concerns about animal welfare, especially in the case of exotic leathers (reptiles, kangaroos, etc.), to the point that important luxury brands have begun to ban them (e.g., Prada eliminated kangaroo leather since 2020).

These issues highlight the need to rethink leather goods in a sustainable way (Range Revolution, 2024).

In this context, the present work aims to transform the concept of leather goods by uniting material innovation and sustainable design. It intends to:

- critically analyze the environmental and social impacts of the traditional leather supply chain
- examine existing solutions in the form of eco-leathers and cruelty-free alternative materials
- present the EcoModular project, a concept of a sustainable modular bag, evaluating its design strategy and potential impacts.

The methodology followed is qualitative in nature and based on secondary research and experimental design. First, a review was conducted of the available literature and data: international reports (FAO, Ellen MacArthur Foundation, 2017), scientific studies (e.g., life cycle analysis – LCA – of new materials), sustainability databases of materials (Higg MSI), and industry articles.

These sources provided the key parameters on environmental impacts (CO_2 emissions, water consumption, land use, chemical pollution) and social impacts (working conditions, animal welfare) associated both with traditional leather and alternative materials.

The quantitative data collected—for example, the carbon footprint per square meter of material or the water consumption per kilogram of product—were used to develop objective comparisons among various options.

In parallel, a design research approach was adopted for the development of the EcoModular project. This included the analysis of case studies of emerging brands in sustainable luxury, the selection of innovative materials (based on low impact and technical performance criteria), and the definition of design principles (modularity, durability, traceability).

The concept was then analyzed through strategic evaluation tools: Key Performance Indicators (KPIs) of product sustainability were identified (e.g., carbon footprint per unit, percentage of renewable materials, % of recycled components, estimated life cycle duration).

2. CONTEXT AND PROBLEMS OF TRADITIONAL LEATHER GOODS

The bovine leather supply chain has a significant ecological footprint throughout its entire life cycle.

At the agricultural stage, cattle breeding for meat and leather entails massive emissions of greenhouse gases (methane produced by ruminant digestion and nitrous oxide from manure). To this is added the deforestation of sensitive ecosystems (for example, the Amazon) to make space for pastures and forage crops, resulting in biodiversity loss and a reduction in CO_2 absorption sinks (Cline, 2012).

The water footprint of leather is extremely high: it is estimated that producing 1 kg of bovine leather requires about 17,000 liters of water, considering the water needed to hydrate the animals and grow their feed. Translated into a finished product, a single leather bag may require over 17,000 L of water—the equivalent of around 450 five-minute showers—making leather one of the most water-intensive materials (Hakansson, 2023).

Moreover, traditional tanning is a highly polluting chemical process: typically, chromium salts (chrome tanning) and other hazardous substances are used. For every kg of tanned leather, up to 2.5 kg of chemical agents (tannins, solvents, dyes) and 250 liters of water are used in the tanning baths, generating liquid waste containing heavy metals (chromium, lead) and solid waste (trimmings, sludge) that are difficult to safely dispose of. If not properly treated, such effluents cause contamination of rivers and soil with toxic effects on aquatic fauna and local communities.

In addition to environmental aspects, traditional leather production presents ethical and social issues.

In recent years, the use of exotic leathers (crocodile, snake, kangaroo, etc.) has raised public outcry due to the cruel methods of acquisition and the risk of driving wild species toward decline.

Growing consumer awareness about animal welfare has led several brands to adopt cruelty-free policies, giving up furs and exotic leathers. As mentioned, Prada announced in 2020 the end of kangaroo leather use, and similar decisions have affected python and crocodile skins in sustainability-conscious fashion houses.

Another aspect is the health and safety of tannery workers, who are often exposed to toxic substances: in some manufacturing areas (e.g., Southeast Asia), the lack of adequate protections leads to high incidence rates of skin and respiratory illnesses among tanners due to contact with chromium and organic solvents.

From a socio-economic perspective, it is important to note that leather is not merely a by-product of the meat industry but a valuable commodity: up to 26% of a slaughterhouse's revenue can derive from the sale of hides. This means that the demand for leather contributes to keeping intensive livestock farming economically viable—with all the associated ethical and environmental implications (Kumar, 2020).

In summary, the issue with conventional leather goods is twofold: on one hand, a high ecological footprint (carbon footprint, water footprint, pollution) that is difficult to reconcile with climate and environmental protection goals; on the other hand, ethical and social issues (animal suffering, labor conditions) that are increasingly less acceptable to consumers and stakeholders. This problem justifies the search for innovative solutions to make leather goods more sustainable in a comprehensive way (D'Adamo, 2024).

3. EXISTING SOLUTIONS: ECO-LEATHERS AND ALTERNATIVE MATERIALS

In light of the issues described above, the fashion industry is exploring two parallel paths towards sustainability: innovating traditional tanning processes to reduce their impact, and developing alternative materials to animal leather that are ecological and cruelty-free.

In the first direction, some cutting-edge tanneries are adopting metal-free tanning processes (without chromium), using vegetable tannins or biobased tanning agents (derived from polyphenols or enzymes) that avoid heavy metals. Systems are implemented to recirculate tanning and finishing baths to reduce water consumption by up to 15%, and treatment plants recover chemical substances (such as chromium) from the waste, returning clean water. Additionally, renewable energy and automation help reduce emissions and waste (e.g., photovoltaics in tanning districts, sensors that accurately dose reagents).

These incremental improvements make the supply chain cleaner but do not eliminate dependence on animal raw materials and their major impacts (breeding and slaughter). This is why the second strategy is emerging strongly: creating alternative eco-leathers, i.e., materials that mimic the properties of leather without being derived from animals, favoring renewable or waste resources and less impactful processes.

Among the new materials developed in recent years, those obtained from plant sources (leaves, fruits, mushrooms) or from biotechnological cultures stand out, often referred to as next-gen biomaterials.(Sean Mowbray, 2024)

Cactus Leather (Desserto)

One of the pioneering materials is cactus leather, developed by the Mexican company Desserto.

It is an innovative fabric obtained from the dried leaves of the nopal cactus, a succulent plant common in arid regions of Mexico that grows spontaneously without the need for irrigation or pesticides.

Desserto's founders developed a process to transform cactus biomass into a flexible and resistant material, similar to leather, suitable for bags, footwear, clothing, and furniture. From an environmental standpoint, the advantages are considerable: as nopal is rain-fed (uses only rainwater), water consumption for producing the material is negligible compared to leather tanning. A preliminary LCA analysis conducted by Desserto shows that cactus leather generates about 5% of the greenhouse gas emissions of equivalent bovine leather and about 77% less than polyurethane faux leather (in other words, it reduces CO_2 emissions by more than 18 times compared to animal leather). (Hakansson, 2022)

Similarly, water use is reduced by 99.5%: cactus cultivation requires no artificial irrigation, so producing 1 m^2 of Desserto requires only fractions of a liter of water, compared to several thousand liters for 1 m^2 of bovine leather. Furthermore, cactus plantations have a carbon sequestration effect: 14 acres of organic nopal cultivation can absorb up to 8100 tons of CO_2/year, with minimal emissions (about 15 tons) from agricultural operations, making the process carbon-negative (climate-positive).

The resulting material is 100% vegan, free of animal components and added toxic substances; about 90% of its composition is plant-based, with a small fraction of synthetic polymers used as a binder, making it partially biodegradable. Functionally, Desserto is very similar to traditional leather in softness and appearance, while having a significantly cleaner production chain.

Thanks to these characteristics, Desserto has received international recognition and has attracted the interest of numerous fashion brands that see cactus leather as a way to reduce the ecological footprint of their collections. (Desserto, 2020)

Pineapple Leather (Piñatex)

Another emblematic example of eco-leather is Piñatex, an innovative material made from pineapple leaf waste. Developed by Spanish designer Carmen Hijosa through research in the Philippines, Piñatex embodies the principles of the circular economy and the cradle-to-cradle approach, transforming agricultural waste into a valuable resource. Each year, pineapple cultivation generates over 25 million tons of leaves that are normally burned or left to decompose in the fields.

Piñatex uses this by-product without requiring dedicated land or crops: "we don't need to use even a hectare of land: no water wasted, no pesticides or fertilizers. We are simply taking a waste and turning it into valuable items," say the creators.

Cellulose fibers are extracted from the leaves through mechanical decortication; the remaining biomass is reused as organic fertilizer or converted into biogas, avoiding waste.

The dried and pressed fibers form a non-woven sheet that serves as the material's base, later finished with a polymer coating to improve resistance and waterproofing. Piñatex is thus composed of 80% pineapple plant fibers and 20% coating resin. The result is a lightweight and flexible fabric, suitable for many uses (clothing, accessories, furniture); the surface can be printed or colored similarly to animal leather.

A major advantage of Piñatex is its biodegradability: being mostly made of natural biomass, at end of life, under appropriate industrial composting conditions, the material is compostable, degrading much faster than tanned leather (which can persist for decades due to treatment) and fully plastic-based synthetic leathers.

Economically, Piñatex has proven competitive: its cost is about €23 per m², comparable or even lower than bovine leather (€25–38 per m²). This has favored its adoption by several international brands, including Puma, Camper, and Hugo Boss, which used it for a line of men's sneakers aimed at ethically conscious consumers.

Piñatex thus represents a successful case of sustainable innovation, capable of combining impact reduction (use of waste, no added resources or water) with technical and economic performance suitable for the market.(Hijosa, 2016)

Mushroom Leather (Mycelium: Mylo)

Among the emerging materials is mushroom leather, obtained from mycelium (the network of filaments that forms the structure of fungi). Mycelium can be cultivated on organic waste substrates (e.g., sawdust) in controlled conditions, growing rapidly in a few days to form a soft and pliable layer, very similar to leather in appearance and feel. One of the leading products in this field is Mylo, the result of biotechnological research by the American startup Bolt Threads. Mylo gained prominence thanks to a collaboration with designer Stella McCartney, a pioneer of cruelty-free fashion, who presented some garments and accessories made of mycelium in 2021. The production process involves growing the mycelium in a nutrient-rich bioreactor derived from organic waste; in about two weeks a fungal material sheet is obtained, which is then stabilized, tanned, and dyed using traditional techniques to give it the desired properties, resulting in Mylo (Bolt Threads., 2021).

The final product is certified bio-based (mainly composed of renewable ingredients) and verified 100% vegan throughout the supply chain. A notable aspect is that, unlike many synthetic leathers, Mylo is plastic-free: its composition does not require polyurethane or PVC, making it fully biodegradable and free of microplastics. Stella McCartney declared: "I'm incredibly proud to be the first fashion house in the world to create garments with Mylo, a luxuriously soft and sustainable mushroom leather... certified bio-based and plastic-free," emphasizing how this alternative aligns with ethical values without compromising luxury and quality (Mc Cartney, 2021).

Apple Leather (AppleSkin)

Another solution born in Italy is apple leather, developed from apple industry waste.

The most well-known example is AppleSkin, a material patented by the South Tyrolean company Frumat.

AppleSkin valorizes the residues from juice and cider production (peels, cores, pulp) that would otherwise be waste. Fully embracing the circular economy model, this material uses a locally abundant resource, turning a disposal problem into a value-added solution. The process involves drying and grinding apple waste into a fibrous powder; this is mixed with binding resins and laminated onto a backing fabric, creating a faux leather sheet. The resulting composition is about 50% apple fibers (organic biomass) and the rest synthetic polymers needed to provide strength and workability.

AppleSkin is 100% vegan and free of toxic substances; thanks to its high renewable biomass content, its ecological footprint is lower than that of fully plastic synthetic leathers.

In terms of properties, "apple leather" has a thickness and consistency comparable to light leather, making it suitable for making bags, coverings, and accessories. It shows good durability, though perhaps not reaching the robustness of thick traditional leather.

A notable advantage is its local origin: in fruit-growing regions like South Tyrol, the raw material is zero-km and transport impact is minimal, creating a virtuous cycle where apple agricultural waste becomes fashion material. AppleSkin shows how innovation, sustainability, and territory can intertwine to generate new low-impact materials with commercial potential.

However, since it is largely composed of polyurethane, AppleSkin is not completely biodegradable; it still represents a significant step forward in reducing the plastic component of conventional vegan leathers and increasing the bio-based material content (Frumat 2022).

Table 1. The table below provides a concise comparison between traditional leather, synthetic PU leather, and the main alternative materials, based on three key indicators: CO_2 emissions, water consumption, and compostability/ biodegradability

Material	Carbon Footprint (kg CO_2e per m²)	Water Consumption (L per m²)	Biodegradability
Bovine leather (hide)	30–40 (very high, due to farming and tanning)	~20,000 (extremely high)	Low (very slow degradation; chrome tanning prevents composting)
Synthetic leather (PU)	~10 (medium, fossil origin)	~1,400 (high, industrial production) (~14x less than leather)	None (not biodegradable, it is plastic)
Cactus leather	~2–3 (very low; ~5% of leather)	~100 (negligible; ~0.5% of leather)	Partial (90% bio-based, partially compostable)
Pineapple leather (Piñatex)	~5 (low; plant waste + 20% resin)	~500 (very low; leaf waste, minimal processing)	High (80% plant-based, compostable at end of life)
Mushroom leather (Mylo)	~2–3 (very low; 8% of leather)	~200 (low; mycelium cultivation, low water requirement)	High (100% bio-based, fully biodegradable)
Apple leather (AppleSkin)	~7–10 (medium; 50% biopolymer)	~1000 (moderate; industrial processes [less than leather and animal origin])	Partial (partly biodegradable, bio-based component; plastic component non-biodegradable)

4. SUSTAINABLE PRODUCTION PROCESSES

Beyond the choice of materials, sustainability in leather goods also depends on production methods.

Let's examine some initiatives and technologies that make the supply chain more sustainable: the adoption of a circular model, blockchain traceability, and technological innovation to mitigate environmental impact.

Circular Economy in Leather Goods

The circular economy aims to transform the linear "produce, use, discard" model into a continuous cycle of reuse and recycling. In leather goods, this primarily means recovering production scraps: offcuts, trimming waste, shavings, and leather residues. Instead of being disposed of as special waste (often with high environmental and economic costs), these scraps can be reused. For example, some tanneries recover over 90% of their scraps and repurpose them for other sectors: fertilizers and animal feed (proteins, collagen from residues), gelatins for the food industry, granules for construction, etc. On average, over 70% of tanning waste is recycled or otherwise valorized (Minney, 2022).

Additionally, creative upcycling initiatives are spreading: in Tuscany, for example, the BisBag project collected premium leather offcuts from luxury production (otherwise considered hazardous waste due to chromium) and transformed them into new leather accessories, giving a "second life" to high-quality materials destined for disposal. This approach not only reduces waste but also creates added value and local economic opportunities (Fletcher, 2016).

Finally, designing products with a circular perspective facilitates repair and recycling: designing bags and accessories in a modular way or with easily separable parts simplifies the replacement of worn components and recycling at end-of-life. The use of mono-materials or compatible materials makes it easier to disassemble a product and recover leather, metals, and fabrics separately. These strategies, along with take-back programs (some brands offer credit to customers who return worn bags, which are then refurbished or recycled), help close the loop in the leather goods sector (Ramchandani, 2020).

Blockchain and Supply Chain Transparency

Blockchain offers a powerful tool to ensure transparency and traceability throughout the leather goods supply chain. This technology allows for the creation of a secure and immutable digital ledger of all transactions and stages in the supply chain. Applied to the sector, it allows each raw material batch or finished product to be associated with a digital "identity card" (Cui, 2020).

In practice, using a QR code or NFC tag on the product, the consumer can access (via app or web) certified information on the origin of raw materials (for example, the farm or the field where a plant material was grown), the processing stages (tannery, manufacturing), and the actors involved. All this data is recorded on blockchain, and therefore unalterable, and can be consulted by anyone with the code. Blockchain supply chain traceability enables transparent monitoring of the origin and quality of products at every stage of the process, minimizing the risk of fraud.

For the consumer, this means being certain of purchasing an authentic product made according to the declared standards (e.g., leather from farms not involved in deforestation, or certified organic plant-based materials). For the company, blockchain reinforces trust and protects the brand from counterfeiting or opaque practices, creating a bond of transparency with customers.

In the future, blockchain could also support the circular economy by tracking, for example, repair or resale cycles of a product and certifying its durability. Although still in early stages in the fashion sector, this technology is rapidly revolutionizing the concept of transparency, making the leather goods supply chain more open and accountable.

Technologies to Reduce Environmental Impact

Technological innovation is making leather production cleaner and more efficient.

Metal-free tanning processes (without chromium) or those using vegetable tannins are being introduced, which are safer for human health and free from heavy metals. New biobased tanning agents (e.g., based on plant polyphenols or enzymes) make it possible to stabilize leather collagen without resorting to chromium salts. Efforts are also being made to reduce water consumption: some tanneries have developed optimized tanning and finishing recipes that enable water and chemical savings of up to 15%, by splitting and reusing processing baths without compromising their effectiveness. Wastewater is treated with advanced systems that recover chemical products (such as chromium) and reduce pollutant loads, returning purified water to the environment. (Ghosh & Saha, 2020)

Organic waste from the process (hairs, fats, proteins) is increasingly being recovered: for example, intact hairs extracted from depilation baths are used to produce fertilizers or agricultural biostimulants, preventing them from ending up as waste.

On the energy front, many tanneries and leather goods facilities are improving energy efficiency by installing photovoltaic systems, recovering heat from thermal processes, and replacing obsolete machinery with more efficient versions.

Finally, digitalization and automation allow for constant monitoring of production parameters (temperature, pH, water consumption) and real-time adjustments to optimize processes, reducing chemical waste and defects. For instance, sensors in drums accurately dose reagents, minimizing excess, and CAD systems for leather cutting maximize the use of each hide, reducing waste.

All these innovations contribute to a more sustainable leather goods supply chain, minimizing emissions, waste, and risks, while maintaining high product quality standards (Hosey, 2012).

5. ECOMODULAR PROJECT

EcoModular is a modular eco-designed bag that embodies many of the solutions described so far, offering a tangible example of what the future of sustainable fashion accessories might look like.

EcoModular is a bag with a contemporary and functional design, conceived to adapt to different user needs and to have a prolonged life cycle.

Concept and Design.

The EcoModular bag was conceived as a composable system rather than a single product. During the design phase, designers deconstructed the bag into its essential parts—main body, handles/straps, decorative outer panels, closures, and pockets—imagining them as interchangeable modules.

The concept is based on the idea of multi-use modularity: with a few gestures, the user can reconfigure the bag to suit different needs. For example, EcoModular can transform from an elegant handbag to a spacious backpack simply by repositioning the handles and adding an extension module to the base. Or, by removing the top part, it becomes a compact crossbody bag for minimalist daily use.

This versatility is made possible by a patented fastening system—composed of quick-release metal clips and hidden guides—which securely connects the modules while maintaining a clean aesthetic. The modular design required numerous tests and prototypes to ensure strength: the junction points were subjected to mechanical stress tests to ensure that the bag, even when composed of multiple assembled segments, withstands loads and tension without failure. Formally, the design is modern and linear, inspired by clean geometries. The essential lines also facilitate modularity, since modules with simple geometric shapes (rectangular, trapezoidal) are easier to combine. EcoModular adopts a minimalist and functional aesthetic, where every zipper and joint has a purpose. The concept also incorporates the "design for disassembly" philosophy: each component can be removed with simple tools, making it easy to replace or recycle individual parts.

Eco-friendly Materials.

In line with its sustainable mission, EcoModular uses materials selected for their low environmental impact and high quality. The main material used for the outer panels is Desserto® cactus leather, chosen for its nappa-like look and excellent durability and flexibility. Cactus leather provides a soft yet robust feel to the touch and is available in various colors without traditional tanning (colors are achieved using low-impact organic dyes).

Some interchangeable modules, designed to add texture and stylistic variety, are instead made of Piñatex® (pineapple leather) with a metallic finish and AppleSkin with saffiano embossing: this shows how EcoModular can also serve as a platform to experiment with different bio-based materials within the same product.

The inner lining is made of GOTS-certified organic cotton fabric, printed with natural dyes; in some versions, there is also an interior made from recycled PET (from recycled bottles) for more sporty variants, thus aiming at the reuse of post-consumer materials.

Even the metal hardware—buckles, hooks, base feet—is chosen with a green approach: the hardware is made of 70% recycled stainless steel and is nickel-free.

Every detail, from snap buttons to zippers, is designed for durability: these components are built to withstand repeated opening cycles without losing effectiveness, ensuring a long product life.

Functionality and Modularity.

EcoModular's distinctive feature is its extreme adaptability. The basic set includes: a main module (which serves as the body of the bag), a removable top module (which can serve as an independent pouch), a lower extension module to increase capacity, two types of shoulder straps (a short one and a long adjustable one for backpack use), and a pair of rigid handles.

The modules are connected by a system of zippers and hidden magnetic fasteners along the horizontal joints: this solution allows for intuitive, almost "automatic" assembly, guided by magnets, while also ensuring secure closure thanks to sturdy side zippers.

Modularity concerns not only shape but also style: EcoModular offers replaceable outer covers. The front and back of the main module are flat panels fastened with internal snap buttons, which can be detached and replaced with covers of different colors or materials. In a few seconds, for example, one can switch from a minimal-looking black cactus leather bag to a silver Piñatex panel bag for a special occasion, without needing to move the contents to another bag. This personalization encourages creative reuse: instead of buying several bags to match different outfits, the customer can update their EcoModular by changing the panels or adding accessory modules.

From a circular economy perspective, this approach reduces overall resource consumption, since a single multifunctional product replaces the purchase of multiple single-use products.

Applied Circular Economy.

EcoModular incorporates principles of circularity at various stages of its life cycle.

During production, standardized modules allow for material cut optimization: the panel patterns for cactus and pineapple leather were designed to minimize waste, and the few offcuts generated (thanks also to the use of rectangular shapes) are recovered to make small coordinated accessories such as keychains and straps.

Standardization also facilitates maintenance and repair: each element can be ordered as a replacement part. The EcoModular brand encourages customers to return the bag for periodic maintenance (for example, replacing the lining if worn, or renewing the surface finish of the plant-based leather after years of use) by offering a low-cost refurbishing service.

The strategy also includes the possibility of launching a subscription program: trend-conscious customers can subscribe to receive seasonal new module kits (new colors, limited-edition covers designed by artists, etc.) to apply to their EcoModular, while sending back any modules they no longer use. Returned modules are sanitized, inspected, and made available again as refurbished or rental products, thus ensuring a virtuous reuse cycle.

6. IMPACT AND KPIS

To evaluate the success of a project like EcoModular, it is necessary to measure both its environmental and socio-economic impact through appropriate Key Performance Indicators (KPIs).

Reduction of Environmental Impact (LCA):

One of the main objectives of EcoModular is to reduce the ecological footprint compared to a traditional bag, and for this purpose, a key KPI is the reduction in CO_2 emissions and resource consumption as shown by the Life Cycle Assessment (LCA).

The production of an EcoModular bag results in about 60% fewer greenhouse gas emissions compared to a high-end equivalent bag made from animal leather. This is thanks to the use of materials such as cactus leather, which is carbon-negative upstream in the supply chain, and the renewable energy used in the assembly workshops. Additionally, there is over 80% water savings during the tanning and processing phase, as plant-based materials do not require the lengthy soaking processes typical of leather.

Product Lifespan and Reuse Rate:

Product longevity is a crucial KPI for sustainability. EcoModular is designed to have an above-average life cycle: the stated goal is at least 10 years of active use. To monitor this, the brand tracks the age of bags in use and collects post-sale feedback. A high reuse rate indicates that modularity and construction quality are effectively translating into extended product life.

Modularity and Accessory Sales:

A specific indicator for EcoModular concerns the actual use of its modularity by consumers. This is measured, for example, by the average number of additional modules or covers purchased per bag sold. A high average value (e.g., 3 extra modules per customer in the first year) indicates that users are taking advantage of the product's personalization and continuous update potential, reducing the need to purchase entirely new bags.

Market Acceptance and Customer Engagement:

On the market side, traditional KPIs such as total sales, annual growth, and market share in the sustainable luxury bag segment are obviously important.

EcoModular also evaluates qualitative parameters: customer satisfaction (measured through post-purchase surveys with the Net Promoter Score) and engagement on social media and brand communities. A high level of engagement—such as customers sharing photos of their EcoModular configurations or participating in co-design workshops—indicates that the product has succeeded in creating a loyal community around the brand's values.

Circular Economy Indicators:

Specific circularity KPIs are defined.

Among these: the percentage of recycled or renewable material used in the bag, the end-of-life recovery rate, and the number of bags regenerated through the refurbishing or second-life program. The latter KPI, though still marginal during the launch phase, is under close observation: the company aims to be able to collect and recirculate at least 50% of the units sold once they reach the end of their first use cycle, whether by reselling them as certified second-hand or by recycling their materials.

7. CONCLUSION

The leather goods sector is undergoing a profound transformation, driven by growing environmental awareness, ethical concerns, and the need for innovation in materials and design. This article has demonstrated how the shift toward sustainable fashion can be supported through a combination of eco-friendly materials, circular design principles, and technological advancements.

By analyzing the environmental and social impacts of traditional leather, and evaluating alternative bio-based and cruelty-free materials—such as cactus leather, Piñatex, Mylo, and AppleSkin—it becomes evident that viable solutions already exist. These materials, although not free of limitations, offer significantly lower carbon footprints, reduced water consumption, and greater biodegradability compared to conventional options.

The EcoModular project illustrates how these principles can be applied in practice. With its modular structure, versatility, and integration of sustainable materials, EcoModular represents a replicable model for designing fashion accessories that are adaptable, repairable, and resource-efficient. The project further embraces the values of circular economy through design for disassembly, low-waste production, and post-use regeneration strategies.

Ultimately, the path to a more sustainable leather goods industry lies in rethinking not only what products are made of, but how they are made, used, and reintroduced into the value chain. Eco-design, transparency through technologies like blockchain, and business models centered on reuse and customization are crucial tools to align fashion with environmental and social responsibility. The case of EcoModular suggests that sustainable innovation in leather goods is not only possible but can be desirable, functional, and aesthetically compelling.

REFERENCES

Bolt Threads., (2021). Mylo™: The leather alternative made from mycelium. Retrieved from https://boltthreads.com

Cline, E. L., (2012). *Overdressed: The Shockingly High Cost of Cheap Fashion.* Penguin.

Cui, H., & Leonas, K. K. , (2020). "Blockchain Technology in Footwear Supply Chain." In *Leather and Footwear Sustainability* (pp. 23–56). Springer.

D'Adamo, I., Gagliarducci, M., Iannilli, M., & Mangani, V. (2024). Fashion Wears Sustainable Leather: A Social and Strategic Analysis Toward Sustainable Production and Consumption Goals. *Sustainability*, 16(22), 9971. https://doi.org/10.3390/su16229971

Desserto., (2020). Environmental Impact Assessment of Cactus Leather Production. Desserto Official Reports. Retrieved from desserto.com.mx

Ellen MacArthur Foundation, (2017). A New Textiles Economy: Redesigning Fashion's Future. Retrieved from https://ellenmacarthurfoundation.org

European Commission., (2020). Circular Economy Action Plan: For a cleaner and more competitive Europe. Retrieved from https://ec.europa.eu

FAO, (2013). Tackling climate change through livestock: A global assessment of emissions and mitigation opportunities. Rome: Food and Agriculture Organization of the United Nations. Retrieved from www.fao.org

Fletcher, K., & Grose, L. (2012). *Fashion and Sustainability: Design for Change*. Laurence King Publishing. Minney, S. (2016). *Slow Fashion: Aesthetics Meets Ethics.* New Internationalist.

Frumat., (2022). AppleSkin: The bio-based leather alternative made from apple waste. Company White Paper. Retrieved from appleskin.com

Ghosh, S., & Saha, S. (2022). "Sustainable supply chain management in the leather industry: A systematic literature review." *International Journal of Logistics Research and Applications*, 25(5), 1–20.

Hakansson, E., (2022). Cactus leather creators release material life cycle assessment –Collective Fashion Justice Retrieved fromCollectiveFashionJustice.org

Hakansson, E., (2023). *Total Ethics Fashion.* Hardie Grant Books.

Higg Materials Sustainability Index (Higg MSI). (n.d.). Sustainable Apparel Coalition. Retrieved from https://msi.higg.org

Hijosa, C., (2016). Piñatex: Sustainable textile innovation from pineapple leaf fibre. Ananas Anam Ltd. Retrieved from softlinebrandpartners.com

Hosey, L., (2012). *The Shape of Green: Aesthetics, Ecology, and Design.* Island Press.

Kumar, P. S., & Joshiba, G. J., (2020). *Leather and Footwear Sustainability: Manufacturing, Supply Chain and Product-Level Sustainability.* Springer.

Louis Vuitton's New Bioclimatic Atelier. (2023). *Architectural Digest.*

McCartney, S., (2021). Stella McCartney launches garments made from Mylo mushroom leather. Press Release.

Minney, S., (2022). *Regenerative Fashion.* Laurence King Publishing.

Ramchandani, M., & Coste-Maniere, I., (2020). "Leather in the Age of Sustainability: A Norm or Merely a Cherry on Top?" In *Leather and Footwear Sustainability* (pp. 11–22). Springer.

Range Revolution. (2024). Embracing Circular Economy: The Future of Leather Goods.

Sean Mowbray, S. (2024, February). Fashioning a Circular Future for Traditional and Alternative Leather. *Mongabay.* https://news.mongabay.com/2024/02/fashioning-a-circular-future-for-traditional-and-alternative-leather/

Softline Brand Partners, (2025). Vegan leather and its benefits Retrieved from SoftlineBrandPartners.com

Somers, C., (2025). *The Nature of Fashion.* Chelsea Green Publishing.

Short Papers

ENHANCING EDUCATIONAL FEEDBACK WITH LLM-POWERED MULTI-AGENT SYSTEMS

Lorenzo Pozzi
Piazza Copernico s.r.l
Via Francesco Gentile, 135, 00173 Roma RM, Italy

ABSTRACT

Large Language Models (LLMs) have the potential to assist teachers in their roles and enhance the student learning process. This work explores a system aimed at two key objectives: to assist teachers in interpreting student exam mistakes and to provide students with more detailed feedback. The workflow includes three agents operating sequentially: the first is responsible for analyzing the student's exam responses; the second agent reconstructs the thought process that led to the incorrect answer; the third synthesizes the findings from the previous agents to produce personalized feedback tailored to the student's learning needs. Hence, unlike prior work focusing on single-agent setups, this study explores the potential of multi-agent frameworks. Using a passage from the TOEFL reading comprehension exam alongside a hand-crafted set of open-ended questions, each paired with a wrong answer and a correct baseline, the system analyzes student responses, grounds its assessments in the passage, and convincingly simulates the reasoning behind student errors. We hope this work fosters further research of LLM-powered multi-agent systems for classroom teaching to enhance both teacher efficiency and student engagement in the learning process.

KEYWORDS

Multi-Agent, Second Language Comprehension, Generative AI

1. INTRODUCTION

The rapid advancement of artificial intelligence has opened new possibilities for enhancing the educational experience, particularly through the use of Large Language Models (LLMs). These models have demonstrated significant potential in assisting teachers with various tasks and improving the learning process for students (Wang et al., 2024). However, most current implementations have centered around using single LLM-based agents for specific educational purposes, such as grading or tutoring. In contrast, this paper investigates multi-agent systems to solve multiple educational tasks simultaneously. The advantage of using this framework is that each agent can be designed to specialize in a particular function, allowing for greater depth in each step of the process while benefiting from the synergy created within the system as a whole. Specifically, the proposed system involves a sequence of three specialized agents, all working toward the common goal of providing constructive feedback for a second-language reading comprehension exam.

The first agent analyzes the student's responses against a correct baseline answer, focusing on factual accuracy and logical consistency. This helps students better understand their mistakes and learn how to avoid them in the future. The second agent attempts to reconstruct the reasoning that may have led to the student's incorrect answers. This simulation provides teachers with valuable insight into students' thought processes, helping them identify patterns of misunderstanding and adjust their instruction accordingly. Finally, the third agent synthesizes the insights from the previous agents to generate personalized feedback based on the student's answers.

To summarize, our key contributions are: (1) exploring a multi-agent workflow for error correction and feedback in second-language reading comprehension exams; (2) performing extensive analysis of the agents' responses to identify emerging mistake patterns; (3) identifying system limitations and areas for improvement in error interpretation and student thought process simulation.

2. LITERATURE REVIEW

Several authors have considered interleaving LLMs and education. (Wang et al., 2024) give a comprehensive overview of the possible applications including Question Solving, where LLMs are tasked to solve questions in a zero-shot setting; Error Correction, which focuses on providing feedback to students' errors; Confusion Helper, which contrary to the previous two avoids giving direct solutions but rather uses LLMs to generate guidance that helps students solve problems themselves. EduChat (Ouellette, 2006) is a notable example of an LLM-based agent tailored specifically for educational purposes. It provides personalized, equitable, and empathetic support to teachers, students, and parents through dialogue. Similarly, FreeText (Matelsky et al., 2023) leverages LLMs to automatically assess student responses to open-ended questions and offer targeted feedback. A key limitation of these systems is their focus on single-task LLMs. Multi-agent systems address this by employing multiple LLMs, referred to as agents, each focused on a specific task but able to collaborate, creating a synergistic workflow. For instance, Boiko et al. (2023) developed agent-based systems to assist in the design, planning, and execution of scientific experiments. Similarly, Math Agents (Swan et al., 2023) help convert equations into LaTeX and Python, supporting mathematical exploration and proof.

In summary, while prior work highlights the educational potential of multi-agent LLMs, most focus on fact-based feedback without examining the reasoning behind student errors. Additionally, to our knowledge, no existing systems target second-language reading comprehension. This study addresses that gap by exploring multi-agent LLM systems in this specific context.

3. METHODOLOGY

3.1 Agents Definition

The developed system comprises three agents built on GPT-4 (version gpt-4o-mini) and it follows the framework proposed by CrewAI[1], where each agent is a specialized entity performing a defined set of actions while communicating with other agents to share relevant information. The agents are defined as follows:

- Answer Analysis Agent: analyses baseline and student answers, referencing the exam passage to confirm or refute claims. Its goal is to provide a factual and content-centric breakdown supported by appropriate evidence.
- Student Perspective Simulator: reconstructs the thought process behind the wrong answer, offering insights into the possible mental steps that led to the mistake.
- Student Feedback Agent: is tasked to write personalized feedback building on the insight from the previous steps. The agent has to adapt to the student's first language.

In this configuration agents operate sequentially, receiving the output from the previous steps in the cascade of tasks. Each is defined by two fields, i.e. goal and a backstory. Additionally, agents are assigned specific tasks or actions during their turn of execution. Each action is defined by two fields, namely description and expected output. All the agent configurations are available on GitHub[2].

3.2 Exam Passage and Query Construction

As input for the multi-agent pipeline, we extracted a passage from the dataset developed in (Tseng et al., 2016). This dataset includes questions from the listening comprehension section of the TOEFL exam, a standard assessment of academic English for non-native speakers. However, unlike the original multiple-choice format, the experiment conducted is shaped as an open-ended reading comprehension exam. Following prior work (Day & Park, 2005) and (Freeman, 2014), we created 22 diverse open-ended questions grouped into four categories:

[1] https://github.com/crewAIInc/crewAI
[2] https://github.com/lopozz/Enhancing-Educational-Feedback-with-LLM-Powered-Multi-Agent-Systems

- Vocabulary in Context (n=4): requires inferring the meaning of unfamiliar words or phrases based on context.
- Inference Questions (n=4): involve drawing logical conclusions beyond stated information.
- Gist Questions (n=4): focus on the student's ability to identify the central idea of larger segments.
- Specific Information Questions (n=7): test recall of directly stated details from the passage.

3.3 Response Analysis

The multi-agent system was prompted with the exam passage and a set of question-answer pairs. Hence, the responses from each agent were individually analyzed using specific evaluation criteria to define the failure or success of the generation and then assigned a code to each result. For Answer Analysis Agent, success was defined as providing logically sound justification for the correctness or incorrectness of each answer. The reasoning had not only to be logically valid but also had to be supported by references to the exam passage. The Student Perspective Simulator was evaluated on its ability to plausibly reconstruct the thought process behind incorrect answers. A failure in this scenario involves lack of plausibility or failing to take into account the knowledge and linguistic capabilities of an exam candidate. For both of these agents, flaws in the student answers and the possible underlying causes were categorized using predefined codes. The Student Feedback Agent was assessed based on its ability to synthesize insights from the previous agents and deliver personalized feedback in the student's first language, Italian, while also adopting a tone that resembled a supportive teacher.

4. RESULTS

4.1 Error Analysis from the Point of View of the Reasoner

One of the objectives of the agent was to understand the flaws present in the exam incorrect answers. A detailed analysis of the generated response revealed several types of errors identified. The most common was the absence of supporting evidence from the exam passage, as highlighted by phrases like "[...] there is no direct reference indicating the cost of producing customized pencils". This type of mistake appeared across all the query types discussed in the previous section. The second most frequent issue involved overly narrow answers, where questions required a comprehensive list or a general synthesis, but responses lacked sufficient breadth or abstraction. Another recurring issue was the presence of contradictions. Unlike the first type of error, which involved a lack of evidence, these mistakes occurred when the candidate's answer directly contradicted explicit information provided in the text. Finally, some errors were linked to flawed inferential reasoning. These mistakes indicated that the given answer lacked logical conclusions from the information provided in the text.

In one instance, the agent failed to fully grasp the nature of the mistake of the answer. Here, the incorrectness was due to a failure to track referential cohesion, particularly in distinguishing between two students mentioned in the passage.

4.2 Reasoner Evidence Analysis

Giving solid evidence when providing effective is crucial. One of the agent's key tasks was to identify, when necessary, specific passages from the text that could support its statements regarding both the baseline and incorrect answer. In this regard, Answer Analysis Agent successfully validated 18 out of 21 cases by quoting verbatim passages from the context to back up its claims. For example, in one instance, it referenced the text directly: "[...] the customer expresses her desire to move away from the campus tutoring center and start her own business, stating, — I want to start doing tutoring work independently —", demonstrating strong alignment between its reasoning and the source material. In the remaining cases, the agent provided references to relevant portions of the passage without directly quoting them. For instance, it stated: "The speaker mentions that a simple batch of standard business cards would cost around two dollars". Although not a verbatim reference, this technique still effectively anchored the explanation in the text.

When evaluating incorrect student answers, grounding was equally critical. The Answer Analysis Agent employed a range of techniques to support its assertions, including direct quotations, e.g. "Phrases such as — I have been working at the campus tutoring center as a math tutor — serve to clarify [...]", paraphrased explanations, e.g. "The conversation explicitly states that the character has been a math tutor but is looking to establish a new tutoring service independently", as well as step-by-step reasoning, such as breaking down arithmetic errors. It also identified cases where responses lacked any textual support, highlighting a common issue among student mistakes.

4.3 Error Analysis Taking the Student Perspective

Interpreting the reasons behind a student's mistakes can be quite complex. However, analyzing the thought process simulations facilitates the identification of plausible "why"s behind many errors. One common cause was the presence of flawed assumptions. In one case, the simulation suggested that the student incorrectly assumed that "the town shop's cards were cheaper because it was a local shop". Another recurring issue was a lack of attention to critical details, which impacted responses to synthesis-based questions, and detail-based ones. These mistakes were often linked to oversimplified responses or incorrect reasoning. The agent also highlighted potential gaps in the student's vocabulary, which contributed to misunderstandings of the semantic meaning of specific words. Lastly, mistakes were attributed to a misunderstanding of the question itself. However, the agent was less effective when analyzing vocabulary-related questions. While it could identify the target word, it often confabulated, offering implausible justifications for the student's mistake. This highlights a clear limitation in vocabulary-specific feedback.

4.4 Downstream Pipeline Errors

A sequential pipeline enhances interpretability and system coherence, but it also introduces vulnerability to error propagation. The Student Feedback Agent was able to generate coherent feedback in Italian, but its effectiveness was compromised when earlier agents failed in their tasks. For instance, the agent wrote "Hai mostrato un buon impegno nella tua risposta, ma ci sono alcuni aspetti che necessitano di chiarimento. Il tuo errore principale è stato interpretare "oggetti fissi" come elementi visibili e riconoscibili, collegandoli erroneamente alla necessità di una migliore visibilità". Although the tone is supportive, an error from the previous step propagated leading to incoherent feedback.

5. CONCLUSION

This study investigated a multi-agent system powered by Larg Language Models (LLMs) to assist teachers in analyzing student exam mistake and generating personalized feedback. The first agent performed well in identifying factual inaccuracies and logical inconsistencies in student responses but struggled with nuanced issues like referential cohesion—limitations that interestingly persisted even with a more advanced model (gpt-4o). The second agent generally produced plausible simulations of students' thought process, which could support teachers understanding common misconceptions and intervene more effectively. Nevertheless, its performance was less reliable when addressing vocabulary-related questions, possibly due to the LLM's advanced linguistic abilities, which may hinder realistic simulation of less proficient learners. For instance, in one case, the model failed to recognize the confusion between the English word "stationery" and the similar-sounding Italian word "stazionari." This highlights the need for further refinement in the agent's handling of semantic nuances, particularly in a second language context. The third showed strong adaptability to the target language and maintained a correct tone in all the generations.

REFERENCES

Boiko, D.A., MacKnight, R. and Gomes, G., (2023). Emergent autonomous scientific research capabilities of large language models. arXiv preprint arXiv:2304.05332.

Dan, Y., Lei, Z., Gu, Y., Li, Y., Yin, J., Lin, J., Ye, L., Tie, Z., Zhou, Y., Wang, Y. and Zhou, A., (2023). Educhat: A large-scale language model-based chatbot system for intelligent education. arXiv preprint arXiv:2308.02773.

Day, R.R. and Park, J.S., (2005). Developing Reading Comprehension Questions. Reading in a foreign language, 17(1), pp. 60-73.

Freeman, D., (2014). Reading comprehension questions: The distribution of different types in global EFL textbooks. In English language teaching textbooks: Content, consumption, production (pp. 72-110). London: Palgrave Macmillan UK.

Grabe, W. and Stoller, F.L., (2019). Teaching and researching reading. Routledge.

Jamieson, J., Jones, S., Kirsch, I., Mosenthal, P. and Taylor, C., (2000). TOEFL 2000 framework. Princeton, NJ: Educational Testing Service.

JOHNSTON, P.H., (1981). Prior knowledge and reading comprehension test bias. University of Illinois at Urbana-Champaign.

Kendeou, P., Van Den Broek, P., Helder, A. and Karlsson, J., (2014). A cognitive view of reading comprehension: Implications for reading difficulties. Learning disabilities research & practice, 29(1), pp. 10-16.

Kern, R.G., (1994). The role of mental translation in second language reading. Studies in second language acquisition, 16(4), pp. 441-461.

Matelsky, J. K., Parodi, F., Liu, T., Lange, R.D. and Kording, K. P., (2023). A large language model-assisted education tool to provide feedback on open-ended responses. arXiv preprint arXiv:2308.02439.

Ouellette, G.P., (2006). What's meaning got to do with it: The role of vocabulary in word reading and reading comprehension. Journal of educational psychology, 98(3), p. 554.

Swan, M., Kido, T., Roland, E. and Santos, R.P.D., (2023). Math agents: Computational infrastructure, mathematical embedding, and genomics. arXiv preprint arXiv:2307.02502.

Tseng, B. H., Shen, S.S., Lee, H. Y. and Lee, L. S., (2016). Towards machine comprehension of spoken content: Initial TOEFL listening comprehension test by machine. arXiv preprint arXiv:1608.06378.

Wang, S., Xu, T., Li, H., Zhang, C., Liang, J., Tang, J., Yu, P.S. and Wen, Q., (2024). Large language models for education: A survey and outlook. arXiv preprint arXiv:2403.18105.

EXPLORING AI DIGITAL TEXTBOOK DESIGN FOR PERSONALIZED DIAGNOSTICS IN SPECIAL EDUCATION

Seol-Hee Lee, Yedana Lee and Ga-Young Lee
Baekseok University, Republic of Korea

ABSTRACT

Amid the rapid digital transformation of the Fourth Industrial Revolution and the acceleration brought on by the COVID-19 pandemic, AI digital textbooks have emerged as a promising tool to support personalized learning for students with disabilities. This study explores the design of an AI digital textbook that enables structured pre-diagnosis of learners with diverse cognitive, motor, and communication characteristics. Based on a developmental research methodology, we identified core design principles through a review of literature and case studies, and developed a prototype that supports both initial and ongoing diagnosis prior to instruction. A usability evaluation was conducted with five experts, including four in special education and one in educational technology, to assess the system's effectiveness and gather feedback.The findings highlight the need for data-driven, differentiated learner modeling as a foundation for AI digital textbook development in special education. This study contributes to the integration of AI in inclusive education and provides a practical foundation for future research on adaptive learning technologies and personalized support systems.

KEYWORDS

AI Digital Textbooks, Personalized Diagnostics, Special Education, Learner Modeling, Usability Evaluation

1. INTRODUCTION

The concept of the 'Fourth Industrial Revolution' emerged at the World Economic Forum in 2016, heralding a digital transformation. The COVID-19 pandemic further accelerated the democratization of online education. Amidst these changes, AI has great potential to promote equality among learners by acting as an intelligent tutor to support personalized education (Ally & Perros, 2022). AI-powered intelligent tutoring systems (ITS) are gaining traction, especially in the field of special education, where individualized education is essential. Previously, special education teachers had to create individualized education plans (IEPs) and conduct learning assessments, which put a heavy burden on teachers (Rashid & Wong, 2023). In response, the South Korean government announced a plan to develop AI digital textbooks in 2023 and plans to introduce them to the educational field in stages from 2025 (Ministry of Education, 2023). The plan is to provide a more inclusive learning environment by reflecting the needs of each type of disability in accordance with UDL (Korea Education and Research Information Service, KERIS, 2023).

However, there is also controversy surrounding the introduction of AI digital textbooks. Some countries (e.g., Sweden and Finland) have opted to revert back to paperback textbooks due to issues of declining literacy and decreased concentration after the introduction of digital textbooks (Hwang, 2024). Accordingly, studies are being conducted in Korea to investigate teachers' perceptions and needs in order to carefully consider the development and introduction of AI digital textbooks. Special education teachers were concerned about issues such as device breakage, lack of big data related to the learning of students with disabilities, and distractions in the learning environment. At the same time, however, they seem to be looking forward to the possibility of using it as a personalized diagnostic and learning tool for students with disabilities with various cognitive, physical, and communication characteristics and learning styles (Kim, 2024; Lee et al., 2024a). Reflecting these teachers' expectations and concerns, Lee et al. (2025) proposed design principles for AI digital textbooks, following the stages of analysis, design, implementation, and evaluation. Based on this design principle, this study aims to present a specific development model (direction) of AI digital textbooks for personalized diagnosis. The research questions are as follows

A. What are the functions of AI digital textbooks for personalized diagnosis of students with disabilities?
B. What is the effectiveness and improvement direction of personalized diagnosis identified through usability evaluation?

2. RESEARCH METHOD

In this study, we aimed to explore the design direction of AI digital textbooks for systematically diagnosing the learning level of students with disabilities. This study was conducted by applying the developmental research method and consisted of the following steps: design, development, and evaluation.

2.1 Design

We conducted a theoretical exploration of AI digital textbooks for students with disabilities and design principles derived based on this. The theoretical exploration centered on the concept and utilization of AI digital textbooks, design principles of AI digital textbooks for students with disabilities, needs analysis, and data-based learner analysis. Based on the results of literature and case analysis, we derived design principles to analyze learners' cognitive characteristics, motor skills, main communication methods, emotions, and interest factors to provide pre-diagnosis of learners using AI digital textbooks.

2.2 Development and Evaluation

To support effective learning through the use of AI digital textbooks, a pre-diagnostic prototype was developed based on the identified design principles. This prototype analyzes learners' cognitive characteristics, motor skills, primary communication methods, and their emotions and interests related to learning. In this study, pre-diagnosis refers to a structured process of collecting individualized learner information prior to instruction. The pre-diagnostic process was designed to include two stages: initial diagnosis and ongoing diagnosis.

For the usability evaluation of the prototype, a total of five experts (four in special education and one in educational technology) were invited to review the system. The usability evaluation was conducted in written format, and open-ended feedback was collected regarding the strengths of the prototype and areas requiring improvement. These responses were then analyzed qualitatively.

3. RESULTS AND DISCUSSION

3.1 Analysis Results

Through the analysis of previous literature and cases, it was possible to derive the implications that learners with disabilities should be diagnosed in consideration of their more complex characteristics than those of ordinary students. According to a study by Lee et al. (2024), it is important to analyze the learner's cognition, motor ability, communication method, and emotional characteristics to design an AI digital textbook for students with physical disabilities, and it is suggested that individualized learning activities should be designed to reflect these characteristics. In addition, Lee et al. (2024) emphasized that differentiated and personalized support should be provided for visually impaired students, considering not only the degree of visual impairment, but also the learner's multiple intelligence and willingness to participate.

3.2 Prototype Development

Pre-diagnosis is primarily categorized into 'initial diagnosis' and 'ongoing diagnosis'. The initial diagnosis begins with the teacher entering the student's information (name, gender, grade, area of disability, and degree of disability). Next, the student's cognitive level is divided into three broad categories (e.g., A, B, C), and a survey is presented to determine the student's literacy level.

Questions are also presented to assess the motor skills required to use the AI digital textbook (see Figure 1). First, ask the student what assistive device they need to use the tablet PC. The responses to this question are broadly divided into direct selection (using fingers and a touch pen, using an eye mouse or head mouse, using a joystick and switches together), and indirect selection (using the scanning function and switches). If the 'direct selection type' was selected, the student was asked to check whether drag-and-drop was possible, whether continuous touch was possible, or whether only one touch was possible in the activity using the tablet. In addition, it will also check whether additional functions such as handshake correction and auto complete are needed for accurate selection. When choosing the 'indirect selection type', the user can also select the scanning speed, the color of the scanning area display, and the degree of darkness.

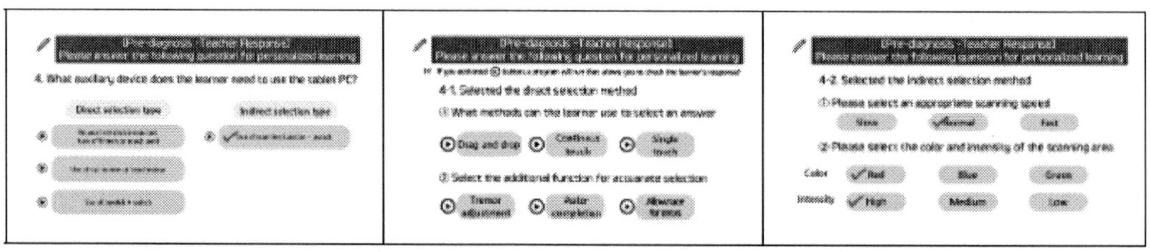

Figure 1. Assessment questions for diagnosing the learner's motor skills

After diagnosing the motor skills, the students' main forms of communication were identified. This section identified whether they could communicate verbally, using written symbols, using pictorial symbols, or using gestures. While learners were asked to respond to specific items (e.g., communication preferences, motor skill demands), these were used for interface calibration and not formal diagnostic validation.

In addition, the learners were asked to evaluate their learning emotions (e.g., they are interested in the class and actively participate in it, their interest varies depending on their physical condition, they are not interested in academics but actively participate in classes in the form of games/play, etc.).

"Regular diagnosis" was created to diagnose according to learning units, sessions, and days of the week and to present individualized learning goals. It was designed to identify the student's mood and interest in the learning content on a day-to-day basis and to suggest individualized learning goals and learning formats (e.g., scanning mode, presentation of game-type activities, provision of picture communication boards, etc.).

3.3 Usability Evaluation

A usability evaluation was conducted involving five experts, including four specializing in special education and one in educational technology, to assess the system's interface and gather feedback on its functionality. One strength of the design is its capacity to facilitate a comprehensive pre-diagnostic process across multiple domains, enabling personalized analysis and customization of assistive settings and content formats. Additionally, it allows for detailed adjustments to learners' expression methods based on the pre-diagnostic results, enhancing usability. However, although the design principles include considerations for learners with severe disabilities who communicate through gestures, the actual prototype does not appear to fully reflect these considerations. The key findings from the usability evaluation are presented in Table 1.

Table 1. Key aspects of usability evaluation and reflected results (partial excerpt)

Key Aspects of Usability Evaluation	Reflected Results
The primary users are expected to be students attending special education schools, but the current prototype appears to be designed for students at a relatively high level.	Since students with disabilities exhibit widely varying learning leve designing a prototype exclusively for students with very low learnin levels would limit the diversity of content and formats. Consequentl this aspect was not incorporated into the design.
Questions such as "How do you feel today?" and "A you interested in the learning content?" alone are insufficient for providing tailored diagnostics based on units, sessions, or days of the week.	The unit (or session) goals were revised to be presented at multiple levels, allowing for selection.
When diagnosing the need for assistive devices to us the AI digital textbook, a survey format where teachers respond is not an accurate evaluation methc	To determine whether "direct selection (finger touch, eye-tracking mouse, joystick + switch)" or "indirect selection (scanning function switch)" is more suitable, a direct student performance-based evaluation (e.g., game-based format) is planned.

4. CONCLUSION

This study aims to develop an AI-based digital textbook that systematically analyzes learners based on cognitive ability, motor function, main communication method, and interest factors for students with disabilities. In order to effectively design and develop an AI-based digital textbook for students with disabilities, learner modeling research for students with disabilities and AI digital textbook design and development research considering specific disability areas should be conducted.

In order to provide personalized learning, it is essential to collect various data from learners and analyze and model it. In learner modeling for general students, clustering techniques are used to classify types of learners using data such as academic achievement, learning engagement, completion rate, and correct answers rate (Shi et al., 2020, Liu et al., 2022). On the other hand, research on learner modeling for students with disabilities is relatively scarce. Therefore, further research is needed to identify what data is particularly important in the learning environment of students with disabilities, as well as unique data that only students with disabilities have, and how to effectively model it.

It is also necessary to outline design directions for AI digital textbooks tailored to different types of disabilities. Students with disabilities have different learning styles and needs depending on the characteristics of their disability, so a personalized learning environment that reflects this is required. By using intelligent technologies such as AI, it is possible to diagnose the media environment, options, assistive device settings, preferred learning methods, topics of interest, and learning methods according to the type of disability of the learner, thereby providing a more user-friendly digital textbook (Lee, et al., 2024b). However, research on digital textbooks suitable for specific areas of disability is still lacking, and more research is needed.

REFERENCES

Ally, M. et al., (2022). Artificial intelligence in the fourth industrial revolution to educate for sustainable development. *Canadian Journal of Learning and Technology*, Vol. 48, No. 4, pp. 1-20. https://doi.org/10.21432/cjlt28287.

Hwang, H. J., (2024). *A data-driven examination of the current state of digital educational innovation: Focusing on AI digital textbooks.* Korean Intelligent Information Society Promotion Agency, Daegu, South Korea.

Kim, J. H., (2024). An analysis of special education teachers' perceptions of the educational use of AI digital textbooks. *Journal of Special Education Research*, Vol. 26, No. 4, pp. 45–67.

Korean Education and Research Information Service, (2023). *AI digital textbook development guidelines.* KERIS, Daegu, South Korea.

Lee, G. Y. et al., (2024a). Special education teachers' needs analysis regarding AI digital textbooks for students with physical disabilities. *Journal of Physical, Multiple, and Health Disabilities*, Vol. 67, No. 4, pp. 143–157.

Lee, G. Y. et al., (2024b). Design principles of digital textbooks for visually impaired students. *Journal of Visual Disabilities*, Vol. 40, No. 1, pp. 53–79.

Lee, S. H. et al., (2025). Development of design principles for AI digital textbooks for students with physical disabilities. *Journal of Physical, Multiple, and Health Disabilities*, Vol. 68, No. 1, pp. 201–220.

Liu, Y. et al., (2022). Exploring factors affecting online learners' intention to continue learning in e-learning: A meta-analysis. *Proceedings of IEEE 2nd International Conference on Educational Technology (ICET)*. IEEE, pp. 76–80. https://doi.org/10.1109/ICET55642.2022.9944532.

Ministry of Education, (2023). *Digital-based educational innovation plan.* https://www.moe.go.kr/

Rashid, S. M. M. et al., (2023). Challenges of implementing the individualized education plan (IEP) for special needs children with learning disabilities: Systematic literature review (SLR). *International Journal of Learning, Teaching and Educational Research*, Vol. 22, No. 1, pp. 15–34. https://doi.org/10.26803/ijlter.22.1.2.

Shi, L., Cristea, A. I., Toda, A. M., and Oliveira, W., (2020). Revealing the Hidden Patterns: A Comparative Study on Profiling Subpopulations of MOOC Students. *Proceedings of the 28th International Conference on Information Systems Development (ISD2019)*. Toulon, France, pp. 123–134. https://doi.org/10.48550/arXiv.2008.05850.

ENHANCING DIGITAL COMPETENCY IN PRE-SERVICE EARLY CHILDHOOD TEACHERS: A PROGRAM DEVELOPMENT STUDY

Minkyung Suh and Wooyong Jeun
Baekseok University, South Korea

ABSTRACT

This study aims to develop and implement an educational program to strengthen the digital competencies of pre-service early childhood teachers. Program covers three core areas: teachers' professional digital competencies, teaching and learning competencies using digital tools, and strategies to support young children's digital competency. Through a systematic development process of literature review, expert validation, and pilot testing, we designed a nine-session program that incorporated tablet PCs and applications (Todo Math, Todo Hangul, Quiver AR, and Rakugaki AR). Implemented with 8 young children and 10 pre-service early childhood teachers, the results showed significant improvements in pre-service early childhood teachers' digital competencies and their ability to integrate digital tools into play-based learning. Participants improved their skills in selecting, creating, and using digital materials, while fostering young children's digital competency. The findings underscore the need for structured training programs to prepare pre-service teachers for the digital transformation era, and provide a practical framework for early childhood education. This study advocates for continuous professional development and curriculum integration of digital technologies to prepare educators and children for a rapidly evolving digital society.

KEYWORDS

Digital Competency, Pre-Service Early Childhood Teacher, Education Program

1. INTRODUCTION

In the era of digital transformation, young children are increasingly exposed to advanced technologies through various devices, applications, and interactive media. With 89.4% of children aged three and above using the internet, their play and learning environments have fundamentally shifted, a trend accelerated by the COVID-19 pandemic (Ministry of Science and ICT, 2020). Young children, often referred to as digital natives, Digitods, or the Alpha generation, are growing up immersed in digital environments (Holloway et al., 2015). They engage with digital devices during play, mimic smartphone usage, and utilize various forms of digital media. Despite the deep integration of digital technology into their lives, perceptions about their use remain divided (Dong, 2022; Lee, 2019; Yang & Yang, 2018). For the Alpha generation, artificial intelligence and robots are no longer objects of fear; they naturally interact with machines, fostering technological coexistence (Lee et al., 2022). To prepare these children for the digital transformation era, developmentally appropriate education that fosters their digital competencies from an early age is essential.

Digital competency is the ability of young children to know the benefits and risks of digital, to use digital media positively in play, everyday life and activities, to express and communicate their thoughts and feelings, and to participate safely and healthily in the digital society (Bae et al., 2023). Early childhood digital education also promotes ‘digital citizenship’, a fundamental skill for future citizens, while helping young children to understand what digital is, to use digital devices, and to develop early digital literacy. So the Ministry of Education in Korea is actively promoting digital media use and implementing policies to cultivate future-ready talents. Key initiatives include the mandatory integration of AI education from elementary school, the development of customized AI contents for young children, and the establishment of a digital-based educational infrastructure (Ministry of Education, 2020). The Fundamental Plan for Early Childhood Education Development (2023) emphasizes enhancing young children's digital competencies through play-based learning and digital ethics education, while also strengthening teachers' digital competencies

(Ministry of Education, 2024). These efforts have led to the increased development and dissemination of diverse digital content, supported by advancements in information and communication technology (ICT). As a result, digital technology is increasingly used to support young children's play and enhance their interest and motivation.

While young children from digital native generations are increasingly exposed to technology, they do not automatically develop critical digital competencies such as information discernment, critical thinking, and digital citizenship. Children require guided educational experiences to develop healthy attitudes toward digital technologies (Chaudron et al., 2018). Due to young children's developmental characteristics and their dependence on adults, teachers must possess strong digital competencies to provide high-quality digital education (Baek & Kim, 2021; Kim, 2021; Kim & Park, 2022; No & Kim, 2022). Despite the growing availability of digital resources, there remains a significant gap in comprehensive programs to enhance teachers' digital competencies. This study addresses this gap by developing an educational program specifically designed to strengthen digital competencies of pre-service early childhood teachers.

2. METHODS

2.1 Participants

The participants of this study included 8 children (7 boys and 1 girl) and 10 pre-service early childhood teachers residing in City C. Parental consent was obtained for each child participating in the study. The mean age of the children was 72-73 months (approximately 6 years old).

2.2 Program Development Framework

This study aimed to develop and implement an educational program to strengthen the digital competencies of pre-service early childhood teachers. The program was developed following the ADDIE model through stages of analysis, design and development, implementation, and evaluation. The process of the study is illustrated in Figure 1.

A comprehensive literature review was conducted based on key studies (Chaudron et al., 2018; Fallon, 2020; Hardersen, 2012; Hong et al., 2016; Iivari et al., 2020; Korthagen, 2010) to establish the theoretical foundation for the program design. From this review, we derived essential program components including educational goals, content structure, digital media selection, teaching methodologies, and assessment approaches. The program framework was organized around three core competency domains:

a. Teachers' digital professional competencies: Understanding of digital literacy concepts and skills
b. Teachers' teaching and learning competencies: Ability to implement technology and comprehend children's digital literacy needs.
c. Children's digital competencies: Capacity to independently perform age-appropriate tasks using digital technology

2.3 Pilot Testing and Main Implementation

A pilot study was conducted twice in August 2023 with a class of 5 children to assess program appropriateness, timing, and implementation feasibility. Results showed children engaged well with digital media by demonstrating the ability to learn and performing basic functions of the tablet PCs, but required the teacher's assistance to select applications, launch them, and use necessary features. The main study was conducted nine times from late August through October 2023. Children participated during various activity periods (play, small-group, and individual times).

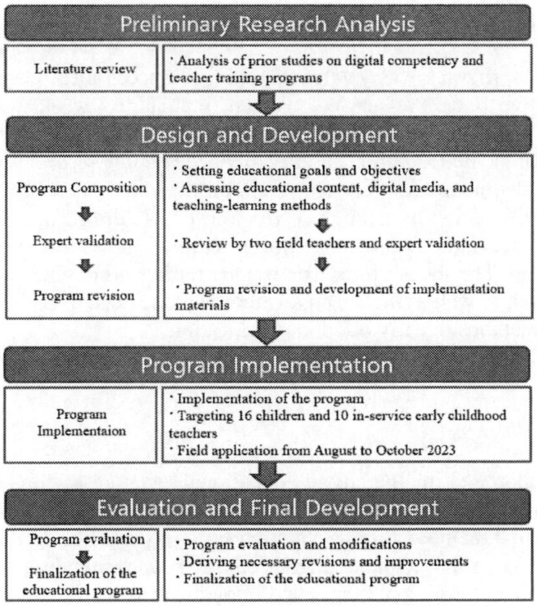

Figure 1. The process of the study

2.4 Program Content and Structure

The program was designed to enhance digital competencies of pre-service early childhood teachers with objectives of strengthening professional competence, developing teaching abilities, and supporting understanding of children's digital needs. The final program incorporated modifications based on validity assessments by experienced early childhood teachers with over two years of digital technology field implementation. As examples of the program structure and content, sessions 1-3 are summarized in Table 1.

Table 1. Program descriptions for sessions 1 to 3

Session	Theme	Digital Activities	Creative Elements	AR Component
1	Environmental protection & recycling	Todo Math/Hangul assessment and activities	Drawing, coloring, cutting butterfly/bee outlines	
2	Making a flowerpot	Todo Math/Hangul activities (line drawing, counting, writing)	Creating plant/animal designs, coloring, cutting	QuiverAR visualization of colored plants/animals; peer assessment with stickers
3	Making dolls	Todo Math/Hangul activities; tablet drawing	Coloring sticks, selecting materials (paper rolls, stickers)	QuiverAR visualization of colored dolls

2.5 Program Implementation

The program was implemented with pre-service early childhood teachers who expressed a strong interest in integrating digital technology into early childhood education. Each participant engaged in hands-on sessions, applied digital tools during interactions with children, and received ongoing guidance to support the practical application of their learning.

2.6 Program Evaluation

The effectiveness of the program was evaluated using a mixed-methods approach that combined both quantitative and qualitative data. Pre-service teachers maintained weekly journals, completed structured self- assessment sheets, and participated in a recorded focus group discussion at the end of the program. A social validity assessment yielded a high feasibility and acceptability score, with an average of 4.8 out of 5.0. Qualitative findings revealed marked improvements in participants' instructional self-efficacy, particularly in using digital tools to modify and individualize curriculum for children with ASD. In their reflective journals, participants emphasized the value of understanding children's individual developmental characteristics for effective digital integration. The focus group discussions further highlighted a growing sense of professional confidence and satisfaction, as well as increased awareness of the need to strike a balance between digital tools and traditional teaching methods in early childhood education.

3. CONCLUSION

This study aimed to develop and implement an educational program to enhance the digital competencies of pre-service early childhood teachers. Based on the findings, several key implications emerged.

First, the study identified the need for systematic training to enhance digital professional competencies of pre-service teachers. Program participants increased their understanding of digital competencies while developing professional skills through planning and implementing technology-based educational activities for young children. This confirms previous findings (Kim & Kim, 2024) that teachers facing the digital transformation era can correctly understand digital competency concepts, recognize development needs, and implement appropriate digital activities for children. These results align with Kim and Park's (2022) work on enhancing teachers' digital competencies. As pre-service teachers belong to the digital native generation, early-stage digital competency education is essential (Momdjian et al., 2025). Consequently, education combining theory and practice is needed to effectively promote teachers' digital competencies for successful classroom performance.

Second, digital technologies and activities in competency programs should respond to the specific needs of both children and pre-service teachers. Since technology constantly evolves, materials must reflect participants' interests and preferences. As pre-service teachers have varying levels of prior digital experience, providing sufficient exploration time enables them to develop appropriate child-centered activities. This research offers practical opportunities for utilizing age-appropriate digital technologies in early childhood education. Teaching digital competencies is essential because current and future educators are digital natives raised in a digital society (Liu et al., 2024; Momdjian et al., 2025). The implementation of digital media should extend beyond basic safety and visual materials to encompass digital literacy and play-based learning approaches (Liu et al., 2024).

In conclusion, the study offers a practical framework for integrating digital competency education into pre-service teacher training. Ongoing professional development is essential to ensure that educators can create inclusive and developmentally appropriate digital learning environments that foster young children's digital literacy, critical thinking, and creativity. These findings contribute to the broader discourse on preparing educators and children for a rapidly changing digital society and lay the foundation for future research and policy development in early childhood digital education.

However, this study has two primary limitations. First, the relatively small sample size and its confinement to a single region may limit the generalizability of the findings. Second, the reliance on self-reported data such as reflective journals and focus group discussions may introduce bias. Future research should involve more diverse participants and include objective evaluation methods to better assess program effectiveness.

REFERENCES

Bae, Y. J., Lim, E. M., Kim, K. R., & Kim, H. J. (2023). *Basic research on how to support digital education for young children* (Custodial Reporting 2023-04). Seoul: Korea Institute of Child Care and Education.

Baik, E. Y., & Kim, H. S. (2021). The current status of early childhood teachers' use of digital media in the early childhood classroom. *Journal of Future Early Childhood Education, 28*(3), 67-96. http://doi.org/10.22155/JFECE.28.3.67.96

Chaudron, S., Di Gioia, R., & Gemo, M. (2018). *Young Children (0-8) and Digital Technology-a qualitative study across Europe*. Publications Office of the European Union.

Dong, P. I. (2022). Do media really harm young children? : Examining young children and media discourses in Korean cultural contexts. *Korean Journal of Early Childhood Education, 42*(5), 29-51. http://doi.org/10.18023/kjece.2022.42.5.002

Fallon, G. (2020). From digital literacy to digital competence: the teacher digital competency (TDC) framework. *Education Tech Research Development, 68*, 2449-2472. https://doi.org/10.1007/s11423-020-09767-4

Hardersen, B. (2012). Digital Competence in the Kindergarten Sector. *Nordic Journal of Digital Literacy, 7*, 227-230. https://doi.org/10.18261/ISSN1891-943X-2012-03-07

Holloway, D. J., Green, L., & Stevenson, K. (2015). Digitods: Toddlers, touch screens and Australian family life. *M/C Journal, 18*(5). 1. https://doi.org/10.5204/mcj.1024

Hong, K., Jo, J., & Park, C. (2016). Development of Software Education Training Program for Early Childhood Teachers. *Korean Association for learner-centered curriculum and instruction, 17*(1), 469-494.

Iivari, N., Sharma, S., & Venta-Olkkonen, L. (2020). Digital transformation of everyday life: How COVID-19 pandemic transformed the basic education of the young generation and why information management research should care? *International Journal of Information Management, 55*, 1-6. https://doi.org/10.1016/j.ijinfomgt.2020.102183

Kim, G. H. (2021). Analysis of kindergarten teacher's needs about teacher's competencies, job performance, and contents of education in the 4th industrial revolution. *The Korea Association of Child Care and Education, 126,* 169-195.

Kim, H., & Kim, K. (2024). The effects of using digital media in art activities on young children's self-concept and drawing representation ability. *The Korean Society For Early Childhood Education, 44*(2), 177-202.

Kim, K. R., & Park, E. H. (2022). Development of an education program to strengthen the digital competency of early childhood teachers. *Early Childhood Education & Care, 17*(3). 103-132.

Korthagen, F. A. (2010). How teacher education can make a difference. *Journal of Education for Teaching, 36*(4), 407-423. https://doi.org/10.1080/02607476.2010.513854

Lee, J., Pyun, Y., & Jeun, W. (2022). *AI education meets Nuri curriculum*. Goyang: Gongdongche.

Lee, Y. M. (2019). The composition of curriculum to improve ICT instructional media competency of early childhood teacher. *Journal of Korea Academia-Industrial cooperation Society, 20*(12), 588-596. http://doi.org/10.5762/KAIS.2019.20.12.588

Liu, S., Reynolds, B. L., Thomas, N., & Soyoof, A. (2024). The Use of Digital Technologies to Develop Young Children's Language and Literacy Skills: A Systematic Review. *Sage Open, 14(1)*. https://doi.org/10.1177/21582440241230850

Ministry of Education (2020). *Artificial intelligence era education policy direction department the key challenge*. Sejong: Ministry of Education.

Ministry of Education (2023). *Third (2023-2027) framework plan for the development of early childhood education*. Sejong: Ministry of Education.

Ministry of Education (2024). *(For teacher-led classroom transformation) How to build capacity for digital-enabled education innovation*. Sejong: Ministry of Education.

Ministry of Science and ICT (2020. 9. 24). *How is our society changing during the Great Transition of the Fourth Industrial Revolution triggered by COVID-19? South Korea Policy Briefing*. Retrieved from https://www.msit.go.kr.

Momdjian, L., Manegre, M., Gutiérrez-Colón, M. (2025). A study of preservice teachers' digital competence development: exploring the role of direct instruction, integrated practice, and modeling, *Evaluation and Program Planing, 109*. https://doi.org/10.1016/j.evalprogplan.2025.102538

No, B. R., & Kim, J. H. (2022). An interview study on the utilization and beliefs of digital media and contents of kindergarten teachers. *Korean Journal of Children's Media, 21*(3), 91-125. https://doi.org/10.21183/kjcm.2022.09.21.3.91

Yang, O. S., & Yang, Y. J. (2018). Developing a measurement scale of technology, pedagogy, and content knowledge(TPACK) for early childhood teachers. *Korean Journal of Early Childhood Education, 38*(3), 29-56. http://doi.org/10.18023/kjece.2018.38.3.002

CONCEPTUAL MODEL OF THE TRIPARTITE APPROACH TO IMPROVING LEARNING OUTCOMES OF THE STUDY PROGRAM

Ljerka Luic and Ana Mulovic Trgovac
University North
Trg dr. Zarka Dolinara 1, Koprivnica, Croatia

ABSTRACT

This research paper examines the perspectives and attitudes of key stakeholders involved in the enhancement of learning outcomes within study programs. These stakeholders include employers, alumni, and current students enrolled in undergraduate, graduate, and postgraduate economics programs. The principal aim is to develop a conceptual model based on a tripartite approach to learning outcomes. Data collection was conducted through field methods and surveys, yielding responses from 171 participants. The collected data were analysed using descriptive statistical tools, which facilitated the identification of critical information dimensions necessary for the formulation of the tripartite model. The results indicate a balanced perspective concerning the influential factors associated with all three dimensions. Nonetheless, future research may benefit from a longitudinal analysis of the evolving perceptions regarding the significance of learning outcomes with contemporary labour market demands. Furthermore, this paper establishes a framework for additional investigation into the informational and communicative elements related to the established dimensions of the model, exploring how the implementation of modernized learning outcomes may influence stakeholder attitudes and public perception.

KEYWORDS

Learning Outcomes, Study Program, Improvement, 3D Information Dimensions, Conceptual Model

1. INTRODUCTION

Digital Europe 2030 serves as a strategic blueprint for the European Union, emphasizing the urgent need to innovate in learning, working, researching, and achieving business objectives amidst the ongoing digital transformation (Wiggberg et al, 2022). While the real sector undergoes substantial evolution, the higher education landscape is notably trailing, unable to keep pace with the rapid changes necessitated by digitalization (Leron, Baconguis, 2021). The McKinsey Global Institute's (Madgavkar at al, 2025) projections indicate that by 2030, automation could displace between 400 to 800 million jobs, impacting roughly 15 to 30% of the global workforce. Furthermore, the International Monetary Fund's World Economic Outlook highlights an unsettling trajectory for the business environment, suggesting that between 75 to 375 million individuals will require reskilling to align with the emerging professions of the future. This landscape necessitates a shift in the essential skills required, moving towards a demand for enhanced cognitive capabilities, psychomotor skills, independence, and accountability in the workplace.

To address these disruptive trends, it is crucial to revise the learning outcomes within academic programs. For instance, disciplines like business economics should align more closely with evolving labor market dynamics, particularly within the GIG economy, where flexibility, adaptability, digital literacy (Eshet-Alkalai, 2004; Eshet-Alkalai, 2012), legal and financial awareness, entrepreneurial competencies, and lifelong learning are paramount. The World Economic Forum's 2020 report further substantiates this by indicating a dual demand for both digital and social competencies among future professions, especially as the green economy gains traction and innovations in data processing, artificial intelligence, engineering, cloud computing, and product development proliferate.

To cultivate a workforce that meets these demands, it is imperative that educational institutions equip students with the requisite knowledge (Alavi, Leidner, 2001; Al-Husseini et al, 2019; Correia et al, 2023), skills (Dijk, 2017; Vasilieva, 2022) and competencies (Mukherjee, 2023).

This includes integrating new technologies (Cózar-Gutiérrez, Sáez-López, 2016) into the curriculum (Škoda, Luić, 2020), particularly around artificial intelligence (Kučić, Luić, 2023) and immersive technologies such as VR/AR/XR, which are increasingly vital in contemporary professional settings. Establishing robust partnerships with professional associations and industry stakeholders is equally essential. Engaging representatives in the educational process through mentorship, internships, and collaborative projects will enrich student development and facilitate a smoother transition into the workforce (Björklund et al, 2019; Fuad et al, 2020).

In summary, enhancing learning outcomes in study programs is critical to transitioning towards a modern educational paradigm that incorporates online and e-learning methodologies (Singh, Thurman, 2019). This approach fosters flexibility and student-centered learning (Kim, 2023; Baričević, Luić, 2023.), thus preparing future professionals to navigate the complexities of digital business environments.

Graduates will be equipped to handle decision-making amid entrepreneurial uncertainty, embrace personal and team accountability in strategic implementations, and uphold social and ethical standards throughout their professional pursuits.

2. METHODS

The objective of this research is to analyze the perceptions and attitudes of vital stakeholders involved in enhancing the learning outcomes of academic programs, specifically focusing on employers, alumni, and current students enrolled in undergraduate, graduate, and postgraduate economics programs at the University North.

Utilizing a *field* research methodology, the study employed an online survey designed through Google Forms, targeting a purposive sample comprising employers, alumni, and current students. In this pilot research, Participants were approached via the official email of the Department of Economics at the University North, from which invitations to participate were dispatched. The survey ensured respondent anonymity. The finalized research sample comprised N = 171 participants, categorically consisting of 28 employers, 65 alumni, and 78 current students across all three academic levels.

The employer questionnaire consisted of 26 items structured into four thematic sets: (1) business organization, activities, and employment of graduates from the University North; (2) competencies acquired; (3) evaluations of academic programs alongside the practical knowledge and skills requisite for contemporary labor market demands; and (4) an assessment of the relevance of current learning outcomes in relation to the needs and expectations of their organizations.

The alumni survey included 23 items, segmented into three categories: (1) socio-demographic data; (2) inquiries regarding the educational experience at the Department of Economics of the University North; and (3) evaluations of the relevance of current learning outcomes concerning their career trajectories and the requirements of the labour market.

For current students, the survey instruments contained 53 questions for undergraduate students, 58 for graduate students, and 37 for specialist (postgraduate) students. These were organized into six categories: (1) socio-demographic factors and study satisfaction; (2) assessment of the significance of learning outcomes related to the study program; (3) evaluation of competencies essential for employability; (4) relevance of existing learning outcomes to labour market demands; (5) perceptions of career development; and (6) student experiences and academic progression throughout their studies.

3. RESULTS AND DISCUSSION

For this research, we develop a conceptual model (Robinson, Karamuftuoglu, 2010; Tuđman, 2014; Bosančić, 2016, Correia et al, 2023) based on a tripartite approach to enhancing learning outcomes. This model identifies three critical stakeholders in the optimization of educational results for study programs: employers, alumni, and current students enrolled in economics programs. These groups are considered pivotal in influencing the perceptions of relevant stakeholders in the academic and professional landscape.

3.1 Employers

In a comprehensive analysis of study programs, respondents identified several key areas for enhancement: 76.9% emphasized the need for increased incorporation of practical tasks and case studies, while 57.7% highlighted a greater emphasis on digital tools and business technologies. Additionally, 53.8% called for stronger connections with industry through collaborative projects and mentoring initiatives, and 50% advocated for the development of entrepreneurial and managerial skills. Notably, 83.7% of participants expressed a willingness to engage in future collaborations with the Department of Economics, particularly in the context of student internships. Of this group, 70.8% showed interest in partnerships involving guest lectures and workshops, and 62.5% were keen on participating in projects and case studies. Employers involved in the undergraduate economics program underscored specific learning outcomes as critical for the existing curricula. The most pivotal outcomes included the ability to "utilize modern information technologies in business" (75%) and the capacity to "distinguish and critically assess key elements of the economic framework relevant to business and management" (67.9%). For graduate studies, the primary learning outcomes identified were the competencies to "plan, organize, manage, and control business operations within an international context" (71.4%) and to "critically evaluate market structures utilizing core concepts of microeconomic theory to develop and implement strategic business plans" (67.9%).

3.2 Alumni

Alumni prioritize several key enhancements in their educational experience, emphasizing the need for more practical teaching methodologies (63.1%) and an increased integration of digital tools and technologies (53.8%). They also highlight the importance of stronger affiliations with industry employers (52.3%) and the cultivation of entrepreneurial competencies (47.7%). The courses deemed most beneficial include a diverse set of disciplines such as Accounting, Finance, Business Communication, Digital Economy, Digital Business Transformation, Innovation and Strategic Development, FinTech, Controlling, Management, Microeconomics, Macroeconomics, Implementation of EU Projects, Marketing, Entrepreneurship, Business Intelligence in Entrepreneurship, Knowledge Management, Global Value Chains, and Strategy and Competitiveness of Companies. Respondents identified several critical skills essential for career advancement that were inadequately developed during their studies. Notably, these include practical skills, digital literacy, financial acumen, leadership abilities, critical and strategic thinking, data analytics, public speaking, and negotiation skills. In terms of the university's engagement with the business sector, alumni rated this connection as satisfactory (43.1%). For improvements, the alumni suggest the introduction of both mandatory and elective courses focusing on contemporary business trends (66.2%), enhanced partnerships with employers and industry mentors (61.5%), increased practical training opportunities (55.4%), and a greater emphasis on international opportunities and exchange programs (33.8%).

3.3 Students

Postgraduate Study: In a survey assessing the relevance of elective courses to labor market demands, 50% of respondents rated their relevance as either very good or excellent on a scale of one to five. Respondents highlighted a need for courses that emphasize practical applications and simulations of EU projects, as well as those focusing on digitalization and artificial intelligence (AI). Notably, 75% indicated that engaging in real-world projects and case studies would greatly enhance their competencies, while 87.5% found value in business process simulations and the development of digital skills utilizing modern tools such as AI, business intelligence (BI), and augmented/virtual reality (AR/VR).

Graduate Study: In the same survey, 66.6% of respondents assessed the relevance of elective courses as good or very good. They pointed to a desire for courses encompassing investments, internet security, and business intelligence. Additionally, there was a consensus on the necessity of integrating project creation with actual companies to foster collaboration between academia and the real sector. Similar to the postgraduate insights, 77.8% of participants believed real project engagement would be beneficial, with 72.2% emphasizing the importance of business process simulations and the cultivation of digital skills through modern technologies. The evaluation of the relevance of each learning outcome associated with the study program is presented in Table 1, using a rating scale from one to five.

Table 1. Evaluation of the Relevance of Learning Outcomes within the Graduate Study Program

Learning outcomes	1	2	3	4	5	6	7	8	9	10	11	12	13	14	15	16	17	18	19	20
Mean Value	**3,88**	**3,88**	4,11	4,33	4,11	4,11	4,27	4,22	4,22	4,27	4,22	4,17	**3,66**	**3,66**	**3,77**	4,39	4,39	4,17	4,67	4,50

Undergraduate Study: For undergraduate respondents, 75% rated the relevance of elective courses as good or very good. They expressed interest in subjects such as artificial intelligence applications in marketing, entrepreneurship, and accounting for non-profit organizations. There was also a call for practical instruction involving information systems and an enhanced focus on EU projects and funding. Importantly, 69.2% felt that participation in real business projects would significantly bolster their competencies. Furthermore, 82.7% expressed a desire for more comprehensive education on entrepreneurship, while an overwhelming 98.1% acknowledged the need for curriculum elements that foster the development of "soft skills." The evaluation of the relevance of individual learning outcomes within the study program is detailed in Table 2 below, rated on a scale from one to five.

Table 2. Evaluation of the Relevance of Learning Outcomes within the Undergraduate Study Program

Learning outcomes	1	2	3	4	5	6	7	8	9	10	11	12	13	14	15
Mean Value	4,17	4,13	**3,29**	4,52	4,24	4,15	4,08	4,23	4,35	4,45	4,38	4,33	4,12	4,40	4,17

4. CONCLUSION

The objective of this research was to develop a conceptual model based on a tripartite approach aimed at enhancing learning outcomes within academic programs. This study explored the perspectives of key stakeholders involved in the improvement of these outcomes, including employers, alumni and current students across undergraduate, graduate, and postgraduate economics programs. The research successfully identified critical information dimensions necessary for constructing the conceptual model, which captures the interplay of factors influencing learning outcomes among these three stakeholder groups. The findings indicate a nuanced equilibrium among the influencing factors across the three dimensions. Utilizing the developed conceptual model, we can delineate the information flows that are essential for structuring the logical framework of communication between functional entities involved in the enhancement of learning outcomes. This information flows can be transmitted and, where necessary, transformed to facilitate effective collaboration among stakeholders. Future research avenues may focus on longitudinal analyses to further examine how perceptions of learning outcomes evolve with the demands of the contemporary labor market. Additionally, this paper lays the groundwork for deeper investigations into the information and communication dynamics integral to the identified dimensions of the model. This could include exploring how modernized learning outcomes influence stakeholder attitudes and opinions, and their subsequent impact on shaping the perceptions of relevant public audiences.

In conclusion, the conceptual model of the tripartite approach emphasizes the critical role of collaboration among employers, alumni, and students in enhancing learning outcomes. By actively involving all stakeholders, this model facilitates the development of more relevant competencies and strengthens the connection between the academic community and the professional sector.

ACKNOWLEDGEMENT

The publication of this paper was made possible through the University North funds, intended to support the scientific research project "Research perspectives of the Design Thinking method in the fields of social sciences and interdisciplinary sciences", to which these authors express their gratitude.

REFERENCES

Alavi, M., Leidner, D.E., (2001). Review: Knowledge Management and Knowledge Management Systems: Conceptual Foundations and Research Issues. *MIS Quarterly*, Vol. 25, No. 1, p. 107.

Al-Husseini, S. et al, (2019). Transformational leadership and innovation: the mediating role of knowledge sharing amongst higher education faculty. *International Journal of Leadership in Education*, Vol. 24, No. 5, pp. 670-693.

Baričević, M., Luić, L., (2023). From active learning to innovative thinking: The Influence of Learning the Design Thinking process Among Students. *Education Sciences*, Vol. 13, No. 5, Article No 455.

Björklund, T. et al, (2019). Learning across silos: design factories as hubs for co-creation. *European Journal of Education*, Vol. 54, No. 4, pp. 552-565.

Bosančić, B., (2016). Proces stjecanja znanja kao problem informacijskih znanosti. *Libellarium: časopis za istraživanja u području informacijskih i srodnih znanosti*, Vol. 9, No. 1, pp. 31-58.

Correia, P.R.M. et al, (2023). Using concept maps to surf semantic waves in the pursuit of powerful knowledge structures. *Knowledge Management & E-Learning: An International Journal*. Vol. 15, No. 3, pp. 381-391.

Cózar-Gutiérrez, R., Sáez-López, J.M., (2016). Game-based learning and gamification in initial teacher training in the social sciences: An experiment with Minecraft. *Edu. Int. J. Educ. Technol. Higher Educ.*, Vol. 13, Article No 2.

Dijk, J.A.G.M., (2017). The relation between 21st-century skills and digital skills: A systematic literature review. *Computers in Human Behavior*, Vol 72. Pp. 577-588.

Eshet-Alkalai, Y., (2004). Digital Literacy: A Conceptual Framework for Survival Skills in the Digital Era. *J. Educ. Multimed.* Hypermedia, Vol. 13, pp. 93-106.

Eshet-Alkalai, Y., (2012). Thinking in the Digital Era: A Revised Model for Digital Literacy. *Issues in Informing Science and Information Technology*, Vol. 9, pp. 267-276).

Fuad, D. et al, (2020). Innovation culture in education: a systematic review of the literature. *Management in Education*, Vol. 36, No. 3, pp. 135-149.

Kim, E., (2023). Factors affecting organizational innovation of universities – focusing on shared vision, student-centered value, trust culture, and organizational positivity. *Journal of Curriculum and Teaching*, Vol. 12, No. 1, Article No. 200.

Kučić, M., Luić L., (2023). Tailorscore: Application of Innovative Teaching and Artificial Intelligence to Encourage Creativity. *Proceedings of 16th annual International Conference of Education, Research and Innovation.* Seville, Spain, pp. 7390-7397.

Leron, P., Baconguis, R., (2021). Exploring the dimensions of innovation culture in the public higher education institutions: towards improved organizational performance in research and development. *Journal of Research Management and Administration*, Vol. 1, No. 1, pp. 36-57.

Madgavkar, A. at al, (2025). *Tough trade-offs: How time and career choices shape the gender pay gap*. McKinsey Global Institute. McKinsey & Company.

Mukherjee, T., (2023). *Youth Transition in the Digital Age: Balancing Digital Competency Demands and Preparing for the Future*. In: Deb, S., Deb, S. (eds) Handbook of Youth Development. Springer, Singapore.

Robinson, L., Karamuftuoglu, M., (2010). The nature of information science: changing models. *Information Research: An International Electronic Journal*, Vol. 15, Article No 4.

Singh, V., Thurman, A., (2019). How Many Ways Can We Define Online Learning? A Systematic Literature Review of Definitions of Online Learning (1988-2018). *American Journal of Distance Education*, Vol. 33, No. 4, pp. 289-306.

Škoda, J., Luić, L., (2020). Challenges of Digital Age Curriculum Convergence. *Proceedings of 13th annual International Conference of Education*. Online Conference, pp. 6448-6454.

Tuđman, M., (2014). *Theory of Information Science*. Third revised edition. Croatian University Press. Croatia, Zagreb.

Vasilieva, E., (2022). *Research on the Level of Soft Skills of the Digital Generation*. In: Zaramenskikh, E., Fedorova, A. (eds) Digitalization of Society, Economics and Management. Lecture Notes in Information Systems and Organisation, Vol. 53. Springer, Cham.

Wiggberg, M. et al., (2022). Defining digital excellence: requisite skills and policy implications for digital transformation. *IEEE Education Society Section*. Vol. 10, pp. 52482-52507.

FROM THE PERSPECTIVE OF 3C-3I INTEGRATION: RESEARCH ON THE CHALLENGES AND TRANSFORMATION STRATEGIES OF DIGITAL TEACHING IN MIDDLE SCHOOLS

Xixi Lu
Nanning Normal University, College of Materials and Chemistry, Major in Chemistry
No. 508 Xinning Road, Wuming District, Nanning City 530100, China

ABSTRACT

The Digitalization of education has become a global irreversible trend, and many countries have formulated policies and plans to promote the transformation of education. This study summarizes the current situation and problems in the process of digital education: 1. Teachers' digital literacy has been improved. 2. The imbalance of the use frequency of teaching technology in classroom teaching; 3. Progress has been uneven across regions. Therefore, based on the theory of meta-cognitive education and "3C + 3i" education concept, combined with the real situation of world education, this paper constructs a DPP digital teaching model to guide education and teaching activities, according to the "3 C + 3 I" education concept, this paper puts forward the solution path for the above situation, hoping to provide substantive help for the digitization of education in various regions.

KEYWORDS

3i Education Idea, Digitization, Education Digital Resources

1. INTRODUCTION

"Accelerating education reform in the information age" is one of the strategic tasks put forward in "China's education modernization 2035"(Central Committee of the Communist Party of China, State Council, 2019). Digital education is an important breakthrough to open up a new track of education development and shape the new advantages of education development. Digital education resources enable every learner to carry out high-quality education, personalized education and precision education, it is an important measure to promote educational equity. In the process of educational digitalization, the upward fitting educational goal is the direction, the digital educational resources are the core, and the teaching process with the participation of information technology is the means to realize the transformation of education. In order to speed up the process of digital education, it is necessary to recognize the current situation and improve the implementation of personalized improvement path.

2. THEORETICAL BASIS

Meta-cognitive learning theory was first put forward by American psychologist J. H. Flavell in 1976, which is very representative. But the origins of metacognitive theory go back much further than 1976. As early as in ancient Greece, the famous philosopher Socrates once expressed his wisdom on the basis of "Knowing oneself": "I know I don't know", which embodied the "Subject" consciousness that was absent at that time, it is the sprout of western philosophy from "Object cognition" to "Subject body cognition". The famous Socratic method is to make students feel their own ignorance of the problem step by step, that is, to find their own "Ignorance", which makes students realize that the subject of the problem is themselves, and then enhance the self-awareness (Jianfeng, 2010). Leading into the teaching process means that the teacher guides the students fully, makes the students give priority to the analysis and exploration of the problems or tasks, the teacher only plays an auxiliary role in guiding, and the relationship between the teacher and the students is subordinate, so as to cultivate students' ability to analyze and solve problems.

In summary, metacognitive theory mainly includes three parts: metacognitive knowledge, metacognitive experience and metacognitive monitoring (Suqin, 2008). Metacognitive knowledge refers to the subject's cognition of his own cognitive process, including procedural knowledge and declarative knowledge. For example, before students learn electrochemical knowledge in middle school, they should learn to understand the process of cognition, understanding their own cognitive characteristics and choosing appropriate and efficient learning strategies belong to the students' metacognitive knowledge. And metacognitive experience is the subject in the cognitive process triggered by the cognitive and emotional (Fenfen, 2008). For example, anticipating failure or success when a cognitive activity is about to be performed. Metacognitive monitoring is a core part, which means that the subject monitors and adjusts the process of cognitive activities in order to complete the cognitive activities efficiently and satisfactorily. For example, teachers in the teaching process constantly reflect: such a teaching method can be accepted by students? Can the student receive the full meaning of what I have said?

2.1 3C+3I Educational Philosophy

In the past two years, China's education adheres to the "3C" concept, steadily advancing the process of education digitization. 3C concept refers to "Connection first, content-based, cooperation to" (Jinpeng, 2024-1-30). The so-called connection refers to the connection of educational resources, educational services and educational ideas in various regions and platforms, and the collection of excellent educational resources from various places into treasured classics, for example, under the National Public Reading Platform for smart education, the platform connects local platforms (e.g. school online, Chinese University MOOC, etc.) in various provinces, cities and regions in China to support people across the country to learn with their thoughts. With the advancement of Technology and the digital transformation of teachers in China, the platform is becoming more and more popular, the educational data stored on the internet has been further expanded, and the National Smart Education Public Service Platform has formed the ability of portal aggregation and switch control, which can strongly support the further deepening of National Digital Education. "Cooperation for the sake of cooperation" refers to the cross-border exchange and cross-regional collaboration of the educational digital service platform. Cross-border exchange refers to cross-border cooperation with governments, enterprises, schools, etc., promoting the digital education resources of the service platform to realize the integration of teaching and learning. Cross-regional collaboration refers to bringing high-quality digital educational resources to less developed areas of the world through "Resource Abroad".

International means broader International cooperation, advocating global digital cooperation, sharing educational resources and reaching educational cooperation to all countries, all regions and all races in the world (General Office of the Ministry of Education, 2024).

2.2 3C+Metacognitive Theory

By combining metacognitive theory with the 3C (Connection First, Content Core, Collaboration Essential) , we can construct a framework that takes into account the principles of cognitive neuroscience, deep knowledge construction, and social learning, the efficient teaching mode of web-based learning. This kind of combination not only strengthens the learning effect, but also cultivates autonomous and strategic learners. In the process of learning, metacognitive strategies help to screen key connections and avoid the interference of invalid information. Monitoring and regulation ensure that core knowledge is processed in depth, not superficially. Reflection sessions help identify "When/where to use this knowledge". The metacognitive skills trained in 3C (such as self-questioning and goal-disassembling) can be applied to any field and have universality.

3. CURRENT SITUATION

3.1 Imbalance of the Use Frequency of Teaching Technology in Classroom Teaching

Survey based on Shang Linlin (Linlin, 2023). Using social science statistical software to analyze the survey data, the results are shown in Table 1. Through the comparative analysis of online teaching and offline teaching, it is found that in the online sample, teachers' use of information-based teaching is mainly presented in the "Knowledge presentation" and "Open question raising" links, however, there is still a big gap in "Solving complex problems", "Evaluation strategy" and "Innovation of teaching mode" supported by digital technology. Digital technology has been actively applied in knowledge presentation, management guidance and teaching feedback in offline teaching, while it has shown a fault-like decline in "Solving complex problems" and "Innovating teaching models". For example, in a middle school chemistry class, teachers' use of intelligent technology to assist classroom teaching mainly focuses on situational introduction and knowledge presentation, which is less involved in student activities, no real use of digital technology will be abstract knowledge visualization, concrete, cannot help students to complete the deeper level of knowledge internalization. In other words, teachers do not develop students' metacognitive ability through digital technology, but only focus on the improvement of students' cognitive ability. It can be seen that whether online or offline teaching, teachers' use of digital means to teach is extremely uneven, in different teaching links in a class, the frequency of relying on digital technology to improve classroom effects is greatly unbalanced, which will lead to serious polarization in the long run and slow down the deep-seated pace of education transformation to digital.

Table 1. Statistics of teaching behavior times of research samples

The act of teaching	Number of online sample teaching behaviors	Number of offline sample teaching behaviors	Category
Classroom lead-in with the support of technology	19	20	Behavior and activity in class
Knowledge presentation supported by technology	109	92	
Collaborative learning supported by technology	24	28	
Ask open-ended questions	67	37	
The resolution of complex problems	3	4	
Summary and promotion supported by technology	42	28	
Outward bound with technical support	26	25	
Evaluation strategy supported by technology	7	31	Tasks and projects after class
Feedback and planning with technical support	43	85	
Feedback and evaluation	3	20	

3.2 The Process of Digitalization Varies from Region to Region

Based on Zhang Huirong (Huirong, Jingyu, Yi & Naiqing, 2021). The results of a survey of teachers' attitudes, proficiency and satisfaction with online teaching are shown in Table 2. Teachers in rural areas do not realize that online education derived from digital technology can promote the digital transformation of national education. Due to the strong connection and cooperation of online education affected by digital technology, teachers in rural areas do not realize that online education can promote the digital transformation of national education, with the characteristics of high-quality content, online teaching can better develop the 3C and 3i concepts in offline conventional teaching. The attitude of teachers to online education determines the extent of online education in the region. If the digital process between urban and rural areas develops in the long term, it is bound to cause great damage to the education, economy and People's happiness of the villages and towns. In general, urban teachers are better than rural teachers in using digital equipment for online education, and the digital coordinated development between different regions still needs to be further improved.

Table 2. Urban-rural differences in teachers' information-based teaching performance during online teaching

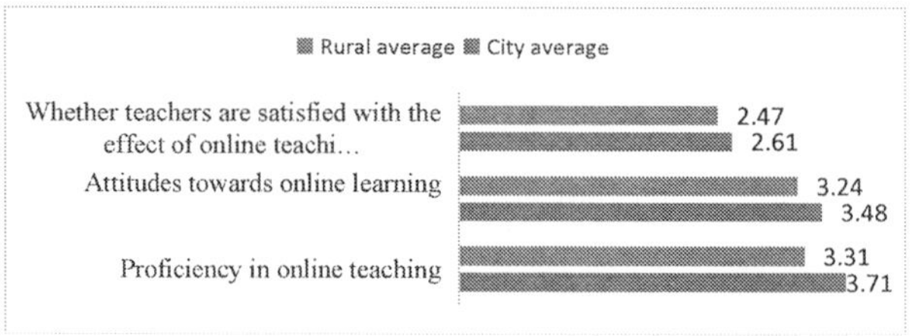

4. STRATEGY

The 3i concept is the further deepening and expansion of the 3C education concept. So how to integrate the concept into secondary education, secondary education to bring more vitality of the change?

4.1 The Embedding of DPP Digital Teaching Model

In the 3i concept, digital education needs to be intelligent, which means that education needs to use artificial intelligence technology for modern learning. In the 3C education concept, it says to be content-based, then from 3C to 3i must be a content as the core, digital information technology as a means to build a personalized teaching model process. With the deepening of digital education, this paper proposes DPP digital teaching model, as shown in Figure 1. In this model, education digitization is the core, and the ultimate goal of teaching activities is to improve students' core literacy, enhance teachers' digital literacy, and be able to carry out intelligent evaluation, the teaching content is expressed in a digital and situational way in the teaching process, and the teaching evaluation is given by completing the project-based inquiry and creation of the teaching process, then by the teaching appraisal personalization feedback student's study degree, thus achieves the closed-loop. From the macro point of view, the DPP digital teaching model aims to promote a deeper level of digital education, and aims to accumulate the energy of digital education in each teaching activity, so as to prepare for the next stage; from the medium point of view, the DPP digital teaching model aims to promote a deeper level of digital education, the educational goal of DPP digital teaching model is to improve both teachers and students. It requires that teaching activities should not only improve the overall quality of students, it also requires the designer of the teaching activity-the teacher-to be able to draw inspiration from progress. For example, the teacher has changed from a model for playing videos using a whiteboard to a creative construction of a virtual situation, and from a tool user to an instructional designer. To achieve a win-win situation; from the microscopic point of view, each teaching activity should include teaching content, teaching process and teaching evaluation, and the whole teaching process can be renovated by reforming the micro-units of teaching activities. For example, in the teaching process, the use of virtual laboratory technology enables students to learn chemical knowledge and complete the experiment by hand, and combines vision, hearing and touch, which is convenient for students to build a knowledge framework faster.

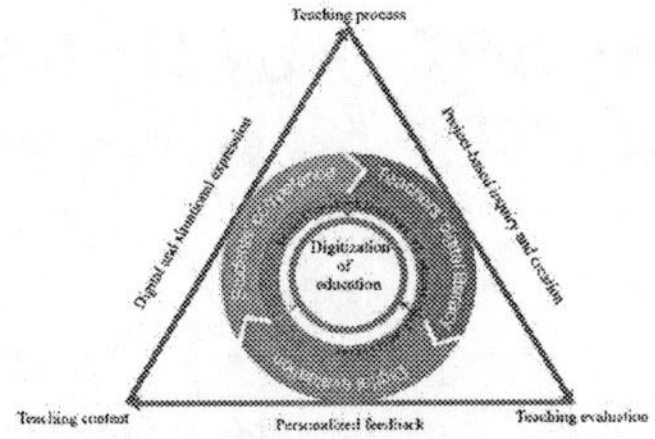

Figure 1. DPP digital teaching model

4.2 Strengthening the Management of Digital Information Security

The biggest challenge lies in resolving the power and responsibility mismatch and structural contradictions caused by the hierarchical system and digital barriers 16. The sharing of digital resources is a double-edged sword. While bringing educational equity and educational resources to developing regions and countries, it also brings the challenges of digital information and personal privacy, when individuals or institutions log on to websites or download apps to use educational resources online, there is a risk of leaking real information to websites and apps. The power of the individual is lost in a two-way vortex, but the energy of the state and government working with local people can stand in the vortex. Follow the digital ethics guidelines and construct a "Two-way drive" governance evaluation mechanism (Jia, Xinyu, Guoxing & Haozhe. 2025). Clear Association rules in different dimensions and multiple levels, establish the connection between network behavior and rule standards and improve the driving sensitivity, timely monitor, capture, analyze, and feedback bad network behavior, to achieve early discovery and early governance, reduce the bad network behavior in the learning of educational digital resources. That is to say, the upper level of the country and the government need to formulate relevant strong policy documents to ensure the information security of users, and the grass-roots people and society also need to raise the awareness of information security, do not do harm to other people's information security behavior, in the global sharing of educational information resources at the same time, global information management is urgently needed, a wide range of global cooperation to establish educational digital resource sharing mechanism, so as to ensure that educational resources fair, just, open, and not by the harmful elements of monopoly.

REFERENCES

Central Committee of the Communist Party of China, State Council. (2019). China's education modernization 2035[EB/OL].

Fenfen, G. (2008). Research on teachers' metacognitive teaching. Chinese adult education, (03), 115-117.

General Office of the Ministry of Education. (2024). China-Africa Cooperative Action Initiative for the protection of educational and cultural heritage.

Huirong, Z., Jingyu, W., Yi, L. & Naiqing, S. (2021). The "Digital divide" between urban and rural basic education: characterization, causes and countermeasures -- an empirical study based on online teaching. Education and the economy, 37(04), 20-28.

Jia, G., Xinyu, K., Guoxing, X., & Haozhe, J. (2025). Easy "Floating" into "Embedding": Technical Dilemma and relief strategy of digital literacy education for college students empowered by GENAI. Modern educational technology, 35(04), 25-34.

Jinpeng, H. (2024-1-30). Work together to promote application, sharing and innovation in digital education-keynote speech at the World Digital Education Congress 2024.

Linlin, S. (2023). Research on secondary school teachers' ability to use information technology from the perspective of teaching behavior analysis (Inner Mongolia Normal University).

Suqin, J. (2008). On the application of metacognition in multimedia-assisted English listening class. Acta Hunan University of Science and Engineering, (01), 214-215.

Zongkai, Y. (2023). Promoting education digitalization under the guidance of Chinese pedagogy. Educational Research, 44(07), 12-16.

RAIL TRAFFIC RESEARCH IN GERMANY: EVALUATION OF GUIDELINES, REGULATIONS AND DESIGN VALUES WITH RESPECT TO CLIMATE CHANGE

Jens Lehnen[1], Michael Eble[1], Carina Herrmann[2] and Frauke von den Driesch[2]
[1]*Dr., valantic*
Bolzstraße 6, 70173 Stuttgart, Germany
[2]*German Centre for Rail Traffic Research at the Federal Railway Authority*
August-Bebel-Straße 10, 01219 Dresden, Germany

ABSTRACT

Climate change leads to essential challenges such as global temperature increases, extreme weather events or ecological and social challenges. Transport infrastructure is increasingly affected by climate change, e.g. through damage to infrastructure caused by floods or landslides. In response to these developments, this ongoing research project takes a systematic approach to evaluate transportation infrastructure regulations for rail and road networks in response to climate change. It analyses how existing guidelines, norms, and design values can be revised to adapt to the effects of rising temperatures, extreme weather events, and other climate-induced risks. The applied methodology is based on a comprehensive literature review, gathering global best practices on regulatory adaptation. This informs a detailed evaluation of selected regulations, where standardized fact sheets highlight outdated assumptions and recommend targeted improvements in safety margins, material specifications, and maintenance protocols. Finally, the research project proposes an implementation strategy that guides relevant authorities and stakeholders through formal revision procedures, coordination efforts, and ongoing monitoring. This multi-phase process combines scientific data, stakeholder input, and structured evaluations to ensure both practicality and credibility. The goal is to support the strengthening of infrastructure resilience and public safety under evolving climatic conditions. By linking research findings with actionable steps, the concept offers a robust foundation for enduring, climate-responsive, and cost-effective regulatory updates. The project began in January 2025 and will conclude in November 2025.

KEYWORDS

Climate Resilience, Infrastructure Adaption, Regulatory Analysis, Transportation Sector, Implementation Strategy, Standards Assessment

1. INTRODUCTION

Rail transport in Germany provides significant value by efficiently and environmentally transporting large volumes of passengers and goods. In passenger transport, it offers a reliable alternative to road traffic, alleviating highway congestion and reducing pollution (Maierhofer et al., 2024). Freight rail transport is essential for the logistics industry as it economically and safely moves large shipments over long distances. The continuous modernization and digitalization of infrastructure further enhance the attractiveness of rail transport, improving sustainability and competitiveness (Pinkhasik and Herrmann, 2021). In addition, digital solutions such as digital twins to enabling real-time monitoring, Predictive Maintenance to monitor equipment and infrastructure or Passenger Information Systems to provide real-time updates and personalized travel information are increasingly being used (Jeschke and Grassmann, 2021). Overall, rail transport in Germany is indispensable for a sustainable mobility and transport strategy.

Climate change intensifies weather events like extreme heat, heavy rain, and storms, straining transportation infrastructure (Kasraei et al., 2024; Nemry & Demirel, 2012). Railways face damage from flooding, storms, and heat, with rising temperatures causing rail expansion and coastal tracks threatened by sea level rise (Garmabaki et al., 2021; Park et al., 2022). Regulations guide infrastructure planning and operation, but evidence shows they may no longer fully address new climatic conditions like heatwaves and heavy rainfall.

Updating these rules is resource-intensive and varies widely across studies (Kind et al., 2021; Siefer et al., 2018).

Against this backdrop, the research project "Evaluation of Guidelines, Regulations, and Design Values with respect to Climate Change", led by the German Centre for Rail Traffic Research at the Federal Railway Authority (DZSF), aims to provide a consistent, transparent, and evidence-based overview of regulations that may require revision. The focus is on identifying how current standards, guidelines, and norms in road and rail infrastructure can be adapted to accommodate changing climatic conditions. The project started in January 2025 and will conclude in November 2025.

2. OBJECTIVES AND APPROACH

The primary purpose of this research project is twofold. First, it aims to develop an overview establishing which regulations and standards merit consideration for climate-related adjustments in the road and rail sectors. Second, it seeks to offer detailed recommendations for a subset of these regulations by assessing current provisions considering shifting climatic conditions such as global temperature rises or changes in precipitation conditions, then proposing specific changes. These insights will be extrapolated into a broader strategy for future or more extensive adaptation efforts. The goal is to produce a systematic approach for evaluating and prioritizing the regulations that may need updating, culminating in a step-by-step strategy that can guide the practical implementation of revisions.

An interdisciplinary project team, whose expertise spans infrastructure, climate studies, data analysis, literature reviews, and consulting in both rail and automotive sectors, carries out the research. The work follows a sequence of interconnected sub-tasks. First, a literature review identifies relevant documents, studies, and methods on climate adaptation in regulations. Next, the team selects and analyzes those regulations that appear most susceptible to climate impacts, based on insights from literature. From these findings, a structured strategy is developed in a third step on how to update or amend the relevant rules, norms, or guidelines. Throughout the process, the team collaborates with the client and a research committee from different institutions and sectors, ensuring alignment with practical needs and validation of the final recommendations.

The project's deliverables include a literature review summarizing studies, methods, and regulations requiring climate-focused revisions, followed by an in-depth analysis of selected regulations, highlighting weaknesses and areas in need of revision. An implementation strategy provides a roadmap for adapting regulations, supported by a concise "handbook" for practitioners. Progress updates, interim reports, and a final comprehensive report synthesize findings, with an expert working group ensuring practical relevance throughout. This study is being funded by the German Federal Ministry Transport (German abbreviation: BMV) in the context of the BMV Research Network. The BMV Research Network is the cross-modal research format in the BMV's departmental research. In 2016, seven departmental research institutions and specialist authorities of the Federal Ministry for Transport (BMV) joined forces to form a network under the guiding principle of "Knowledge - Ability - Action". The aim is to research urgent transport issues of the future in a number of areas, including climate change, environmental protection, reliable infrastructure, digitalisation, renewable energies and transport economic analyses.

2.1 Sub-Task 1: Literature Review of International Studies and Regulations

In today's fast-paced environment, where dissemination of knowledge has accelerated tremendously, literature research is still crucial for scientific purposes as it allows researchers to build upon existing knowledge by identifying gaps, trends, and established findings within a particular field (Tang et al. 2022). It provides a comprehensive understanding of the rapidly evolving research landscape, enabling scholars to design studies that are both relevant and innovative amidst the constant influx of new information. Thus, we undertook a comprehensive literature analysis using a scientifically rigorous approach (Snyder, 2019; Rowley and Slack, 2004), reviewing a total of 81 documents.

The aim of sub-task 1 is to compile and analyze a broad selection of documents related to climate adaptation. The review focuses on three main domains. First, it reviews climate adaptation in regulations, highlighting updated standards and methodologies for revisions. Second, it addresses transportation-specific climate adaptation, highlighting any projects in the road or rail sectors that have assessed or revised relevant technical guidelines. Third, it explores the processes and frameworks employed by other organizations to pinpoint essential climate-related changes. The team uses databases like IEEE Xplore, Google Scholar, and Web of Science, reviews non-academic reports, and consults climate, engineering, and risk assessment experts to address gaps. All sources are filtered according to their relevance, publication date (with emphasis on material published from 2010 onward), and credibility, including peer review and scientific acceptance. This thorough approach ensures that potentially relevant case studies and research findings are included. Table 1 gives an overview of different search categories and search results based on 81 documents analyzed so far. The content given in the table below is sorted by the studies' sources and the number of their publications.

Table 1. Sources of analyzed documents by number (top 10; exemplary)

Source	Category	Number of documents
Federal Ministry for the Environment, Nature Conservation, Nuclear Safety and Consumer Protection (BMUV)	Federal authority	7
Deutsches Institut für Normung e. V.	International association	6
European Union	Association of states	4
Federal Ministry for Digital and Transport	Federal authority	4
Forschungsgesellschaft für Straßen- und Verkehrswesen	Research institute	3
International Organization for Standardization (ISO)	International association	3
Federal Office for the Environment	Federal authority	2
Federal Ministry for Climate Action, Environment, Energy, Mobility, Innovation and Technology (BMK)	Federal authority	2
Cerema (Centre for Studies on Risks, the Environment, Mobility and Urban Planning)	Research institute	2
Potsdam Institute for Climate Impact Research (PIK)	Research institute	2

The documents were analysed in detail and evaluated concerning three specific questions. First, we analysed the documents regarding specific changes that were made due to the effects of climate change:

- **Integration of Climate Scenarios:** More and more take climate change into account, ensuring more resilient infrastructure planning and development.
- **Adaptation of Construction and Planning Standards:** Standards for structures such as roads, bridges, and tunnels have been adjusted to integrate better materials and construction methods that can withstand extreme weather conditions.
- **Introduction of Early Warning Systems and Risk Management:** New systems now enable early detection and response to extreme climatic events in infrastructure management.
- **Changes in Emission Certificate Allocation and Trading:** Procedures for the allocation of emission certificates have been revised to create incentives for emission reductions and to stabilize the market.
- **Strengthening the Resilience of Transportation Infrastructures:** Many countries enhance transportation resilience by reinforcing structures and improving drainage.

These adaptations reflect the growing importance of considering climate change impacts in infrastructure planning and development. In a second step, we analysed the reasons for the adaptions. The main aspects were:

- **Increase in Extreme Weather Events:** Climate change intensifies extreme weather events, straining and damaging transportation infrastructure.
- **Enhancing Resilience:** The adaptations aim to increase the resilience of infrastructure against the impacts of climate change to prevent failures and ensure safe usage.
- **Compliance with Legal and International Obligations:** National and international climate protection commitments require that infrastructure is designed sustainable and climate-friendly to reduce emissions and meet climate goals.
- **Long-Term Durability and Cost Reduction:** Proactive adaptation measures can reduce long-term costs by decreasing the need for expensive emergency repairs and reconstruction after extreme weather events and extending the lifespan of infrastructure.
- **Protection of Public Safety:** Adapting infrastructure to climate change reduces accident risks, preventing failures and ensuring public safety, especially on heavily trafficked roads and bridges.

These reasons emphasize the urgency and importance of adapting regulations in the transportation infrastructure sector to effectively confront the challenges posed by climate change (Moretti 2018, Colin 2016).

Finally, we examined in detail how the process of adapting the regulations is implemented. Typically, the six essential steps for changing regulations in transportation infrastructure due to climate change are:

1) **Recognition and Analysis:** The process begins with identifying the need for adaptations based on scientific studies, reports, and the increasing frequency of extreme weather events. This leads to a comprehensive risk analysis of existing infrastructure.
2) **Stakeholder Engagement and Consultation:** Formation of working groups and networks that involve relevant stakeholders such as government agencies, experts, scientists, and the public to gain a broad perspective and gather feedback.
3) **Development of Strategies and Measures:** Based on the analysis, concrete adaptation strategies and technical standards are developed to address identified risks and vulnerabilities.
4) **Pilot Projects and Testing:** Conducting pilot projects to test the effectiveness of proposed adaptations and to gather practical experience.
5) **Implementation and Monitoring:** Execution of new regulations and guidelines, accompanied by continuous monitoring and evaluation of measures to ensure their effectiveness.
6) **Regular Evaluation and Updating:** Continuous updates to strategies and regulations ensure infrastructure resilience amid evolving science, experience, and climate conditions.

This iterative and participative process ensures that the changes are both effective and practical, increasing resilience to the impacts of climate change.

The main outcomes of sub-task 1 include a catalog of relevant documents and data sources on climate adaptation in the transportation field, a preliminary listing of regulations (covering roads, rail, and other pertinent domains) that warrant a closer look, and a summary of known adaptation strategies or frameworks to guide subsequent analyses.

2.2 Sub-Task 2: Detailed Analysis of Selected Regulations

In sub-task 2, the project team narrows down a set of regulations for in-depth study, basing its choices on the output of the literature review. The selection process considers whether regulations might be subject to serious climate-related stresses (such as extreme heat damaging rail lines or heavy rainfall undermining road embankments) and strives to achieve a representative sample covering a variety of regulatory formats, authors, and technical domains. The analysis itself begins by confirming the availability of official texts, then developing standardized fact sheets that document each regulation's scope, climate sensitivity, and existing safety or design margins. The team determines whether the documents reference relevant climate conditions or rely on outdated assumptions about weather extremes. It suggests new threshold values, safety factors, or inspection frequencies that might improve resilience. These fact sheets typically cover background information, climate relevance, existing provisions, potential adaptations, and the feasibility of implementing updates, as well as the responsible organizations (Blackwood et al. 2022, Fankhauser 2016).

Regular communication with subject matter experts ensures that the chosen regulations retain their significance. If the analysis reveals minimal adaptation needs, resources can be redirected to regulations that demand closer scrutiny. By the end of sub-task 2, the project team has compiled a collection of standardized documents summarizing climate adaptation challenges for each examined regulation, laying the groundwork for the forthcoming implementation strategy. One recommendation for regularities could be, for example, to focus predictions rather on forecasts and climate predictions than on past climate scenarios (Hewitt and Moufouma-Okia, 2023).

2.3 Sub-Task 3: Implementation Strategy for Adapting Rail and Road Transport Regulations to Climate Change

Sub-task 3 develops a strategy to update regulations for climate demands, aiming for replicable, scalable procedures. It prioritizes adaptation needs, considering safety, weather event likelihood, and costs. The strategy outlines steps for revisions, including drafting changes, reviews, consultations, publishing updates, and integration into practice. Stakeholder involvement is crucial, identifying key participants and resolving conflicts over costs or interests. Regular monitoring and feedback loops ensure regulations evolve with climate projections. A handbook supports users, addressing obstacles and solutions. The goal is efficient, enforceable regulatory changes, engaging stakeholders throughout.

3. CONCLUSION

The intensified challenges of climate change necessitate that infrastructure withstand both incremental shifts, such as rising average temperatures, and acute incidents like storms or floods. Effective adaptation of design values, safety margins, and maintenance routines is vital to protect passengers, staff, and physical assets.

A significant outcome of this project is its transparent overview of precisely which climatic variables affect particular regulations and how these regulations can be updated. The fact-sheet approach synthesizes both brief and comprehensive regulatory documents into clear, comparable references, a process that benefits decision-makers in setting priorities. By establishing consistent methods for assessing climate impacts and identifying suitable modifications, the project encourages a more coordinated adaptation process that can be extended to other rules or infrastructure sectors, thereby enhancing collaboration among authorities, operators, and researchers over time. While the focus remains on the German and, in part, European context, the project's methodology can be adapted worldwide, given that climate resilience constitutes a global challenge. The recommendation to engage with stakeholders and address potential obstacles—ranging from lengthy revision cycles to competing budgetary concerns—further underlines the need for realism in any regulatory update initiative. Additionally, this study should be conducted on a regular basis to consider further climate changes and scenarios such as insufficient reductions in GHG.

Certain topics lie outside the scope of the project. The primary concern is adaptation to climate effects, not greenhouse gas mitigation. Other infrastructure systems, such as waterways, are not covered in detail, and broader climate protection strategies remain beyond the project's focus. The work centers specifically on adapting road and rail infrastructure standards to evolving climate-related demands.

REFERENCES

Blackwood, L. et al., (2022). Nature-based solutions as climate change adaptation measures for rail infrastructure. *Nature-Based Solutions*, Vol. 2.

Colin, M. et al., (2016). Adaptation of Transport Infrastructures and Networks to Climate Change, *Transportation Research Procedia*, Vol. 14, pp. 86-95.

Fankhauser, S., (2016). Adaptation to Climate Change, *Annual Reviews of Resource Economics, pp. 209-230.*

Garmabaki, A.H.S., et al. (2021): Adapting Railway Maintenance to Climate Change, *Sustainability 2021*, 13, no. 24.

Hewitt, C. and Moufouma-Okia, W. (2023). Climate Services Based on Climate Predictions and Projections, *World Meteorological Organization Bulletin,* Vol. 72 (2).

Jeschke, S. And Grassmann, R., (2021). Development of a Generic Implementation Strategy of Digital Twins in Logistics Systems under Consideration of the German Rail Transport. *Applied Sciences* 11, no. 21.

Kasraei, A. et al, (2024). Climate change impacts assessment on railway infrastructure in urban environments, *Sustainable Cities and Society*, Vol. 101.

Kind, C. et al, (2021). Adaptation Standard: Analyse bestehender Normen auf Anpassungsbedarfe bezüglich Folgen des Klimawandels, *Climate Change*, 56/2021.

Maierhofer, J. et al., (2024). Digitalization and Reliability of Railway Vehicles and Tracks—Condition Monitoring and Condition-based Maintenance. *Berg Huettenmaenn Monatshefte 169*, pp. 264–268.

Moretti, L., (2018). Climate Change and Transport Infrastructures: State of the Art. *Sustainability*, 2018.

Nemry, F. and Demirel, H., (2012). Impacts of Climate Change on transport: a focus on road and rail transport infrastructures. *Publications Office of the European Union*, JRC72217.

Park, C. et al., (2023). Advanced Representative Rail Temperature Measurement Point Considering Rail Deformation by Meteorological Conditions and Rail Orientation. *Int. J. of Precision Engineering and Manufacturing* 24, pp. 239–249.

Pinkhasik, P. and Herrmann, P., (2021). Learning from external stakeholders: Evidence from two railway projects in Germany, *Project Leadership and Society*, Vol. 2.

Rowley, J. and Slack, F., (2004). Conducting a literature review, *Management Research News*, Vol. 27 No. 6, pp. 31-39.

Siefer, T. et al, (2018). Untersuchung der Regelwerke für den Bahnbetrieb auf Schwachstellen hinsichtlich des zu erwartenden Klimawandels, *EBA-Forschungsbericht,* 2018-08.

Snyder, H., (2019). Literature review as a research methodology: An overview and guidelines, *Journal of Business Research*, Vol. 104, November 2019, pp. 333-339.

Tang, R. Et al., (2022). A literature review of Artificial Intelligence applications in railway systems, *Transportation Research Part C: Emerging Technologies*, Vol. 140.

FEASIBILITY STUDY ON THE PROCESSING OF RECYCLED HDPE/PLA MIXED FILAMENT FOR FDM 3D PRINTING

Peng-Ching Chiu[1], Jui-Hung Cheng[2] and Chang-Zhe Lin[2]
[1]*Department of Chemical and Materials Engineering. National Kaohsiung University of Science and Technology, Taiwan*
[2]*Department of Mold and Die Engineering, National Kaohsiung University of Science and Technology, Taiwan*

ABSTRACT

This study investigates the feasibility of producing FDM (Fused Deposition Modeling) 3D printing filaments by blending recycled high-density polyethylene (rHDPE) and recycled polylactic acid (rPLA). Through melt blending and extrusion, composite filaments were fabricated and evaluated for their printability and mechanical performance. A range of rHDPE/rPLA ratios were tested using tensile strength measurements and visual assessments of printed specimens. The findings indicate that rHDPE exhibits excellent process stability, while the inclusion of rPLA improves surface quality and interlayer adhesion. However, rPLA's lower thermal stability presents challenges under high-temperature conditions. This research contributes to sustainable additive manufacturing by demonstrating the potential of using recycled polymer blends in filament production and identifying optimal blending ratios for balanced mechanical and processing performance.

KEYWORDS

Recycled PLA (rPLA), Recycled HDPE (rHDPE), Sustainable 3D Printing, FDM Filament, Melt Blending, Mechanical Properties

1. INTRODUCTION

Amid growing global concern for environmental sustainability, the reuse of plastic waste has emerged as a pressing issue within materials science and manufacturing engineering. Additive manufacturing, particularly fused deposition modeling (FDM), predominantly utilizes virgin polymers such as high-density polyethylene (HDPE) and polylactic acid (PLA). However, these materials are derived from non-renewable sources, involve energy-intensive production processes, and contribute significantly to environmental degradation due to by-product generation and post-consumer waste (Mejia et al., 2020; Tesfaw et al., 2022). In contrast, the development of recycled alternatives offers a pathway toward circular material use and reduced ecological impact. This study explores the feasibility of using recycled HDPE (rHDPE) and recycled PLA (rPLA) to fabricate composite FDM filaments through melt blending and extrusion techniques. Five blend ratios of rHDPE/rPLA were prepared and tested to evaluate their extrusion behavior, printability, and mechanical properties. The research focuses on identifying the optimal blend ratio that balances mechanical strength, thermal stability, and ease of processing. By addressing the technical challenges associated with recycling thermoplastics for 3D printing, this study contributes to the development of sustainable filament alternatives and aligns with recent findings advocating distributed mechanical recycling in additive manufacturing (Beltrán et al., 2021).

2. METHOD AND RESULTS

2.1 Methodology and Materials

This study examined five rHDPE/rPLA blend ratios - 100/0, 90/10, 50/50, 10/90, and 0/100 - (Figure 1) to investigate their potential for producing recycled filaments suitable for FDM 3D printing (Figure 2). Recycled HDPE and PLA were first dried to remove moisture, then processed using a single-screw extruder. The extrusion temperature varied from 160°C to 200°C depending on the blend ratio. The extruded filaments were drawn and cooled to achieve a consistent diameter of approximately 1.75 mm. Each batch was visually inspected to ensure uniformity in surface finish and dimensional accuracy. For printability evaluation, tensile test specimens were fabricated using a desktop FDM 3D printer. Printing parameters were standardized across all samples: nozzle temperature was set at 210°C, bed temperature at 60°C, layer height at 0.2 mm, and print speed at 50 mm/s. These conditions minimized processing variation and allowed accurate assessment of material behavior. The fabrication of specimens followed ASTM D638 Type V standards to ensure comparability (Figure 3). This consistent methodology ensured that differences in print quality and mechanical performance could be attributed to the material composition rather than to external processing variables (Hasan et al., 2024).

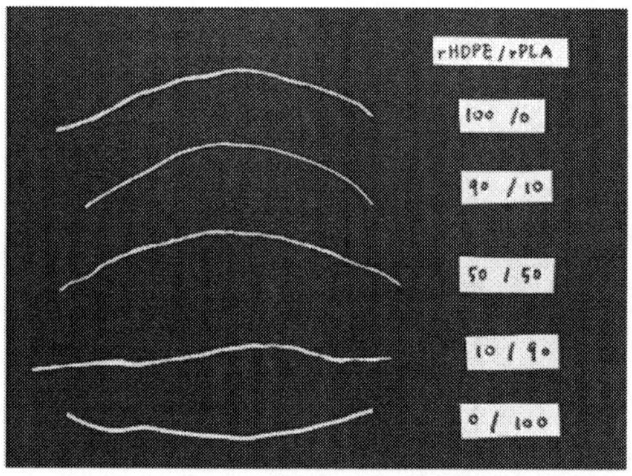

Figure 1. Filament samples with five rHDPE/rPLA blend ratios

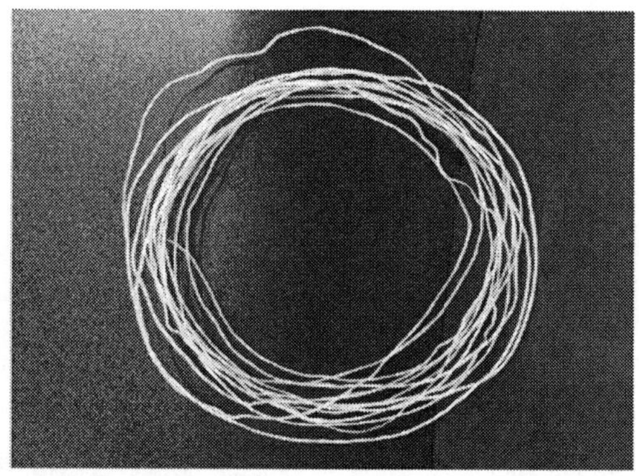

Figure 2. Visual appearance of extruded filaments

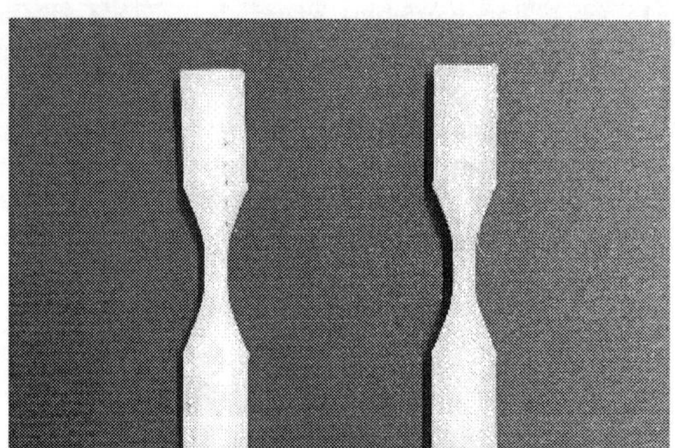

Figure 3. Tensile test specimens following ASTM D638 standard

2.2 Production Process

The extrusion and printability of the rHDPE/rPLA blends were evaluated in terms of process stability, visual appearance, and interlayer bonding. The 100% rHDPE filament displayed excellent melt flow, a smooth surface, and a consistent diameter within ±0.05 mm tolerance, confirming its suitability for filament production. In contrast, the 90/10 rHDPE/rPLA blend produced slightly rougher filaments, likely due to differences in melt viscosity, although extrusion remained stable with minimal parameter adjustment (Figure 4). During FDM printing, the 100% rHDPE specimens (Figure 5) exhibited accurate geometry and reliable bonding between layers. The 90/10 sample (Figure 6) showed improved surface detail and more refined contours, attributed to the flow characteristics of PLA. However, the addition of PLA also introduced minor warping, indicating reduced thermal stability. All prints were produced on the same equipment to ensure consistency, and results were documented visually. This comparative analysis of extrusion behavior and printing outcomes across blends provided a strong foundation for evaluating material feasibility. The observations corroborate previous findings that rHDPE is highly compatible with melt-extrusion and FDM processes (Mejia et al., 2020).

Figure 4. Comparison of rHDPE/rPLA blend specimens

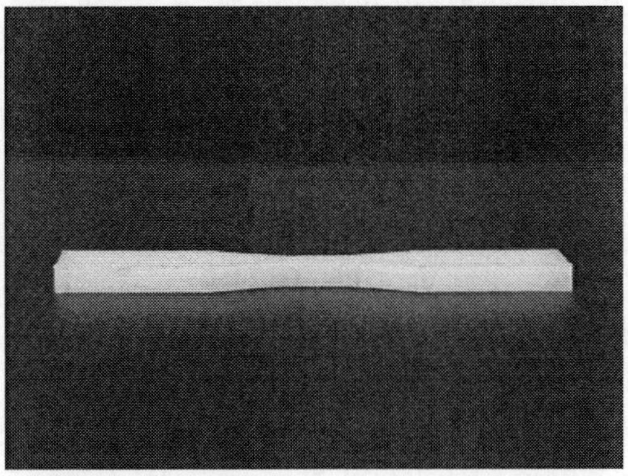

Figure 5. Surface and geometric characteristics of 100% rHDPE print

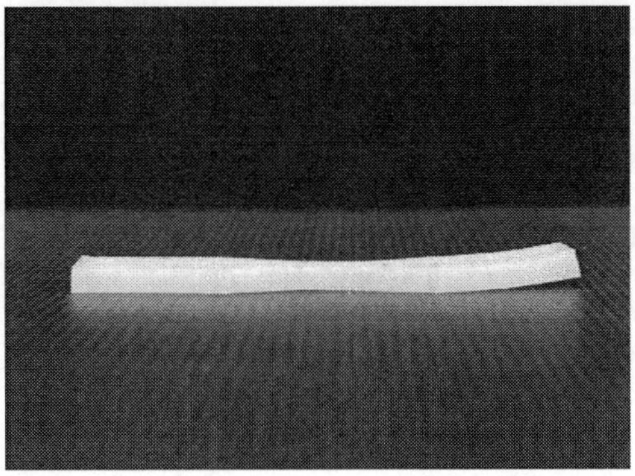

Figure 6. Surface quality and warping in 90% rHDPE + 10% rPLA print

2.3 Planned Test Items

To further validate the performance of rHDPE/rPLA composite filaments, two key evaluations are planned: tensile testing and heat deflection temperature (HDT) analysis. Tensile testing, based on ASTM D638 Type V, will measure tensile strength, Young's modulus, and elongation at break. These metrics are critical for understanding mechanical behavior and suitability for structural applications. HDT testing will assess the material's thermal resistance by determining the temperature at which samples deform under a specified load. This test is particularly important for PLA-rich blends, which tend to soften at lower temperatures. The combination of these two methods will offer insights into both structural durability and heat tolerance, essential for practical FDM use. The test results will be compared across all blend ratios to determine the most balanced formulation. Furthermore, correlations between extrusion behavior, printing quality, and mechanical performance will help refine the criteria for sustainable filament development. This approach follows established practices in evaluating recycled bio-based materials for additive manufacturing applications (Sanyang & Sapuan, 2023).

2.4 Expected Outcomes

It is anticipated that the mechanical and thermal behavior of rHDPE/rPLA filaments will vary significantly depending on blend composition. Blends with higher rHDPE content are expected to exhibit better toughness and thermal processing stability, while potentially lacking optimal interlayer adhesion. Conversely, blends with higher rPLA content are likely to show improved print appearance and bonding characteristics but reduced heat resistance and dimensional stability under thermal stress. Among the tested formulations, the 50/50 blend is hypothesized to provide the best trade-off between structural performance, printability, and ease of processing. This balance makes it a strong candidate for general-purpose applications in FDM printing. The upcoming tensile and HDT tests will validate these assumptions and may lead to refined recommendations for material formulation. Future studies may also explore the incorporation of compatibilizers to improve phase interaction, as well as fiber reinforcements to enhance mechanical properties. These modifications would expand the range of applications and address current limitations, in line with material improvement strategies reported in previous additive manufacturing research (Tümér & Erbil, 2021).

3. CONCLUSION

This study demonstrates the potential of recycled HDPE and PLA blends for use in FDM 3D printing applications. Through controlled melt blending and extrusion, filaments were successfully fabricated using five different rHDPE/rPLA ratios and evaluated for their extrusion behavior, printability, and preliminary mechanical characteristics. The results suggest that rHDPE contributes to processing stability and toughness, while rPLA enhances surface quality and layer adhesion. Nonetheless, rPLA's relatively low thermal stability poses limitations for certain applications, especially those involving elevated printing temperatures. Based on initial assessments, the 50/50 blend shows particular promise in balancing mechanical performance with aesthetic and process stability. This suggests that such a formulation may serve as a viable eco-friendly alternative to virgin filament materials. Moving forward, further mechanical testing, including tensile strength and HDT analysis, will be conducted to confirm the suitability of these materials in broader functional applications. Additionally, the integration of compatibilizers and reinforcing additives could further enhance performance and reliability. Overall, this work contributes to the advancement of sustainable additive manufacturing by demonstrating the feasibility of upcycled thermoplastics in FDM filament production and their potential for adoption in industry and education alike.

REFERENCES

Beltrán, F. R., et al. (2021). Evaluation of the technical viability of distributed mechanical recycling of PLA 3D printing wastes. *Polymers, 13*(8), 1247.

Hasan, M. R., et al. (2024). Impact of process parameters on improving the performance of 3D printed recycled polylactic acid (rPLA) components. *The International Journal of Advanced Manufacturing Technology, 131*(7), 3751–3779.

Mejia, E. B., et al. (2020). Upcycling of HDPE waste using additive manufacturing: Feasibility and challenges. In *2020 Advances in Science and Engineering Technology International Conferences (ASET)* (pp. 1–6). IEEE.

Sanyang, M. L., & Sapuan, S. M. (2023). Effect of mechanical recycling on the mechanical properties of PLA-based natural fiber-reinforced composites. *Journal of Composites Science, 7*(4), 141.

Tesfaw, S., et al. (2022). Evaluation of tensile and flexural strength properties of virgin and recycled high-density polyethylene (HDPE) for pipe fitting application. *Materials Today: Proceedings, 62*, 3103–3113.

Tümer, E. H., & Erbil, H. Y. (2021). Extrusion-based 3D printing applications of PLA composites: A review. *Coatings, 11*(4), 390.

IDENTIFICATION AND ANALYSIS OF DATA SOURCES ON THE IMPACT OF RIVER FLOODING ON TRANSPORT INFRASTRUCTURE IN GERMANY

Michael Eble[1], Jens Lehnen[1], Frauke von den Driesch[2] and Carina Herrmann[2]
[1]Dr., valantic
Bolzstraße 6, 70173 Stuttgart, Germany
2German Centre for Rail Traffic Research at the Federal Railway Authority
August-Bebel-Straße 10, 01219 Dresden, Germany

ABSTRACT

River flooding poses significant challenges to public transport infrastructure, particularly in the context of climate change. This paper presents an ongoing research project concerning both identification and exemplary analysis of data sources relevant to understanding and mitigating the impacts of river flooding on public transport systems in Germany. By examining various data types and sources, the study aims to build a basis that could then later on be used to inform strategies for enhancing river flood management in Germany. Methodologically, a comprehensive literature and geodata review is carried out to identify and analyze relevant sources. Specific selection criteria are defined to determine quality and relevance of the data, and geodata is analyzed to identify geographical areas at risk. The study emphasizes the importance of intensive interdisciplinary cooperation and highlights the significance of modern technological solutions for future warning systems. Five priority areas in Germany have been identified and the vulnerability of specific regions to flood damage is being analyzed. The report emphasizes the need for continuous adaptation and increased cooperation to meet challenges posed by climate change-induced river flood events. The research relies on systematic literature review as well as qualitative stakeholder research to consider practical aspects of river flood and crisis management. The project began in January 2025 and will conclude in November 2025.

KEYWORDS

Resilience, Infrastructure, Transportation Sector, Crisis Management, River Flooding, Data Analysis

1. INTRODUCTION

Rail transport in Germany provides significant value by efficiently and environmentally transporting large volumes of passengers and goods. In passenger transport, it offers a reliable alternative to road traffic, alleviating highway congestion and reducing environmental pollution (Maierhofer et al., 2024). The continuous modernization and digitalization of its infrastructure further enhance the attractiveness of rail transport, improving its sustainability and competitiveness (Pinkhasik and Herrmann, 2021).

At the same time, climate change is driving more frequent and intense weather events, including extreme temperatures, heavy rainfall, and storms, all of which affect transportation infrastructures in negative ways (Kasraei, A. et al, 2024, Nemry and Demirel 2012). Those climate change effects already have significant impact on rail infrastructure, e.g., they can cause damage to track and signals, whether through track flooding, storm damage, or heat and humidity (Garmabaki et al., 2021; Mitropoulos et al., 2025). Systematic literature review shows that the number of studies concerning effects of flood on transport infrastructure has been increasing strongly since the mid 2000's (Watson and Ahn, 2022).

Especially the increasing frequency and intensity of river flooding events have underscored the vulnerability of transport infrastructure. In Germany, incidents such as the 2021 floods in the Ahr Valley have demonstrated the devastating effects on bridges, roads, and railways (Szymczak et al., 2022). Understanding the impact of such events requires comprehensive data analysis to inform adaptation and mitigation strategies.

Against this backdrop, the research project “Exemplary analysis of data sources on the impact of river flooding on transport infrastructure”, led by the German Centre for Rail Traffic Research at the Federal Railway Authority (DZSF), aims to provide a systematic overview of data sources that might be helpful to support future cross-modal analyses of impacts on the transport system in the event of flooding. The project started in January 2025 and will conclude in November 2025.

2. OBJECTIVES AND APPROACH

The research project aims to address the challenges of flood and crisis management with respect to transportation infrastructure, particularly focusing on the impacts of river floods. The overarching goal is to provide insights that could be used for cross-modal impact assessment on transport axes and associated restrictions in the event of flooding. The project seeks to develop a comprehensive understanding of the current state of transportation systems and management practices when confronted with river flood events. This should create a basis for communicating expected climate impacts and their specific consequences and associated needs for action.

To achieve this objective, the project identifies and evaluates a broad range of data sources. It aims to compile a structured, cross-modal overview of the management of transport infrastructure in the event of river flooding in Germany through a detailed analysis of legal frameworks and methodologies. The project emphasizes the importance of interdisciplinary collaboration, leveraging the prospective communication of expected effects of flooding and the associated options for action. The investigations are supported by flood-related geodata in order to spatially concretize the compilation and analysis of the river flooding management. The project also focuses on creating a robust overview that categorizes literature according to its relevance for river flood management. An important focus of the project is the creation of a comprehensive compilation and analysis of the impact and management practices in the event of flooding in selected priority areas in Germany that are associated with potential flood damage to the transport system.

The project aspires to support research and practice, promoting a continuous adaptation process to tackle the challenges posed by climate change-related flooding. It underscores the necessity for ongoing collaboration among stakeholders to effectively safeguard transportation infrastructure.

This study is being funded by the German Federal Ministry for Transport (German abbreviation: BMV) in the context of the BMV Research Network. The BMV Research Network is the cross-modal research format in the BMV's departmental research. In 2016, seven departmental research institutions and specialist authorities of the Federal Ministry for Transport (BMV) joined forces to form a network under the guiding principle of "Knowledge - Ability - Action". The aim is to research urgent transport issues of the future in several areas, including climate change, environmental protection, reliable infrastructure, digitalisation, renewable energies and transport economic analyses.

2.1 Sub-Task 1: Literature Review and Collection of Data Sources

In today's fast-paced environment, where dissemination of knowledge has accelerated tremendously, literature research is still crucial for scientific purposes as it allows researchers to build upon existing knowledge by identifying gaps, trends, and established findings within a particular field (Tang et al. 2022). It provides a comprehensive understanding of the rapidly evolving research landscape, enabling scholars to design studies that are both relevant and innovative amidst the constant influx of new information.

The aim of sub-task 1 is to compile and analyze a broad selection of documents related to river flooding, including research reports, regulatory texts, academic journal articles, and studies published by government agencies or industry associations. Therefore, the approach within sub-task 1 of the research project encompasses several methodological steps for systematically identifying and evaluating data sources related to river flood management. Thus, the research team undertook a comprehensive literature analysis using a scientific approach (Snyder, 2019; Callahan 2014; Rowley and Slack, 2004). Additional geodata research complements this by systematically searching public and governmental databases to collect hydrological and topographical data concerning river areas.

The identified data is structured and assessed to gather information on river flooding, its impact on transport infrastructure and the associated current practice of incident management. A thematic analysis is conducted with a focus on existing framework conditions and flood management requirements in Germany, e.g. protective measures, alarm plans, and recovery processes. To provide first examples for priority areas, GIS tools (Geographic Information Systems) are used to visually represent spatial data and map affected areas cartographically. A multi-step procedure integrates heterogeneous data sources, with the data being checked for consistency through manual validations.

A scoring system is employed for quality assessment of data sources, where subject matter experts evaluate the significance of the data sources. Questionable datasets are validated through supplementary sources or expert opinions. Document analyses and interviews with specialists supplement the research to ensure the consistency and applicability of the findings. Finally, the data is consolidated into a comprehensive overview, serving as the basis for evaluating river flood management measures and preparing action recommendations. This structured evaluation aims to enhance the reliability of the results and identify research gaps in the field of river flood management for transport infrastructure.

So far, the main outcomes of sub-task 1 include a catalog of 100+ relevant documents on river flooding and in the context of transportation published between 2007 and 2025, including a preliminary listing of regulations (covering roads, rail, and other pertinent domains) that warrant a closer look, a summary of known adaptation strategies or frameworks to guide subsequent analyses as well as 30+ geo data sources. Stemming on this, the primary concern is the lack of concrete implementation plans and sparse information on the results and evaluation of measures already implemented, which makes it difficult to achieve a comprehensive overview and understanding of flood management and the resulting potential impact on transport infrastructure.

2.2 Sub-Task 2: Data and Information Acquisition

In sub-task 2, the project team builds on the results achieved in sub-task 1 and focuses on the systematic acquisition of additional data and information in selected priority areas to analyze the flood management with a focus on transportation infrastructure. The priority areas range from heavily populated urban regions to adjacent rural areas (Rhine-Main region, Ruhr area, Elbe region around Dresden, Danube region in Bavaria, Black Forest and Upper Rhine). The selection of the five priority areas is based on their high relevance for the population, economy and transport infrastructure as well as their different hydrological profiles.

First, an acquisition concept is developed, providing an overview of the information and data to be collected, including hydrological data, transportation infrastructure details, and institutional information. This concept also outlines the methods for data collection, such as surveys, interviews, and workshops, to obtain high-quality and region-specific information. Standardized documents, such as request letters and interview guidelines, are also used. Secondly, the data acquisition process involves proven activities, including contacting relevant experts and institutions, as well as collecting data through surveys and interviews. If required, workshops shall also be held to identify practical solutions.

In each of the five priority areas examined, responsibility for different aspects of fluvial flood management lies with different levels of disaster control authorities ("Katastrophenschutzbehörden"), which are organized at state level, regional level, and local level, respectively. For each of the levels of authority, different focus topics have been defined, and specific questionnaires have been developed. For example, a local firefighter organization answers very detailed questions regarding operational management plans, while a regional authority provides insights into overarching approaches to coordinating organizations that act on the local level.

The information collected will then be structured and compiled. After acquisition, the gathered data is analyzed to assess the vulnerability of transportation infrastructure in priority areas. This analysis considers different data for various transportation modes and focuses on main transport routes, such as railway tracks and federal highways. Finally, a structured collection of the acquired information and data is created to allow for a comprehensive analysis of management practices regarding river floods.

2.3 Sub-Task 3: Data Preparation and Data Analysis

With solid data from the previous phase, sub-task 3 focuses on the impact of river flooding on transportation infrastructure in the selected priority areas and its management. Protective measures, alert plans, and effects of river flooding events are analyzed. The study considers three modes of transport: rail, road, and waterways. In the road sector, federal highways as well as federal roads and state roads are considered, as they are crucial for passenger and freight transport.

The collected data is systematically structured to ensure focused and thematic analysis. GIS tools shall be used to create visual representations of the affected areas and transportation routes, enabling the identification of flood risk zones and infrastructural vulnerabilities. A comparison among the different priority areas is conducted to assess regional differences and similarities. Here, stakeholder engagement is another key element, with the strategy describing how to identify those who are involved in the practice of flood management (e.g., public organizations, or the private sector).

The emphasis is on a detailed and comprehensive overview of management practices focusing on transport systems when confronted with river flood events in the priority areas. The robustness of the results is ensured through extensive validation and comparison with official reports. The findings will be documented in a data catalog so that future projects can build on top of the results.

3. CONCLUSION

Analyzing the impact of river flooding on transport infrastructure requires a multifaceted approach, integrating various data sources. Addressing the challenges in data integration and promoting collaborative efforts are essential steps toward building resilient transport systems in the face of increasing flood risks. So far, the ongoing research project underpins the importance of an integrated approach to river flood management, focusing on crisis response and management practices.

However, there are gaps in integrating knowledge and geodata into infrastructure and river flood management. For instance, specific maps for river flood risks affecting transport infrastructure are lacking. Interactive maps and GIS systems are essential for risk assessments and emergency planning. More precise geodata could improve risk models and planning processes. Flexible and regional emergency plans, as well as joint exercises, enhance preparedness.

In the long term, climate change demands a strategic alignment as it increases the intensity and frequency of river floods. Advances in climate analyses and hydrodynamic models offer insights for future scenarios. Strengthened cooperation between state and municipal levels, as well as interdisciplinary approaches, could foster sustainable solutions. Research, practice, and legislation need to be better interconnected to enhance infrastructure resilience. Funding mechanisms are necessary to secure resilience against extreme events. Ultimately, continuous data collection and political support are essential for effective river flood management measures.

REFERENCES

Callahan, J.L., (2014). Writing literature reviews: A reprise and update. *Human resource development review*, 13, no. 3, pp. 271-275.

Garmabaki, A.H.S., et al., (2021): Adapting Railway Maintenance to Climate Change, *Sustainability 2021*, 13, no. 24.

Kasraei, A. et al, (2024). Climate change impacts assessment on railway infrastructure in urban environments, *Sustainable Cities and Society*, Vol. 101.

Maierhofer, J. et al., (2024). Digitalization and Reliability of Railway Vehicles and Tracks—Condition Monitoring and Condition-based Maintenance. *Berg Huettenmaenn Monatshefte 169*, pp. 264–268.

Mitropoulos, L. et al., (2025). Road network accessibility assessment during flood events by using infrastructure facility-based indices. *International Journal of Disaster Risk Reduction* 122, pp. 1–20.

Nemry, F. and Demirel, H., (2012). Impacts of Climate Change on transport: a focus on road and rail transport infrastructures. *Publications Office of the European Union*, JRC72217.

Pinkhasik, P. and Herrmann, P., (2021). Learning from external stakeholders: Evidence from two railway projects in Germany, *Project Leadership and Society*, Vol. 2.

Rowley, J. and Slack, F., (2004). Conducting a literature review, *Management Research News*, Vol. 27 No. 6, pp. 31-39.

Snyder, H., (2019). Literature review as a research methodology: An overview and guidelines, *Journal of Business Research*, Vol. 104, November 2019, pp. 333-339.

Szymczak, S., et al., (2022). Impacts of Heavy and Persistent Precipitation on Railroad Infrastructure in July 2021: A Case Study from the Ahr Valley, Rhineland-Platinate, Germany. *Atmosphere*, 13(7), 1118.

Tang, R. et al., (2022). A literature review of Artificial Intelligence applications in railway systems, *Transportation Research Part C: Emerging Technologies*, Vol. 140.

Watson, G. and Ahn, J.E., (2022). A Systematic Review: To Increase Transportation Infrastructure Resilience to Flooding Events. *Appl. Sci.* 12, 12331, pp. 1–27.

CULTURAL HERITAGE SUSTAINABILITY IN THE DESTINATION BRANDING OF SRI LANKA AND THE MALDIVES

Adriaan De Man

United Arab Emirates University

PO Box 15551 Al Ain, UAE

ABSTRACT

Destination branding is structured on a successful commodification of cultural heritage. While this applies to all sorts of regions and places, tourism-intensive island nations such as Sri Lanka and the Maldives depend fundamentally on unique cultural and natural experiences as key components in their branding strategies. This paper explores the procedures through which heritage elements are integrated into the destination branding of Sri Lanka and the Maldives, by assessing how each of these Indian Ocean countries promotes its cultural assets for international tourism competitiveness. The comparative significance of cultural heritage in brand positioning, followed by subsequent marketing strategies, as a pillar for tourism development is a permanent concern for public and private stakeholders.

KEYWORDS

SDGs, Indian Ocean, Tourism Marketing, Brand Positioning

1. INTRODUCTION

The use of cultural heritage for tourism branding is a common practice, namely for island nations with monetizable or stageable traditions intertwined with local identity. The tourism industry of both Sri Lanka and the Maldives builds on cultural practices and natural landscapes, which are building blocks to a sense of uniqueness and authenticity (Chandrapala, 2021, Maaz, 2021). Given its larger geographical extension, Sri Lanka more readily accentuates religious expressions and cultural landmarks, while the heritage of the Maldives stems mainly from maritime history associated with Islamic traditions and the authenticity of craftsmanship. The two neighboring destinations are economically and logistically intertwined yet, in addition to their specific tourism propositions, the corresponding branding practices are dissimilar and set each one apart within the global tourism market. Cultural heritage plays a major part in this effort and can be analyzed comparatively.

2. COMPARATIVE ANALYSIS

While both destinations are greatly structured on resort offerings, the tourism appeal of Sri Lanka is extensively connected with cultural heritage. The country counts on multiple UNESCO World Heritage sites, including urban features and other tangibles such as forts and temples, enhancing its attractiveness as a cultural tourism destination (Widodo, 2023). Local heritage is further structured in the context of the colonial past, in the form of architecture or cultural intangibles, often related to the history of Buddhism, and other expressions of art.

The 2019 “So Sri Lanka” campaign (Godamunne, 2023) captured this diversity as a branded strength by concentrating on the perception of authenticity. The resulting brand identity is based on the conservation and promotion of heritage sites such as Sigiriya, Anuradhapura, and Kandy (Gamage, 2022), and on visits to sacred sites, traditional performances, and festivals like Vesak or Sinhala and Tamil New Year (Bindloss et al., 2022), which are expected to increase the cultural density of an often beach-oriented stay in terms of international tourism.

This is precisely the case in the Maldives, best known for an abundance of atoll-built luxury resorts and sea-related leisure tourism. The country, however, has a distinctive but often overlooked cultural heritage identity, not very promoted by mainstream tourism marketing. This heritage has been shaped by the maritime past and a blend of influences from Indian, Arab, and African origins. Unique expressions of mosque architecture and consuetudinary art forms such as mat weaving and lacquerware reflect a characteristic cultural individuality (Cui and Li, 2025). The Maldives has been making efforts to integrate cultural and natural heritage, as the latter has become a lucrative yet monothematic source of revenue to the local economy. Examples include the ethnography of fishing communities and certain tangibles, e.g., the Hukuru Miskiy (Pradines and Topan, 2023) or Friday Mosque, so that visitors are invited to explore indigenous elements of what is portrayed as traditional Maldivian life.

The brand positionings of Sri Lanka and the Maldives are comparable in the general sense that cultural heritage is used for similar purposes, that is, for generating tourist destination attractiveness. However, the practical assimilation of heritage into branded capacity building occurs differently in both cases. In the case of Sri Lanka, the emphasis on comprehensive cultural landscapes, comprising Buddhist heritage, places the destination in contact with consumers seeking history, spirituality, and cultural engagement, and in line with the most recent indicators (Sumanapala et al., 2024). Supply on the ground meets such expectations and marketing strategies focus on the cultural tours offering access to sites, religious experiences, and festivals as part of a comparatively more diversified approach to cultural heritage. This is not linearly replicable in the Maldives, chiefly for geographical reasons, and the Ministry of Tourism together with private stakeholders exhibits cultural heritage equally as a primary component and an extension of luxury tourism. In comparison to Sri Lanka, the Maldives has insisted on its natural resources for a very long time, while cultural heritage has plausibly been subordinated to resort tourism. In recent years, a branding narrative based on indigenous island life, however staged the matching experience may turn out to be, has been gaining traction in tourism campaigns. This is in line with the evolution of the Maldives "One island, one resort" tourism policy over the last three decades (Sae-Tang and Yamchuti, 2021).

In fact, branding efforts have allowed for a more refined marketing and promotion. Campaigns and online presence have successfully integrated cultural heritage into the public messaging of Sri Lanka and the Maldives (Bandara et al., 2023, Jayaraman and Makun, 2022). As far as Sri Lanka is concerned, this translates into visuals of ancient sites, religious rituals, and cultural festivals, promoting the island as a destination where history and culture are to be experienced through everyday life. The Sri Lanka Tourism Development Authority (SLTDA) makes use of different marketing channels (Nuskiya, 2018), that is, social media, digital content, and influencer partnerships, in an effort to add cultural heritage to the nation-brand. More recently, the Maldives started engaging in strategic branding diversification by clearly including cultural heritage as a standalone element, following shifting preferences (Sathiendrakumar et al. 2021). Promotional materials emphasize traditions, local cuisine, and architectural uniqueness as an additional layer to luxury resorts. Unsurprisingly, the key marketing investment of the Maldives continues to insist on natural heritage. Nevertheless, there is an increasing emphasis on the cultural and social experience found on the islands. Local people and community-based experiences are now commonly featured in promotional campaigns about the Maldives.

3. HERITAGE BRANDING AND SUSTAINABILITY

Destination branding is intrinsically conditioned by complex production factors, a reason why Sri Lanka and the Maldives encounter recurrent challenges in integrating cultural heritage. The former faces the risk of over-commercialization with regard to heritage attractions, which leads to serious issues, one of which is the loss of perceived authenticity, hence of visitor experience quality. Economic instability in general and infrastructure challenges in particular can also limit physical access to heritage visitor attractions. As is the case in all commodified destinations, cultural appropriation and misuse are a permanent risk to the integrity of communities and their lifestyles (Sugathapala, 2024). In turn, the cultural heritage branding of the Maldives needs to balance a consumer preconception of luxury experience with the reality of unique cultural heritage (Abdulla et al., 2022). As mentioned above, the prominence of high-end resorts can affect the expression of local traditions. Heritage may not be fully showcased to visitors due to the almost total dependence on international tourism on a seascape of scattered islands and coral reefs. The very conservation of this expression of cultural heritage while balancing hospitality performance with sustainable tourism is a fundamental challenge.

Given that the tourism identities of both island nations have been shaped by cultural heritage resources, destination branding is required to pay close attention to the Sustainable Development Goals (SDGs), especially the ones directly connected with local tangibles and practices (Khizar et al., 2023, Pathirana et al., 2024). In particular Goals 8, 11, and 12 (Decent Work and Economic Growth; Sustainable Cities and Communities; Responsible Consumption and Production, respectively) can be identified in this context.

In the case of Sri Lanka, the longstanding community-based and preservation initiatives are structured in arrangement with government branding. One fine example is the "Island of the Gods" campaign, indeed based on cultural heritage and sustainable tourism. Sri Lanka is also fostering ethical heritage conservation for tourism as outlined in the Sri Lanka Tourism Sustainability Code. In this particular situation, some evident SDG were incorporated. Communities are expected to benefit from direct and induced tourism impacts, in particular by increased and better employment in positions related to hospitality, handicrafts, and tour guiding (SDG 8). Also, the preservation of cultural tangibles and of traditions, for the benefit of future generations is present.

The Maldives is equally pursuing SDG-aligned strategies to guarantee that communities and individuals, not only corporations, may benefit from tourism and that heritage remains a core resource. As such, economic start-up opportunities are ensured to local SME and entrepreneurs, as well as cultural performers and artisans (SDG 8). Maldivian traditional architecture and cultural expressions are supported as an ingredient for identity and therefore also for quality tourism (SDG 11). Circular economy practices and other sustainable forms of management at the resorts include waste-reducing and energy-efficient solutions, which directly link to SDG 12.

The application of several other SDGs demonstrates that cultural heritage sustainability is a strategic aim to both Sri Lanka and the Maldives, which directly impacts destination branding. While several operational features contribute to the branded outcome, two main risk factors can be pointed out. First, climate change disturbs the conservation of heritage sites and this is especially critical regarding small island and beach destinations. Second, overtourism can induce cultural dilution, which decreases the brand for residents and visitors alike. Canonical responses to heritage conservation and overtourism are often convergent but do not work automatically for tropical monsoon island-nations with major tourism-driven growth ambitions. Destination branding elements countering these risks are regulatory, hence depend on government involvement, yet depend on full stakeholder collaboration in the form of solutions such as digital distribution platforms, eco-tourism markets, and other public-private partnerships, for instance in the field of heritage conservation. An additional factor affecting both nations but to which the Maldives is dramatically exposed to is coastal vulnerability to extreme events, as became evident during the 2004 Indian Ocean tsunami and its aftermath, namely in terms of tourism redevelopment (Carlsen, 2006; Pardasani, 2006).

4. CONCLUSION

One caveat is to be mentioned here: not all heritage leisure attractions should be developed, hence branded for tourism, according to the same abstract procedures. The ones mentioned in this text make sense in their Indian Ocean setting and their heritage production, supply and consumption should be based on community configurations. Some alternative, non-Western perspectives have been pointed out precisely for this region (Timothy, 2021). Regardless, the destination branding objectives of Sri Lanka and the Maldives are based on similar principles, practices, and sustainability goals. When analyzed through the lens of cultural heritage tourism, though, they differ due to national specificities. While the branding of Sri Lanka strongly highlights excursions and trips to heritage visitor attractions, with particular attention to a cultural and spiritual immersion, the Maldives has for a long time established the image of atoll-centered resort luxury and is currently incorporating cultural heritage into its branded national identity. As a concluding remark, this comparison identifies cultural heritage as a major differentiator in the current tourism industry. Integrating sustainability at all levels of heritage tourism development is vital not only to the inherent quality of the experience but also to maintain international competitiveness, which is not only dependent on pricing, safety, infrastructures, and so forth but also on the cultural uniqueness comprised in the tourism mix.

Recommendations for destination branding are dependent on tangibles, most of which are fundamental production factors that cannot be manipulated and provide uniqueness to the destination in the first place. What can be addressed with increasing sophistication is the branding of a diverse experience not entirely reliant on luxury resorts and open to other forms of experiential (in this case, cultural heritage-based) tourism. This can focus strictly on hospitality through the integration of community elements and mid-range guesthouses that adapt to but do not replace the existing industry and overall tourism supply chain. At the broader level of nation-branding, such elements need to be merged within a development strategy and focus on the notion of collaborative stakeholder sustainability. Straightforward tools for doing so include influencer collaboration, immersive marketing, and social media engagement. From an explicit heritage-oriented branding angle, however, future research is to concentrate on cultural heritage as a brandable variable for both Sri Lanka and the Maldives.

ACKNOWLEDGEMENT

Research was funded by the UAEU College of Humanities and Social Sciences.

REFERENCES

Abdulla, A. and Schmidt di Friedberg, M., (2022). Textiles as Heritage in the Maldives. *Textile and Fashion Education Internationalization: A Promising Discipline from South Asia*. Springer, Singapore, pp. 145-174.

Bandara, K. M. C. S., Premasuriya, P. K., Rathnasiri, E. W. M., Karunarathna, W. R. N. S., Sampath, H. C. P., and Thushara, S. C., (2023). Promotional Strategy for Sri Lankan Tourism: A Case Study of the Sri Lanka Tourism Promotion Bureau, 2023. *Business Cases 2*, University of Kelaniya, Colombo, Sri Lanka, pp. 32-40.

Bindloss, J., Butler, S., Mayhew, B., and Walker, J., (2022). *Sri Lanka*. Lonely Planet, Dublin, Ireland.

Carlsen, J. (2006). Post-Tsunami Tourism Strategies for the Maldives. *Tourism Review International*, Vol. 10, No. 1-2, pp. 69-79.

Chandrapala, M. D. S. H., (2021). Tourism Industry in Sri Lanka (With Reference to Cultural Tourism). *The Journal of Archaeology & Heritage Studies*, Vol. 8m No. 1, pp. 11-15.

Cui, J. and Li, T., (2025). Religion, Power, and National Identity: The Dual Role of Islam in the History and Modernization of the Maldives. *Religions*, Vol. 16, No. 2, p. 201.

Gamage, K.P., (2022). Activity Economy and Cultural Heritage Tourism in Sri Lanka (Special Reference in Galle and Kandy), Trivalent: Journal of Archaeology. *Tourism & Anthropology*, Vol. 3, No. 1, pp. 54-70.

Godamunne, V., (2023). (Un) Making Travel Narratives of Sri Lanka from Ceylon to the "New" Sri Lanka. *Téoros. Revue de Recherche en Tourisme*, Vol. 42, p. 2.

Jayaraman, T. K., and Makun, K., (2022). Tourism and Economic Growth Nexus in Maldives: Asymmetric Analysis and Role of ICT. *International Journal of Business and Society*, Vol. 23, No. 2, pp. 1086-1105.

Khizar, H. M. U., Younas, A., Kumar, S., Akbar, A., and Poulova, P., (2023). The progression of sustainable development goals in tourism: A systematic literature review of past achievements and future promises. *Journal of Innovation & Knowledge*, Vol. 8, No. 4), p. 100442.

Maaz, M., (2021). *Making the Maldives: Exploring how the Body holds Cultural Knowledge*. University of Wales Trinity Saint David, Lampeter, UK.

Nuskiya, A. F., (2018). Impact of Digital Marketing in Tourism Industry of Sri Lanka. *Information and Knowledge Management*, Vol. 8, No. 9, pp. 59-67.

Pardasani, M. (2006). Tsunami Reconstruction and Redevelopment in the Maldives: A Case Study of Community Participation and Social Action. *Disaster Prevention and Management: An International Journal*, Vol. 15, No. 1, pp. 79-91.

Pathirana, M. T., Scott, N., and Sarah, G., (2024). Tourism Development in Sri Lanka: A Case Study of Sustainable Practices in a Small Island Developing State (SIDS). *Tourism Cases*, tourism202400079.

Pradines, S., and Topan, F., (2023). *Muslim Cultures of the Indian Ocean: Diversity and Pluralism, Past and Present*. University Press, Edinburgh, UK.

Sae-Tang, N. and Yamchuti, H., (2021). Relationship between Sustainable Tourism and Environment: Case studies of Maldives and Amphawa Floating Market, Thailand. *Journal of Humanities and Social Sciences Thonburi University*, Vol. 15, No. 1, pp. 149-160.

Sathiendrakumar, R., Jaffur, Z. K., and Seetanah, B. (2021). *The Role of International Tourism in the Maldives. Tourism Planning and Development in South Asia*. CABI, Wallingford, UK, pp. 132-148.

Sugathapala, M.A.S.H., (2024). Harmonizing Tradition and Tourism: A Comprehensive Examination of the Influence of Tourism on the Traditional Lifestyles of Indigenous Communities, *International Journal of Advanced Research and Innovative Ideas in Education*, Vol. 10, No. 1, pp. 660-667.

Sumanapala, D., Timothy, D. J., and Wolf, I. D., (2024). An Agenda for Heritage Tourism Research in Sri Lanka: Current Knowledge and Future Directions. *Journal of Hospitality & Tourism*, Vol. 22, No. 2, pp. 1-24.

Timothy, D.J., (2021). Heritage and Tourism: Alternative Perspectives from South Asia. *South Asian Journal of Tourism and Hospitality*, Vol. 1, No. 1, pp. 177-182.

Widodo, J., (2023). UNESCO World Heritage Branding: Learning from some Cases in Asia. *Journal of City: Branding and Authenticity*, Vol. 1, No. 1, pp. 12-23.

TRAINING SKILLS FOR SUSTAINABILITY: THE 3RPLAY METHODOLOGY

Sara Nosari
University of Turin
Via Sant'Ottavio 20, Turin, 10124, Italy

ABSTRACT

The OECD PISA 2022 data show that a low percentage of students in Italian schools are able to generate original ideas and develop a good level of problem solving. The survey data is highly relevant to Goal 4 (Quality Education) of the 2030 Agenda for Sustainable Development: it highlights a lack of those non-cognitive skills on which the development of critical-creative thinking, necessary for the exercise of active and responsible citizenship, depends. Two needs arise from this: the first, which is cultural, requires the effective recognition of non-cognitive skills as necessary competences and the consequent acceptance of the need for their specific training; the second, which is operational, requires a widespread training action aimed at students and aimed at exercising these skills.

This paper aims to address these needs by promoting the training of non-cognitive skills through the methodology of 3Rplay educational practices. This is a revised version of traditional educational-reflexive practices which includes the methodological device of pretext and an evaluation system. This will be supported by the presentation of the initial findings from its application with primary school teachers and coaches from different sports.

KEYWORDS

Sustainable Change, Critical-Creative Thinking, Problematization, Pretext Device, Evaluation System

1. INTRODUCTION

The OECD PISA 2022 data show that only 5.7 percent of students in Italian schools are able to generate original ideas and develop a good level of problem solving (OECD 2023). The survey data highlights a lack of those non-cognitive skills on which the development of critical-creative thinking, necessary for the exercise of active and responsible citizenship, depends.

Taking into account Goal 4 (Quality Education) of the 2030 Agenda for Sustainable Development, this lack is worrying (Jeronen 2021). In fact, the quality of education is a necessary condition for promoting lifestyles that offer everyone the opportunity to live in an environmentally, socially and economically sustainable world. It is a necessary condition because the accelerated change that characterizes contemporary progress is understood and experienced as a constant overcoming (Rosa 2013). As such, while it opens up possibilities once thought impossible (Marletto 2022), it also forces a constant reconfiguration of the space-time boundaries of our existence in the world. Consequently, this change requires skills that can search the situation for the solution that most authentically maintains and sustains a "human balance" (Morin 2022).

The skills needed for sustainable change include reflection, critical thinking, problematization, judgement, imagination, understanding and creativity (Liverano 2024). These are non-cognitive skills, i.e. skills that cannot be simply measured or acquired. Rather, they are skills that are exercised through reflective educational practice.

The sustainability of humanly balanced change requires therefore, firstly, the effective recognition of non-cognitive skills as necessary skills and the consequent acceptance of the need for their specific education. At the same time, it requires widespread training aimed at developing these skills in everyone, especially the younger generation. These two actions converge in a methodological proposal: that of 3RPlay educational practices (Nosari and Guarcello 2023, 2024), designed and developed by the "REFLECT research group" of the University of Turin.

2. THE 3RPLAY EDUCATIONAL PRACTICES

The methodology of the 3RPlay - *Reflecting, Researching, Replying* - involves setting up group activities that encourage the development of the habit of asking questions (Reflecting), investigating the situation in which one finds oneself (Researching) and developing possible answers to emerging problems (Replying). This methodology is based on the theoretical and methodological framework of the Schönian matrix (1983, 1995) and is characterised by two innovative elements:

- the use of the methodological device of the pretext;
- the use of an evaluation system.

These two elements are epistemologically "general", meaning the methodology can be used with recipients of different ages (10 years and older), in different contexts (formal, non-formal and informal) and to address any issue.

2.1 The Methodological Device of Pretext

In order to initiate and implement the development of the main capacities that characterize non-cognitive skills (problematisation, reflection, understanding, confrontation, judgement, deliberation, ...), 3RPlay practices use the methodological device of the pretext.

A pretext is an image or a sequence of images that enacts (in the form of vignettes) an open-ended situation designed to engage the participants in a problematization. The pretext must be situational and plausible. Indeed, it is essential that it presents a possible experience capable of intriguing and involving.

The proposed situation is designed with suspense. It must be "pending" because only in this way can it raise questions (such as the problem of justice, distribution of resources, access to culture, social welfare, etc.). But it must be a real suspension, i.e. it must raise a question that has not yet been resolved. The pretext is not a ploy to present an already decided situation in an interrogative form.

It is only in this way that the situation staged by the pretext is able to elicit different points of view from the participants, thus activating a process of collective research (Nosari 2023).

Figure 1. Example of pretext on sustainability

The device of the pretext involves participants in three specific actions of non-cognitive skills: *noticing, dwelling on, taking a stand* (Nosari and Guarcello 2023).

First, by introducing a situation, the pretext forces participants to notice something. Non-cognitive skills require attention to detail. It is not just a matter of observing and recording what is there. The attention needed to sustain human balance involves interest, care and concern (Bollow 2020). To have attention then means to become aware. The 3RPlay methodology aims to train this attention. Suspending the situation trains sensitivity to details and the possible connections between them.

The second action of the pretext makes the participants dwell on something. The reflection, judgement and understanding that are characteristic of non-cognitive skills require patience and confrontation with others (Bassot 2023). The 3RPlay methodology exercises with this device aim to develop the habit of not being satisfied with first impressions. It is an approach that only works with dialogue to open up different points of view.

The third action of the pretext encourages the participants to react and to exit the staged situation. The methodological device of 3RPlay forces participants to take a stand. If it initially forces participants not to respond, it is only because the response that sustains human balance cannot be superficial, nor can it be drawn from previous experience. Responsible and sustainable action must be appropriate and meaningful for the specific situation (Critchley 2008). Consequently, only at the end the 3RPlay practices prompts participants to respond. Non-cognitive skills are in fact those that enable the individual to have the courage to expose themselves. Reflection must necessarily lead to an act of judgment that results in a decision (Nosari 2023).

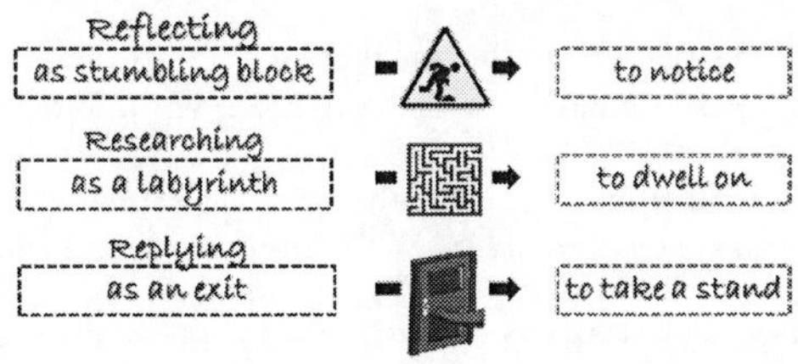

Figure 2. The three actions of pretext

The pretext device motivates innovative participation because its aim is not to find a solution as quickly as possible, but to examine the situation. As such, it engages participants in a kind of competition to read the situation from multiple perspectives.

In this way, participants are not competing to find the right answer. If this were the case, participation could easily be hindered by personal insecurities or biases towards those who are considered better. On the contrary, the methodological device of 3RPlay practices encourages the exercise of non-cognitive skills; since there is no right or wrong question (all questions can contribute to reading the situation!), no participant can feel inadequate.

Consequently, this device makes the 3Rplay methodology suitable for every target group: it facilitates the participation of the insecure teenager; it promotes the involvement of the adult regardless of profession or social position; indeed, it creates a space for discussion that allows everyone to have their say because it puts everyone in the same position.

The effectiveness of pretext was tested in two projects funded by the Fondazione della Cassa di Risparmio of Turin (CRT19, CRT22). The first project, SOCCER - School Football: Educational Roles and Tasks (2020-2021), made it possible to refine the pretext device and collected more than 2,000 questions submitted by the 80 coaches of football schools in the Piedmont region who took part in the 12 sessions organised by the project. The second project, STEREO: Sports Educators for New Societies (2023–24) involved 35 participants, half of whom were primary and secondary school teachers, and the other half were sports coaches in the Pinerolo area. Around 800 questions were asked (Nosari and Guarcello, 2024).

2.2 The Evaluation System

The proposal of the 3RPlay methodology also includes an evaluation system. As reflective practices, the traditional classification would have placed them under informal practices, i.e. practices without a monitoring and evaluation facility. Instead, the 3RPlay methodology designed and tested a monitoring system to record changes in the development and improvement of the basic non-cognitive skill: the problematisation. Indeed, the innovative idea was to monitor and evaluate the data collected during the activities: participants' questions (Nosari and Guarcello 2024).

While the premise of the 3RPlay methodology is that all questions contribute to understanding the situation (there are no right or wrong questions), the questions can be different. The questions may in fact relate to different objects and, depending on the object, read the situation from a point of view that "filters" the situation, including some aspects and excluding others.

The more than 2,000 questions collected during the SOCCER project enabled the research team to identify 4 different typologies of questions and their areas of investigation: descriptive, operational, causal and hypothetical (Nosari 2023; Nosari and Guarcello 2023). The 800 questions collected by the STEREO project then made it possible to refine the evaluation system by identifying a fifth typology: the interpretive one. (Nosari and Guarcello 2024).

Descriptive questions explore those aspects that allow a description of the situation to be returned. They are the questions that filter the situation to sift through the what? To whom? How much? When? Where? For how long? Since when? ...

Operational questions examine the procedural and instrumental aspects of the situation. They are the questions that sift through the methods and tools that led to the situation or that can change it: How does it work? How does it work? With what instruments?

Causal questions examine the situation, on the one hand to reconstruct the causal order that brought it about, and on the other to identify the reason that led to that situation: why? What are the causes? Because of what? For what?

Interpretive questions, on the other hand, explore the possible meanings of the situation. They are questions that "measure" the situation from the subjectivity of the questioner and thus seek its possible meaning: is it right? What is its value?

Finally, hypothetical questions explore the situation by opening up possible hypotheses for development that are not implicit or predictable consequences: What would happen if?

Table 1. The questions typologies

TYPOLOGY	FIELD OF INVESTIGATION
descriptive questions	explore the descriptive aspects of the situation
operational questions	explore the procedural arrangements explore the tools of operation
causal questions	explore efficient causes explore final cause
hypothetical questions	explore possible different hypotheses for the development of the situation
interpretive questions	explore the meaning that the situation may have

From the classification of the questions collected (the classification is currently being revised to identify possible subtypes), the REFLECT research group developed a qualitative evaluation system articulated as:

- evaluation in progress: classification of the collected questions according to the identified types in order to identify and improve the participants' approach to problematising;

- final evaluation: classification of collected questions to record the change in each participant's problematising approach;

- impact assessment: 4/6 months after the end of the course, participants will be interviewed to assess the impact of the non-cognitive skills developed.

The questions collected during the SOCCER project, once classified according to the 4 categories identified, showed that 70% operational questions; 20% causal questions; 9% descriptive questions; 1% hypothetical questions. The final evaluation, in confirming the prevalence of profiles with an operational problematisation, recorded two significant findings for the validity of the methodology. Firstly, the progressive increase in the number of questions: if in the first session some participants asked only one question, at the end of the course all contributed with at least four questions. The second significant result was the broadening of the typologies of questions: if at the beginning of the course some participants asked questions in only one typology, all participants eventually asked questions in at least two typologies.

Instead, the classification of the questions collected during the STEREO project shows a balance between the operational and causal types (40% and 40% respectively). The new typology of interpretative questions collected 10%. The hypothetical typology remains the least common, although it is on the increase (3%). The final evaluation shows that all participants kept the number of questions asked constant (at least 3 per session). Regarding the broadening of the type of questions, the most significant finding is that the teachers, who asked mainly causal questions at the beginning of the course, increased the number of operational questions at the end of the course. The sports coaches, on the other hand, made the opposite change: at the beginning there was a predominance of operational questions, at the end there was an increase in causal questions. The result suggests that comparison between participants from different professional backgrounds and with different relational experiences facilitates the possibility of reading the situation from different points of view.

The results obtained from the application of the 3RPlay methodology show that this version of reflective practices is an effective educational strategy for training and improving non-cognitive skills, which are condition for the promotion of sustainable development.

The evaluation of the SOCCER project's impact (that of the STEREO project is ongoing) shows that participants have become more aware of the complexity of individual situations. The problematising approach increased reflection time before decision-making in 80% of participants, and encouraged confrontation with others in 60%.

3. CONCLUSION

The effectiveness of the 3RPlay methodology essentially depends on the fact that it involves participants equally in matters to which they all feel they can contribute. This methodology can then enhance the formation of non-cognitive skills, which are considered a factor in the quality of education, precisely because the pretext situation initiates a process of participation.

The results obtained encouraged the REFLECT research group to continue in this vein. Several 3RPlay activities have indeed been initiated to train teachers and educators in the didactic use of this methodology. To support this, the research team is detailing the different profiles of the problematizing approach that emerged from the classification of questions.

At the beginning of 2025, a new project based on the 3RPlay methodology was also launched: MANDRAKE: Human and Artificial Thinking Players. Aiming to improve engagement activities and refine question types through the use of artificial intelligence, this new project is intended for students in the final two years of primary school and the first three years of secondary school. The intention is to monitor changes in problematizing approaches at a critical developmental stage for acquiring the skills necessary for a sustainable mindset. The first results will be available at the end of 2025.

REFERENCES

Bassot, B., (2023). *The Reflective Practice Guide. An Interdisciplinary Approach to Critical Reflection.* Routledge, London.

Bollnow, O.F., (2020). *Human Space*. Milano, Mimesis International.

Critchley, S., (2008). *Infinitely Demanding: Ethics of Commitment, Politics of Resistance*. Verso Books, London.

Güneş, F., & Söylemez, Y., (2018). *The Skill Approach in Education: From Theory to Practice*. Cambridge Scholars Publishing, Cambridge.

Jeronen, S., (Ed.), (2021). *Transitioning to Quality Education.* MDPI, Basilea.

Liverano, G., (2024). Developing Character Skills, Values and Virtues through Experience: A Strategy to Promote Sustainable Development. *Formazione&Insegnamento*, 1, 56-67.

Marletto, C., (2022). *The Science of the Possible and the Impossible*. Penguin, London.

Morin, E., (2022). *The Challenge of Complexity*. LUP, Liverpool.

Nosari, S. and Guarcello, E., (2023), The 3 R-Play's educational practices for adult education: the case of football coaches' training, in L. Formenti et al. (Eds.), *New seeds for a world to come. Policies, practices and lives in adult education and learning*. Ledizioni, Milan, pp. 297-305.

Nosari, S. and Guarcello, E., (2024). Promoting Problematizing Competence Through the 3RPlay System: The Case of the STEREO Project. *QTimes – webmagazine*, 3, 44-57.

Nosari, S., (2023). The Methodological Device of "Pretext": 3RPlay's Educational Practices for The Convergence of School and Sports Targets. *Proceedings of WRS International Conference*. Reykjavik, Iceland, pp. 33-37.

OECD, (2023). *PISA 2022 Results (Volume I): The State of Learning and Equity in Education*, PISA, OECD Publishing, Paris, https://doi.org/10.1787/53f23881-en.

Rosa, H., (2013). *Social Acceleration: New Theory on Modernity*. Columbia University Press, New York.

Schön, D., (1983). *The reflective practitioner: how professionals think in action*. Basic Books, New York.

Schön, D., (1995). *Frame Reflection: Toward the Resolution of Intractable Policy Controversies*. Basic Books, New York.

Reflection Papers

THE VALUE, APPLICATION AND CHALLENGES OF ARTIFICIAL INTELLIGENCE IN INFORMATION TECHNOLOGY COURSES IN PRIMARY AND SECONDARY SCHOOLS

Zhiqi Miao
Nanning Normal University
Nanning, Guangxi, China

ABSTRACT

This article mainly introduces the educational value, application methods and challenges faced by artificial intelligence based on mobile terminals in information technology courses in primary and secondary schools. Artificial intelligence can accurately analyze individual differences among students, promote personalized learning, and at the same time exercise students' ways of thinking and promote the development of creativity. In the information technology courses of primary and secondary schools, artificial intelligence can help students master basic learning tools, cultivate problem-solving abilities, and guide students to pay attention to social reality. However, artificial intelligence in information technology courses in primary and secondary schools also faces challenges such as weak professional knowledge of teachers, insufficient equipment and resources. Therefore, it is necessary to enhance teacher training, increase equipment investment and optimize resource allocation. In the future, the integration of artificial intelligence with information technology courses in primary and secondary schools will be deeper and broader, and more emphasis will be placed on practice and application.

KEYWORDS

Information Technology Courses, Artificial Intelligence, Intelligent Literacy

1. INTRODUCTION

As an interdisciplinary field that simulates human intelligence, artificial intelligence is gradually reshaping the educational ecosystem of primary and secondary schools by endowing machines with learning and reasoning capabilities (Wen, 2024). It integrates into information technology courses through intelligent application systems, building new practical platforms. The General Office of the Ministry of Education has issued the "Notice on Strengthening Artificial Intelligence Education in Primary and Secondary Schools", proposing a series of specific tasks and measures to promote the popularization of artificial intelligence education in primary and secondary schools. Against this backdrop, it is particularly important for educators to explore the value, application and challenges of artificial intelligence in information technology courses in primary and secondary schools.

2. THE EDUCATIONAL VALUE OF ARTIFICIAL INTELLIGENCE IN INFORMATION TECHNOLOGY COURSES IN PRIMARY AND SECONDARY SCHOOLS

2.1 Accurate Analysis of Individual Differences to Promote Personalized Learning

In traditional information technology courses in primary and secondary schools, teachers find it difficult to fully consider the learning situations and mastery levels of all students. Due to factors such as class size, teaching resources, and teachers' energy, teaching often focuses on the majority of students and cannot provide precise guidance for each individual. This "one-size-fits-all" teaching model not only fails to meet the personalized needs of students but may also suppress their learning interest and enthusiasm. Artificial intelligence, through big data analysis and learning algorithms, can accurately identify the differentiated needs of students in the process of learning information technology. It can construct a personalized learning profile for each student based on multi-dimensional information such as their learning progress, comprehension ability, interest preferences, and learning habits, simulate behaviors, diagnose states, predict trends, and empower learners for self-improvement and teachers for precise teaching optimization (Ji, 2024).

2.2 Exercise the Way of Thinking and Promote the Development of Creativity

The application of artificial intelligence in information technology courses can cultivate students' creativity and innovative thinking in the educational process. By providing logical problems and challenges, artificial intelligence helps students exercise their logical thinking ability. These problems often involve areas such as logical reasoning and mathematical calculations, effectively training students' logical thinking and fostering their habit of rigorous logical thinking. Moreover, artificial intelligence can stimulate students' imagination and innovative spirit through simulation and interaction. For instance, in programming learning, students can create their own applications with the help of artificial intelligence tools, enhancing their understanding and application of technology.

3. THE APPLICATION OF ARTIFICIAL INTELLIGENCE IN INFORMATION TECHNOLOGY COURSES IN PRIMARY AND SECONDARY SCHOOLS

3.1 Teaching Objectives: from Using Tools to Solving Practical Problems

In the current information technology courses in primary and secondary schools, although the teaching goals of most teachers are set to guide students to master the usage methods of information technology-related tools, hoping that these skills can bring convenience to students' daily lives, the actual effect is not satisfactory. When facing problems in real life, they often have difficulty flexibly applying these skills to find solutions. With the assistance of artificial intelligence, teachers can design teaching cases and scenario simulations that are closer to real life, allowing students to learn and apply information technology in the process of solving real problems. For instance, by leveraging artificial intelligence platforms, students can participate in the design and debugging of smart home systems, experiencing firsthand how information technology is transforming family lifestyles. Artificial intelligence prompts us to re-examine and define the teaching objectives of information technology, promoting the transformation of information technology courses from merely teaching tool usage to solving practical problems and cultivating innovation capabilities.

3.2 Course Content: from Uniformity to Personality Development

In the past information technology courses in primary and secondary schools, teachers often held an idealized expectation that a standardized curriculum system could meet the learning needs of all students. Furthermore, the uniformity of the curriculum leads to a high degree of consistency in the content of exercises and homework, lacking innovation and pertinence, making it difficult to stimulate students' interest and enthusiasm for learning. Artificial intelligence technology analyzes students' learning data to accurately grasp each student's skill level and interest preferences, thereby tailoring personalized teaching plans for each student. This personalized teaching model not only better meets students' learning needs but also effectively improves their learning outcomes. Specifically, artificial intelligence can provide personalized learning packages and resource packages based on students' mastery levels and learning styles. These resources not only include practice exercises and extended knowledge suitable for students' current levels, but also incorporate a rich variety of multimedia materials and interactive elements to stimulate students' interest and motivation in learning.

3.3 Education Methods: from Conventional Methods to Improve Quality and Efficiency

In regular information technology courses, teachers often tend to adopt traditional teaching methods, such as the lecture method, the exercise method and the assignment method. These methods play an important role in imparting basic knowledge and training skills. However, with the rapid development of artificial intelligence technology, its application in the field of education has brought revolutionary changes to the teaching of information technology courses. By introducing artificial intelligence technology, teaching methods have been improved, achieving the goal of enhancing quality and efficiency, and greatly promoting the improvement of students' information technology level. The introduction of artificial intelligence technology enables information technology courses to pay more attention to personalized teaching. With the help of an intelligent analysis system, teachers can accurately grasp each student's learning progress, interests and difficulties, and thus tailor learning content and difficulty for them, ensuring that each student can achieve the best learning results at a pace that suits them.

3.4 Academic Evaluation: from a Single Dimension to a Multiple Evaluation

In traditional information technology courses, the sole use of final exam scores as the evaluation criterion can directly reflect the mastery of knowledge and skills, but it neglects the cultivation of key qualities such as learning attitude, innovation ability, and teamwork. However, artificial intelligence, while assessing knowledge mastery through intelligent testing systems, can also leverage big data analysis and machine learning technologies to deeply explore process data such as classroom interactions, homework trajectories, and online learning behaviors, thereby achieving intelligent and dynamic evaluations of comprehensive qualities like learning attitude, autonomy, and innovative thinking. This not only provides objective and comprehensive assessment feedback but also enables real-time optimization of teaching strategies and supports personalized learning guidance, promoting the transformation and upgrading of the evaluation system from a "result-oriented" approach to a "quality-oriented" one.

4. ARTIFICIAL CHALLENGES AND COUNTERMEASURES OF ARTIFICIAL INTELLIGENCE IN INFORMATION TECHNOLOGY COURSES IN PRIMARY AND SECONDARY SCHOOLS

4.1 Weak and Solution of Teachers Professional Knowledge

The professional quality of information technology teachers is directly related to the quality and effect of artificial intelligence education. However, in information technology education in primary and secondary schools, the weakness of teachers' professional knowledge has become a bottleneck restricting the development of education. Many information technology teachers in primary and secondary schools do not have a deep understanding of the basic principles and applications of artificial intelligence, which limits their teaching ability and effectiveness in the classroom. Therefore, enhancing the professional quality of information technology teachers, strengthening the training of artificial intelligence-related knowledge and the accumulation of practical experience are the keys to improving the quality of information technology education in primary and secondary schools. Schools should establish long-term teacher training programs, including regular training courses, seminars and online learning resources. These activities aim to provide teachers with a continuous learning platform, helping them constantly update their knowledge structure and improve their teaching proficiency.

4.2 Insufficiency and Mitigation of Equipment and Resources

Artificial intelligence education in primary and secondary schools is confronted with the core predicament of shortage of equipment and resources. At the hardware level, it is manifested as insufficient quantity and high cost of teaching equipment such as intelligent robots and high-performance computers. At the software level, there is a lack of suitable intelligent teaching platforms, which makes it difficult for most students to obtain practical experience (Chen, 2024). The solution path requires the construction of a "demand-oriented resource optimization system", by increasing financial investment to build inclusive artificial intelligence laboratories, adopting lightweight solutions such as cloud computing power sharing and virtual simulation platforms, and establishing an inter-school equipment circulation mechanism within the region. Under the premise of ensuring basic hardware coverage, "low-cost and high-perception" teaching scene innovation can be achieved through open-source toolchains and modular teaching aids.

5. CONCLUSION

Artificial intelligence education is both a trend and an inevitable requirement for development. When implementing artificial intelligence education in primary and secondary schools, it is necessary to start from the different cognitive abilities and knowledge acquisition levels of students at different age stages, and rationally consider the course content. The learning content for each stage should be designed to gradually deepen students' understanding of artificial intelligence-related knowledge through five stages: recognition, comprehension, perception, practice, and innovation, laying a solid foundation for their future in-depth study of artificial intelligence (Qin, 2024).

REFERENCES

Guoqiang, C. (2024). Research on innovation of public PE education based on artificial intelligence [J]. *Knowledge Library*, 2024, 40 (22): 107-110.

Min, J. (2024). Research on Teaching Practice of Digital-enabled Adult Continuing Education [A] *Proceedings of 2024 Annual Conference of Henan Private Education Association (Volume 1)* [C] Henan Provincial Private Education Association, Henan Provincial Private Education Association, 2024:3.

Qing, W. (2024). The Technology and the Future [J]. *Research on Open Education*, 2024,30 (06): 2.

Xufeng, Q. (2024). Practice and exploration of artificial Intelligence Education in primary school Information technology curriculum [J]. *Information System Engineering*, 2024, (11): 52-55.

AN OVERVIEW OF THE VILLAGE PROJECT AND INSTRUCTIONAL DESIGN MODELS OPEN ACCESS DATABASE

Fırat Sarsar[1], Özge Andiç Çakır[2*], Christos Vasilakis[3], Stella Tsovou[3], Stavros Theocharis[3], Miša Mojca Cajnko[4], Matic Pavlin[4], Blaž Likozar[4], Saqib Rasool Chaudhry[5], Linda Pavitola[6], Valdis Priedols[7], Nida Şensoy Mercan[8], Berk Deniz[8], Sima Mammadova[9] and Urfan Taghiyev[10]

[1]*Ege University*
Department of Computer Education and Instructional Technologies, İzmir, Turkey
[2]*Ege University*
Department of Civil Engineering, İzmir, Turkey
[3]*Squaredev*
Department of Research & Innovation, Athens, Greece
[4]*National Institute of Chemistry*
Department of Catalysis and Chemical Reaction Engineering, Ljubljana, Slovenia
[5]*Munster Technological University*
Department of Computer Science, Cork, Ireland
[6]*Riga Technical University*
Centre for Pedagogy and Social Work, Riga, Latvia
[7]*Riga Technical University*
Centre for Nature Sciences and Engineering, Riga, Latvia
[8]*EMODA*
Department of Research & Innovation, İzmir, Turkey
[9]*Azerbaijan State Agricultural University*
Department of Information Technology, Ganja, Azerbaijan
[10]*Azerbaijan State Agricultural University*
Department of Agricultural Machinery and Maintenance, Ganja, Azerbaijan

ABSTRACT

In this paper, the general structure of the Virtual Innovative Learning Laboratories for Global Engineering Education (VILLAGE) project (no. 101086464) supported by the European Union in the HORIZON-MSCA-2021-SE-01 call and the "Instructional Design Models open access database" developed within the scope of the project are presented. The instructional design models in the literature were researched in depth and a total of 60 instructional design models were listed on database. The paper summarises the completed deliverables of the project and provides resources for researchers working on instructional design by sharing the structure of the database that can form the basis for scientific publications.

KEYWORDS

Instructional Design, Virtual Lab, VR Training, Engineering Education

1. INTRODUCTION

The Covid-19 pandemic started around the world in early 2020 and had an enormous impact on and changed people's lives all around the world. The situation affected educational institutions, and because of the Covid-19 epidemic, they shifted from in-person to remote learning all around the world during the spring term of the 2019–2020 academic year. One important thing to consider throughout this transition is the importance of having a well-organised planning process based on theory. Effective distance education requires convenient technological infrastructure and pedagogical frameworks that support both teaching and learning processes.

*Corresponding author

The pandemic made it clear that schools and universities needed to improve their readiness for digital deployment (Ayega, 2020; Maisano et al., 2024). Distance learning, on the other hand, is a complex process that requires careful planning based on theoretical foundations. People at all levels of educational administration learnt a lot about how online education works, which helped them respond quickly and plan for future problems (Dedeilia et al., 2020). Peculea (2023) studied undergraduate engineering students and suggested that further research be conducted in online learning environments to examine students' academic performance, factors that motivate and interest them, and the impact of online student engagement and satisfaction during and after the pandemic.

The implementation of virtual reality (VR) in education enables the development of comprehensive virtual environments that accurately replicate complex systems and processes, providing students with access to high-quality information regardless of their physical location. Studies show that VR can significantly enhance educational experiences by making them more engaging and interactive, thereby transforming traditional learning methods (Lvov & Popova, 2019).

VR can be used to create fully equipped, state-of-the-art labs that students and instructors in rural or low-income areas can utilise, thereby enabling inclusive education. VR-based laboratories can serve as significant supplements to engineering education by providing accessible tools for learning.

The objective of VILLAGE project is to reassess the post-pandemic pedagogical methodologies and requirements of engineering educators and students concerning laboratory practice within a multidimensional, multicultural, and multidisciplinary framework. It aims to establish roadmaps for innovative instructional design models that align with stakeholder perspectives, to formulate effective and sustainable teaching models, and to develop a pilot virtual laboratory and its accompanying learning materials to evaluate the proposed model. The project analyses the needs of instructors and students using a multidimensional and multicultural approach, based on data and experience from the seven partner countries (Türkiye, Greece, Ireland, Latvia, Slovenia, Spain, and Azerbaijan) (cordis.europa.eu/project/id/101086464). The themes, such as the role of virtual laboratories in education and the advantages and limitations of VR-based learning environments, were examined with a comprehensive literature review. Qualitative data from focus group interviews and questionnaires conducted with trainers within the project's scope were integrated (D1.1 Global State of the Art Report, VILLAGE Project Consortium, 2024).

Desk research has been conducted to identify (i) infrastructure access, (ii) instructional design methodologies, (iii) content development and lecturing procedures, and (iv) assessment and evaluation processes. This investigation enables us to ascertain the enquiries to be addressed in a secondary analysis part. Data have been collected under the following headings: (i) roles of instructors and students, (ii) instructional design and teaching strategies in the new normal, (iii) instructor-student interaction, (iv) new learning environments, (v) new digital course materials, (vi) applied training, and (vii) problems, solutions, and suggestions. During the desk research, 201 scientific papers were analysed and documented. Researchers organised focus group discussions within their country and ensured that the final report embodied a multicultural and multidisciplinary perspective. A focus group was gathered with 40 members from seven partner nations: Türkiye, Greece, Ireland, Latvia, Slovenia, Spain, and Azerbaijan.

The focus group participants concur that virtual laboratories facilitate the application of theoretical knowledge by providing experience with 3D visualisation, emphasising the foundational facts of the theory, and mitigating risks in practising. Consequently, they concur that educational models must emphasise the visualisation and interaction with authentic engineering systems, offering students immersive experiences through VR technology. For the successful implementation of VR technology in the engineering educational processes, it is of vital importance that instructors obtain pedagogical and technical support, create courses compatible with VR technology, undergo training on VR technology, and acquire skills in VR content development.

Ultimately, VR laboratories allow students to engage in remote learning at their own speed, free from temporal limitations. Consequently, instructional models should strive to enhance technology-facilitated learning and student engagement.

2. INSTRUCTIONAL DESIGN MODELS DATABASE

The open-access instructional design model database, one of the project's deliverables, can be accessed at www.thevillageproject.eu/idmodelsdatabase. This database systematically presents 60 models suitable for engineering education, containing the name, description, design principles, and resource information for each model. The streamlit-based user interface allows users to filter, compare, and access detailed information between models (D1.2 Open Access Database of Instructional Design Models, VILLAGE Project Consortium, 2024).

These models were explained under the following headings: (i) name of the model, (ii) short background of the model, (iii) description of the model, (iv) design (scheme and source) of the model, and (v) sources and other remarks. The dataset is also published as open access on the Zenodo platform (doi.org/10.5281/zenodo.10900874). The instructional design model database provides a user-friendly and functional interface for educators and researchers. As illustrated in Figure 1, users can filter models by name and access the full dataset via the Zenodo platform. Each model entry (such as ADDIE, depicted in the figure) includes a brief background, a detailed description, the model's design (including scheme and source), an expert-evaluated ranking score for VR lab applications in engineering education, and direct links to supplementary resources. The screenshot of the open-access IDMs database from the VILLAGE project website is presented in Figure 1 below:

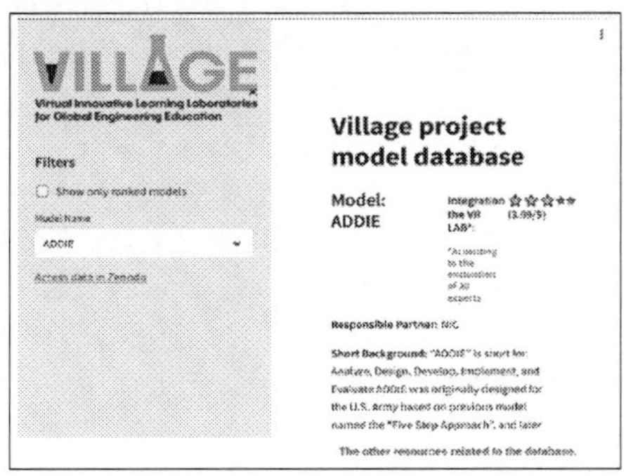

Figure 1. Representational view of the instructional design models database interface

3. CONCLUSION

This paper presents the structure and objectives of the VILLAGE project, summarises its main outputs, and particularly focuses on the database of instructional design models. These 60 instructional design models were prepared by collecting the relevant data from the appropriate dataset. The collection of instructional design models, which is planned to be one of the largest collections, has been converted into a database in the most effective way possible. The project provides a strong foundation for future virtual lab-based teaching models and contributes to open-access academic resources in engineering education.

ACKNOWLEDGEMENT

This study is supported by the VILLAGE project, Virtual Innovative Learning Labs for Global Engineering Education, funded by the European Union under the Horizon-MSCA-2021-SE program (Grant Agreement No. 101086464). The authors would like to thank all project partners for their collaboration and contributions to developing sustainable and inclusive virtual learning environments in engineering education. All project deliverables with a dissemination level of 'PU-Public' have been shared openly on the project website (https://www.thevillageproject.eu/) and the Zenodo platform (zenodo.org/communities/villageproject).

REFERENCES

Ayega, D., (2020). Pandemics and education in Sub-Saharan Africa: Invest in education technology. *American Journal of Educational Research*, 8(8), pp.581–586. https://doi.org/10.12691/education-8-8-10

Dedeilia, A. et al., (2020). Medical and surgical education challenges and innovations in the COVID-19 era: A systematic review. *In Vivo*, 34(3 Suppl), pp.1603–1611. https://doi.org/10.21873/invivo.11950

Lvov, M.S. and Popova, H.V., (2019). Simulation technologies of virtual reality usage in the training of future ship navigators. *Educational Dimension*, 1, pp.159–180. https://doi.org/10.31812/educdim.v53i1.3840

Maisano, D.A. et al., (2024). Remote STEM education in the post-pandemic period: Challenges from the perspective of students and faculty. *International Journal of Educational Technology in Higher Education*, 21, p. 64. https://doi.org/10.1186/s41239-024-00497-8

Peculea, L., (2023). *Online engagement and satisfaction of undergraduate engineering students during the COVID-19 pandemic*. Proceedings of EDU WORLD 2022, vol. 5. Craiova, Romania, pp.177–188. European Publisher. https://doi.org/10.15405/epes.23045.18

VILLAGE Project Consortium, (2024a). *D1.1 Global state of the art report*. Horizon Europe Project No: 101086464. https://doi.org/10.5281/zenodo.15288449

VILLAGE Project Consortium, (2024e). *D1.2 Open access database of instructional design models*. Horizon Europe Project No: 101086464. https://doi.org/10.5281/zenodo.15288506

Posters

PIL@B – INTERACTIVE LABORATORY PROGRAM

Alice Melo Ribeiro[1], Jamila Santos Khalifa[2], Samara Pereira Brito[1]
and Anna Luiza Moreira Lana de Almeida[1]
[1]Universidade de Brasília – UnB, Brazil
[2]Instituto Federal Goiano – IFGO, Brazil

ABSTRACT

PIL@B is a science outreach project that develops a remote experimentation platform to promote interactive and accessible Science teaching using digital technologies. The initiative focuses on creating online experiments using microcontrollers, producing educational content, and engaging in science communication. Through collaboration with universities and schools, the project integrates technology, education, and extension to innovate Science education at different levels.

KEYWORDS

Remote Experimentation, Science Communication, Educational Technology

1. INTRODUCTION

PIL@B is an outreach project dedicated to the development and promotion of remote scientific experiments, utilizing Digital Information and Communication Technologies (DICT). Its primary goal is to foster access to real, interactive scientific experimentation, creating new teaching and learning spaces (Khalifa et al., 2023). By integrating these technologies, students can become familiar with knowledge production in an active and innovative way, as highlighted by Geraldi (2016).

The project emerged in 2020, initially as The project emerged in 2020, initially as Coordination of Scientific Development (CDC) PILAB, within the Institute of Biological Sciences at a university. It was conceived to address the lack of physical laboratories in public schools, aiming to democratize access to experimentation through remote laboratories accessible via the internet . This initiative is particularly relevant for under-resourced schools, as DICT have the potential to facilitate teaching and enhance the pedagogical process in contexts with limited infrastructure, such as many Brazilian public schools (Silva, 2020). The lack of infrastructure, in particular, is one of the greatest barriers to biology education in these institutions (SEAD DF, 2022).

Subsequently, PIL@B integrated the World Pendulum Alliance (Erasmus+), establishing the WP@E-lab sub-network (Figure 1). This international collaboration boosted the project's expansion, leading to partnerships with schools, research centers, and other universities, such as UFSC's RexLab. These partnerships resulted in joint meetings focused on accessibility and active experimentation (Amarante-Segundo et al., 2022). Currently, PIL@B has expanded its activities beyond offering remote laboratories. The project now encompasses the development of open-access experiments using microcontrollers like Raspberry Pi and Arduino, in addition to investing in the training of future teachers and the dissemination of scientific knowledge across different educational levels, from primary to higher education.

Figure 1. QR Code to access the remote experimentation network

2. OBJECTIVES

The project aims to expand access to scientific experimentation via the internet, especially for public school students and teachers who lack adequate laboratory infrastructure. To achieve this, the project is developing its own remote experimentation platform. It also seeks to produce and disseminate high-quality educational content that integrates Digital Information and Communication Technologies (DICT) into Science teaching. Additionally, PIL@B works to support teacher training by promoting innovative teaching practices based on remote experimentation and active learning methodologies.

3. ACTIVITIES

3.1 Development of Educational Technologies

Creation of a new remote experimentation platform, accessible via web browser, featuring experiments such as Newton's Disc, Automated Irrigation, Bee Monitoring, and Remote Microscope. The system is developed in Python and made publicly accessible, accompanied by tutorials, videos, and lesson plans. All resources are co-created with the participation of extension students.

3.2 Science Communication on Social Media

Weekly posts focused on science popularization are produced by the extension team. The project maintains an active presence in science fairs, events, and academic conferences, promoting accessible and engaging scientific content.

4. RESULTS

The PIL@B project has made significant strides in science education outreach. Our Instagram profile, @PIL@B UnB, has garnered over 623 followers and features more than 107 posts, significantly boosting our online science communication efforts. We've also strengthened our role at the intersection of outreach and scientific production through research conducted and papers published. The project benefits from the active participation of two professors and fifteen undergraduate students, who contribute to both technical development and educational dissemination. So far, one remote experiment has been completed, with three more currently in development. The implementation of PIL@B has led to observable improvements in student engagement with scientific content, especially in schools lacking access to traditional laboratories. Teachers involved in our pilot activities have reported greater interest and participation during lessons that incorporated remote experiments. Beyond technical development, PIL@B plays a crucial role in reducing educational inequalities by offering free, scalable resources that are adaptable to diverse teaching contexts across Brazil. The initiative also fosters collaboration through partnerships with other projects in the same field, thereby expanding its impact and network. Looking ahead, PIL@B plans to expand its experiment library, improve the user interface of the platform, and enhance training materials for educators. Our ultimate goal is to scale this initiative to a national level, further democratizing access to quality science education.

REFERENCES

Amarante-Segundo, G. S., Amarante, B. G., Cruz, J. M. R., Ribeiro, A. M., Amador, C. H. S., & Fernandes, H. (2022). World Pendulum Alliance: student's first impressions in Brazil [World Pendulum Alliance: primeiras impressões de estudantes no Brasil]. 8th International Engineering, Sciences and Technology Conference (IESTEC), 8(1), 360–363.

Geraldi, L. M. A., & Bizelli, J. L. (2017). Tecnologias da informação e comunicação na educação: conceitos e definições [Information and communication technologies in education: concepts and definitions]. Revista On-line de Política e Gestão Educacional, Araraquara, (18). Recuperado de https://link.ufms.br/tUyoK

Khalifa, J. S., Cruz, J. M. R., Amarante-Segundo, G. S., & Ribeiro, A. M. (2023). World Pendulum Alliance: experimentação remota [World Pendulum Alliance: remote experimentation]. Revista do EDICC, 9, 1–X. ISSN 2317-3815.

SAEDF. (2022, 1 de agosto). Falta de infraestrutura e de recursos é o maior desafio em sala de aula, aponta pesquisa [Lack of infrastructure and resources is the biggest classroom challenge, survey shows]. SAEDF – Sindicato dos Trabalhadores em Escolas Públicas no DF. Recuperado de https://www.saedf.org.br/index.php/cnte/falta-de-infraestrutura-e-de-recursos-e-o-maior-desafio-em-sala-de-aula-aponta-pesquisa/

Silva, J. B., Bilessimo, S. M. S., Scheffer, G. R., & Silva, I. N. (2020). Laboratórios remotos como alternativa para atividades práticas em cursos na modalidade EAD [Remote laboratories as an alternative for practical activities in E-learning courses]. EaD em Foco, 10(2).

STRATEGIC OPTIONS ON STANDARDS ISSUES FOR ACHIEVING BUSINESS GOALS

Byoungho Jun
College of General Education, Seoul Women's University
621 Hwarangro, Nowon-Gu, Seoul, 01797, South Korea

ABSTRACT

Standards offer numerous opportunities and benefits not only to provide effective and efficient way for business goals internally, but also to increase company's competitiveness externally. Standards and standardization have emerged as strategic management tools, and companies take strategic options on standards for achieving their business goals based on their specific circumstances. This study is to examine how strategic options on standards issues can be taken by companies for business goals based on Miles and Snow typology. It can provide guidelines on standard strategies for business goals.

KEYWORDS

Standards, Strategic Options, Prospector, Defenders, Analyzers, Reactors

1. INTRODUCTION

Standards are used every day by businesses, manufacturers, public bodies, and other organizations as tools for managing vital issues such as trade, regulation, quality, health and safety, innovation, and competitive advantage. Especially Standards offer numerous opportunities and benefits to companies, not only by providing effective and efficient ways to achieve business goals internally, but also by enhancing a company's competitiveness externally (Jun, 2017).

Advancing globalization is affecting more companies than ever before and increasingly involves numerous standard-related issues such as product standard, process standard, and conformity assessments. The importance of standards in a global economy is well understood by leading global companies. It is evident that organizations engaged in standardization initiatives tend to perform better than those that are not (Richard, 2003). Standards turn the concept of "sustainability" into actionable, measurable, and comparable practices. By embedding standards into strategy, organizations not only manage risks and demonstrate compliance but also unlock new opportunities for innovation, market differentiation and meaningful progress toward the SDGs.

Standards and standardization have emerged as strategic management tools, and companies take strategic options on standards for achieving their business goals based on their specific circumstances. The purpose of this paper is to examine how strategic options on standards issues can be taken by companies for business goals.

2. STANDARDS AND STRATEGIC OPTIONS

2.1 Standards in Company

Standards and standardization have now become a fact of life across all industries globally. Economic studies have demonstrated that standards enhance labor productivity and innovation, contributing substantially to overall economic growth (Henk de vries, 2006). Standardization within a company is aimed at maximizing overall operational efficiency and economic performance. Standards can be leveraged externally to achieve competitive advantage in the marketplace as well. Companies actively participating in standardization activities can reduce costs, enhance operational efficiency, facilitate and maintain market access, and gain greater competitive advantage compared to those not engaging in such activities. Standardization issues are inherently connected to all aspects of a company's activities, thus offering numerous benefits. Therefore, companies should take strategic options on standard issues for business goals based on their specific circumstances.

Standards strategies should be aligned with the overall business strategy, implying that strategic options on standards can be implemented across the business levels, strategic level, management level, and operational level (Anthony, 1965). Miles and Snow also addressed three fundamental problems facing the companies, entrepreneurial, engineering (or operational), and administrative problems (Miles and Snow, 1978). The entrepreneurial problem pertains to how a firm establishes its market position, thereby defining specific products or services within the market while the engineering problem involves selecting suitable technologies for the production and distribution processes to effectively address the entrepreneurial problem. The administrative problem relates to the rationalization and stabilization of organizational processes to resolve both entrepreneurial and engineering challenges.

Standards issue also involves those three fundamental problems across the business hierarchy (Kang and Jun, 2012). For example, when a company wants to lead the standardization of a product in an emerging market, it must carefully consider various strategic options that could influence its long-term competitive position. Decisions related to these strategic options constitute a critical entrepreneurial problem confronting the company. A company may need to establish systematic processes aimed at enhancing operational efficiency and reducing costs, such as implementing a quality management system—an approach analogous to addressing the engineering problem. Standardized routine tasks and services within a company facilitate standards management at both the managerial and strategic levels, a practice associated with addressing the administrative problem.

2.2 Strategic Options on Standard Issues

Miles and Snow (1978) identified four general strategic types of organizations—prospector, defender, analyzer, and reactor—based on how firms address three fundamental organizational challenges: entrepreneurial, engineering (or operational), and administrative problems. Prospectors are organizations that seek to pioneer new market opportunities, typically introducing novel product lines in response to emerging environmental trends. Defenders are organizations characterized by a limited range of products, focusing primarily on efficiency enhancement and process improvement within a confined operational domain, rather than actively seeking external opportunities. Analyzers represent an intermediate organizational type situated between prospectors and defenders, typically employing a 'second-but-better' strategy (DeSarbo et al., 2005). They exhibit characteristics blending aspects of both prospectors and defenders (Hambrick, 1983). Reactors are organizations characterized by the absence of a consistent strategic orientation and an inability to effectively respond to technological innovation, typically exhibiting passive reactions to environmental pressures.

A company's strategic options on standard issues can be inferred from the typology by Miles and Snow (1978) as Table 1.

Table 1. Strategic options on standard issues

Types	Strategic options
Prospectors	· a first mover of a market (McDaniel and Kolari, 1987; Conant et.al., 1990). · seek market dominance by introducing innovative technologies and establishing them as market standards · must effectively manage their strategic assets to maintain market competitiveness
Defenders	· pursue cooperative relationships with leading standard-setting firms and aim to play significant roles in standardization competition by leveraging their existing capabilities. · secure market stability through cooperation with leading standard-setting firms
Analyzers	· position themselves variably in the market, either as passive followers of an existing market standard or as active contenders promoting competing technologies to achieve a 'fast-second' advantage, depending upon prevailing market conditions
Reactors	· do not employ a systematic approach to strategy formulation and consequently tend to exhibit lower performance levels

3. CONCLUSION AND FURTHER RESEARCH

This paper demonstrated that strategic options on standard issues could be effectively analyzed through Miles and Snow's (1978) typology, as standard issues encompass three fundamental management tasks: entrepreneurial, engineering, and administrative. Although this model is somewhat dated theories, Miles and Snow's typology has been widely cited across a variety of fields (Beatriz Forés, 2019; Yannick et al., 2024).

In the case of smartphones, Apple has dominated the market as a first mover. As an innovator, Apple, a prospector, has enjoyed the first mover advantages, in which an installed based could be established. Apple has a brand name premium in standardization war, which gives customers with convince that prospectors take all. Many companies in developing countries, as defenders of standardization, adopt dominating market standards and survive a competitive market by utilizing their highly efficient manufacturing capability. Samsung is a good analyzer in standardization wars. As a fast-second company in IT market, Samsung has reputation in a market and a capability to control over the installed base to compete against market leaders. Reactors are supposed to be very low performers, so the importance of standards should be educated to them.

This paper has originality and practical implications in that it provides a guideline on how strategic options on standards issues can be taken by companies for business goals. More case analysis will provide a clearer understanding of strategic options on standards for business goals.

REFERENCES

Anthony, R. (1965). *Planning and Control Systems: A Framework for Analysis.* Boston, MA: Division of Research, Graduate School of Business Administration, Harvard University.

Conant, J.S., Mokwa, M.P., and Varadarajan, P.R., (1990). Strategic Types, Distinctive Marketing Competencies, and Organizational Performance: a Multiple-Measures-Based Study, *Strategic Management Journal*, 11(5), 365-383.

De Vries, H. (2006). *Standards for business – How companies benefit from participating in international standards setting*, IEC.

DeSarbo, W.S., Di Benedetto, C.A., Song, M., and Sinha, I., (2005). Revisiting the Miles and Snow Strategic Framework: Uncovering Interrelationships between Strategic Types, Capabilities, Environmental Uncertainty, And Firm Performance, *Strategic Management Journal*, 26, 47-74.

Forés, B. (2019). Beyond Gathering the 'Low-Hanging Fruit' of Green Technology for Improved Environmental Performance: an Empirical Examination of the Moderating Effects of Proactive Environmental Management and Business Strategies, Sustainability, 11(22), 6299.

Hambrick, D.C., (1983). Some Test of the Effectiveness and Functional Attributes of Miles and Snow's Strategic Types, *Academy of Management Journal*, 26, 5-25.

Jun, B. H. (2017). Case Study on Standards Management: Focused on Small and Mid-sized Businesses, *Journal of the Korea Convergence Society,* 8(8), 239-246.

Kang, B. G., and Jun, B. H. (2012). Understanding standardization strategy based on Miles and Snow Strategic Framework, I- ESA'12 Conference.

McDaniel, S. S., and Kolari J. W., (1987). Marketing Strategy Implications of the Miles and Snow Strategic Typology, *Journal of Marketing,* 51(4), 19-30.

Mies, Y. A. A., Hausberg, J. P., & Packmohr, S. (2024), Digitising Miles and Snow: using cluster analysis to empirically derive digital business strategy types. *Technology Analysis & Strategic Management*, 36(11), 3922–3935.

Miles, R. E., Snow, C. C., (1978). *Organizational Strategy, Structure, and Process*, New York: McGraw-Hill.

Richard, J. (2003). *The need for effective standardization management, Standardization News*, ASTM.

AUTHOR INDEX

Aderemi, A. 199
Akao, K. 137
Alvarez, P. 89
Arai, S. 175
Barot, T. 152
Bellini, C. 29
Benhaoua, K. 37
Beutner, M. 71
Boccuzzi, G. 129
Bonanzinga, V. 48
Boudia, C. 37
Brito, S. 281
Butrime, E. 183
Cajnko, M. 275
Çakır, Ö. 275
Carvalhais, M. 167
Cha, H. 145
Chaudhry, S. 275
Cheng, J.-H. 249
Chicana Huanca, S. 3, 81
Chiu, P.-C. 249
Cimpan, S. 105
Cormican, K. 89
Das, S. 191
de Almeida, A. 281
De Man, A. 258
De Santis, A. 29
Deniz, B. 275
Duche Pérez, A. 3, 81
Eble, M. 244, 253
Ferrer, R. 105
Gallico, D. 208
Gemrot, Z. 63
Gonçalves, B. 159
Green, D. 89
Guedes, K. 121
Guimarães, M. 121
Gutierrez Aguilar, A. 3, 81
Gutierrez Aguilar, O. 3, 81
Hasnine, M. 137
Herrmann, C. 244, 253
Hlatka, L. 152
Houari, A. 37
Iwendi, C. 199
Jeun, W. 229
Jun, B. 284
Khalifa, J. 281
Kohjiba, Y. 175
Korenova, L. 63, 152
Lee, G.-Y. 224
Lee, S.-H. 224
Lee, Y. 224
Lehnen, J. 244, 253
Likozar, B. 275
Lin, C.-Z. 249
Lu, X. 239
Luic, L. 234
Lüttke, J.-P. 71
Mammadova, S. 275
Manganello, F. 129
Marques, M. 97
Martín, M. 29
Matcha, W. 11
Mercan, N. 275
Merica, M. 152
Méxas, J. 121
Miao, Z. 271
Minerva, T. 29
Molinari, A. 113
Monteiro, J. 97, 167
Morais, C. 167
Nakamura, H. 175
Nico, A. 129
Nosari, S. 263
Ojo, O. 199
Pavitola, L. 275
Pavlin, M. 275
Pierluigi, A. 29
Pombo, L. 97
Pozzi, L. 219
Priedols, V. 275
Ramakrishnan, M. 191
Ribeiro, A. 281
Sampaio, S. 89
Sarsar, F. 275
Schmid, A. 63
Shibayama, T. 175
Suh, M. 229
Taghiyev, U. 275
Tavares, R. 121
Tenhunen, M.-L. 55
Theocharis, S. 275
Trgovac, A. 234
Tsovou, S. 275
Tuomaite, V. 183
Ueda, H. 137
Vasilakis, C. 275
von den Driesch, F. 244, 253

Wong, W. 191
Yamada, M. 137
Yaseen, M. 22
Yokoyama, H. 105
Zilkova, K. 152